W0081927

THE VOICE OF SCIENCE

SCIENCE AND CULTURE IN THE NINETEENTH CENTURY
BERNARD LIGHTMAN, EDITOR

THE VOICE
OF SCIENCE

BRITISH SCIENTISTS
ON THE LECTURE CIRCUIT
IN GILDED AGE AMERICA

DIARMID A. FINNEGAN

UNIVERSITY OF PITTSBURGH PRESS

Published by the University of Pittsburgh Press, Pittsburgh, Pa., 15260
Copyright © 2021, University of Pittsburgh Press
All rights reserved
Manufactured in the United States of America
Printed on acid-free paper
10 9 8 7 6 5 4 3 2 1

Cataloging-in-Publication data is available from the Library of Congress

ISBN 13: 978-0-8229-4681-6
ISBN 10: 0-8229-4681-5

Cover art: *Daily Graphic*, September 27, 1876, 1, detail. By permission of the New York Historical Society.
Cover design: Melissa Dias-Mandoly

To my father and mother

CONTENTS

PREFACE AND ACKNOWLEDGMENTS

Early in 1891, Arthur Conan Doyle sent a short story to the *Strand Magazine* entitled "The Voice of Science." The narrative begins with a wealthy society heiress, Rose Esdaile, and her mother, a widow and a "lady of quite remarkable scientific talents," furnishing their house with a host of objects in preparation for a conversazione. After all is ready for this scientific soirée, the discussion between mother and daughter turns to marriage. About this, Rose is in a state of anxiety. Captain Beesly, a man she admires but knows only a little, is expected to propose to her at the soirée. While her mother admits Beesly has many qualities, she had rather hoped Rose would marry Professor Stares, only thirty years old and already a fellow of the Royal Society. At that moment, Rupert, Rose's brother, arrives and attempts to warn Rose not to form a serious attachment to the apparently disreputable Beesly. Rose will not listen and walks away. Rupert then spies a phonograph—the centerpiece of both the conversazione and the story. After listening to it play back the "funny little tinkling voice" of a "celebrated scientist" delivering a lecture on the life cycle of the common barnacle, Rupert takes the "garrulous piece of wood" to his room and records incriminations against Beesly. During the conversazione, just as Beesly declares his undying love for Rose, Rupert sets up the phonograph and replays, in a "squeaky little voice," accusations against the captain of philandering and cheating at cards. Beesly, suitably shamed, flees the party, while Rupert, apologizing, changes the cylinder on the phonograph and plays back the barnacle lecture. Rose marries Professor Stares, with his intellect and "rising fame," but, now and then, remembers the handsome Beesly and "marvels at the strange and sudden manner in which he deserted her."[1]

Whatever else might be going on in Doyle's story, one thing that is not finally clear is which of the many voices described is truly the voice of science. Mrs. Esdaile's fluent discussions of arcane scientific topics in her own home and at the "ladies branch of the local eclectic society" make her voice a strong contender. Against that, however, stand "bitter feminine whispers" of Mrs. Esdaile's "brilliant speeches written out in some masculine hand . . . committed to memory" and of her muddled "geological harangues" and lectures on entomology. Nevertheless, it is Mrs. Esdaile who, with a word, brings back harmony when "thin, ascetic materialists from the University upheld the importance of this life against round, comfortable champions of orthodoxy from the Cathedral Close." A "rattle over the keys" by the pretty Rose also helps.

Perhaps, however, it is the disembodied voice of the phonograph, which says "nothing except what is said to it," that speaks the most scientifically in this narrative.[2] By 1891 the phonograph had become an icon of scientific progress. What was more, the science lectures it recorded were free of the encumbrance or distractions of a physical body or the mistakes or modifications of scribbling and sometimes unscrupulous stenographers. And yet the sound it produced altered the quality of the voice—sonorous scientific speech was transformed into "high piping tones, thin but distinct."[3] The content and cadence remained, but the tone and temper were undeniably distorted. Was anything of importance lost in this remediation? By the cultural measures of the period, it certainly seemed so. After all, the scientists in Doyle's story accumulate fame, speak in authoritative masculine tones (the exceptional Mrs. Esdaile notwithstanding) and exhibit through their very bodies a self-sacrificial commitment to science. These are personal and social attributes and actions that could not be captured by the phonograph.

Doyle's satirical description of the voice of science, draped over a clichéd romance narrative, points to a question that lies behind this book. Amid myriad clashing and clamoring voices, how was public speech mobilized and modulated by those intent on securing the moral meaning and cultural authority of science during the late nineteenth century? The anxiety surrounding this question had a longer history. In 1838 Charles Lyell had written to Charles Darwin urging support for the efforts of the British Association for the Advancement of Science to agitate for science in the public sphere. As he noted to the young naturalist, they lived "in a country where, as Tom Moore justly complained, a most exaggerated importance is attached to the faculty of thinking on your legs, and where, as Dan O'Connell well knows, nothing is to be got in the way of homage or influence, or even a fair share of power, without agitation."[4] Lyell expressed a typical sentiment: learning how to speak effectively in public was a necessary evil if science was to prosper and if society was to be changed according to science's light.

The five dramatis personae in this book—John Tyndall, Thomas Henry Huxley, Richard Proctor, Alfred Russel Wallace and Henry Drummond—

were all keenly aware of the difficulties as well as the importance of speaking in public about science. They all shared a commitment to using their voices to lobby for science in public and were convinced that this was a critical task if science was to garner public support and prosper. They were all also convinced that live speech was a vital medium for persuading the public of science's moral relevance and cultural significance in a period when traditional beliefs were increasingly being challenged and transformed. All five, armed with these convictions, embarked on lecture tours of the United States. Tyndall was first, in 1872–1873 and Drummond last in 1893. Tyndall articulated the commitment of all five when he described the aim of his own tour in a letter to Bence Jones, secretary of the Royal Institution: "I want to do much more than make experiments before the American people. I want to impress upon them the means by which science flourishes, and indeed to cast a certain amount of discredit upon the very work which they have called upon me to do. I want them and England to feel in their heart of hearts the importance of the original investigator and this requires work and thought of a totally different kind from that brought into play in the arrangement of experiment." [5]

Tyndall was not going to America primarily to inform his audiences about the physics of light (though that was his chosen subject) or to dazzle them with dramatic experiments (he relied more on diagrams projected by limelight than live demonstrations). Like the other touring lecturers, Tyndall doubted whether the matter of public science lectures was ever really comprehended by audiences, never mind lodged in their memories. He did, however, believe that through his live lectures he could powerfully connect with their "heart of hearts" and transform their attitude toward science. He wanted to move audiences and to embody for them the importance, moral as much anything else, of science.

All five lecturers vigorously pursued broadly the same agenda. The interest each took in understanding and developing the power of the living voice to effect cultural change and the considerable fuss and attention their tours generated have left rich seams of historical documentation for exploring how "the voice of science" took form, was heard and misheard, and was copied and countered in an age drenched with public speech. This book explores both the performance and the often ambivalent, controversial and contested meanings of speech about science in the late nineteenth century. It is not designed to be a standard history of science lectures. Neither does it seek to provide a systematic and detailed analysis of the composition of particular audiences or the private reactions of individual auditors who kept diaries or wrote letters about lectures attended, valuable though such studies are. Instead, it contextualizes science lectures of a particular kind—those designed to attract the levels of public interest at least sufficient for a successful tour—within the lively and agonistic field of Anglo-American lecture culture. This was science voiced in public, applauded in lecture halls, defended and denounced in pulpits, copiously recorded and reacted to in newspapers and celebrated or scorned in periodicals. All of this and more made up the manufactory of the public voice of science.

The germination and growth of this book owe much to others. It was my colleague David Livingstone's observation, made in passing, that it was surely noteworthy that some of the major controversies in science and religion in the nineteenth century were strongly associated with speech events that first hooked me into the subject. Wilberforce versus Huxley, Tyndall's Belfast address, heresy trials and more all generated, in David's term, "flashpoints" that sparked intense conflict and controversy and generated lasting myths about triumphant science or recalcitrant religion. This encouraged me to try out the thought that there was something about live public speech that raised the cultural temperature, ignited and inflamed preexisting debates and attracted intense public interest. This was never meant to imply that speech was a sufficient or necessary cause for producing any of these things per se. But further investigation suggested that the importance of live talk in several significant instances had been overlooked. I am grateful to responsive audiences in Edinburgh, London and Exeter for allowing me to talk about some early attempts to work this argument out. I have appreciated, too, the anonymous readers and editors (in particular Joe Kember, Simon Naylor, John Plunkett and Charles Withers) of the printed essays based on those early efforts. Martin Hewitt's probing work on "platform culture" in Victorian Britain altered how I approached, and contextualized, science lectures. The superb accounts of lecture culture and political speech in the nineteenth and early twentieth centuries by Tom F. Wright and Jeremy C. Young provided further inspiration and new perspectives to work with.

As the book project emerged and took shape, several colleagues and friends read chapters, offering advice and probing questions. In particular, I would like to thank Nuala Johnson, Oliver Dunnett, David Livingstone and Juliana Adelman, along with my former doctoral students Max Meulendijks and Emma Swain, for reading chapters and providing encouragement. Ciaran Toal, another former PhD student, made his own instructive forays into this territory as his excellent papers in the *British Journal for the History of Science*, *Endeavour* and *Isis* attest. Colm Lavery, Tanya O'Sullivan, William Ward and Mark Wood all offered further inspiration along the way through their own doctoral research on subjects sharing a close family resemblance to what I've tried to do here. The three anonymous reviewers who read, and reread, the entire manuscript also deserve my warmest thanks. The often maligned "reviewer 2" in particular provided detailed, critical feedback, along with information about the precise whereabouts of overlooked archival material. I am most grateful for the time they took to do this.

There are some debts that just keep on accumulating and here I must highlight once again the preternatural ability of Bernie Lightman to see book projects of many kinds through to final completion. Bernie's productivity never ceases to astonish and is evidenced by a career of outstanding service and superlative scholarship. I've leaned heavily on both in writing this book. Bernie's invitation to edit, along with the brilliant Sir Roland Jackson and Nanna Kaalund, volume 7 of the John Tyndall Correspondence Project also helped keep me in the right groove. Among other things, it provided a welcome distraction from the burdens of university administration. Roland and Nanna shared their own expertise, often gently correcting the free-floating speculations of a cultural-historical geographer. Michael Barton, also of Tyndall Project fame, deserves mention here for generously sharing with me his own perceptive interpretation of Tyndall's North American tour.

No book is remotely possible, either, without the dedication of editors, archivists and illustrators. I owe particular thanks to Eleanor Gillers of the New York Historical Society, Kath Stevenson in Special Collections at Queen's University Belfast and the National Library of Scotland's Imaging Services team. I am also indebted to Libby Mulqueeny for her painstaking efforts to reconstruct and map Richard Proctor's second tour. And what can anyone say to properly acknowledge the hard work and enthusiasm of Abby Collier, editorial director at UPP? I suspect I barely know less than a tenth of the efforts she has made to see this and other work that has passed over her editorial desk into print. The scholarly community owes much to her outstanding commitment to academic publishing.

I have witnessed in a much closer way, of course, the constant support, patience and love of my family. When all books are done, academic reputations made or unmade, what remains is something much more important. My wife Susie and my three children—Lydia, Adam and Liam—are a constant and precious reminder of this. And they'll be as quick as any to laugh at these rather pious sentiments. Last but certainly not least, this book is dedicated to my parents. Large sections were written in my father's study. Mum's soup and wheaten bread fueled paragraphs written in the afternoons. These were reminders of much larger sacrifices. This dedication is a small token of my heartfelt thanks.

THE VOICE OF SCIENCE

INTRODUCTION

SCIENCE LECTURES IN AN AGE OF ORATORY

Writing a little over a year before his death in 1895, Thomas Henry Huxley offered his readers reflections on the value of the popular lecture. This form of communication had, over the course of his career, "taxed such scientific and literary faculty as I possessed to the uttermost." Yet, as Huxley confessed, this enormous expenditure of intellectual energy had, from one point of view, largely failed. It was doubtful whether "more than one in ten of an average audience carries away an accurate notion of what the speaker was driving at." Why bother, then, clearing one's throat to address an audience? Huxley explained that the worth of public speaking lay not in imparting knowledge but in its power, altogether independent of the "intellectual worth" of what was spoken, to form and influence the attitudes and affections of the audience. It was on this account that those committed to speaking to the public were in fact "wise in their generation" and "justified by results."[1]

For Huxley, it was the affective power of speech that, in large measure, made the considerable cost of performing popular science lectures over several decades worthwhile. In delivering public lectures, Huxley had, as he put it, "turned to account the peculiarities of human nature" and powerfully "awaken[ed] . . . a sympathy for abstract truth."[2] Scientific instruction had for him been a secondary concern, a still vital task reserved for the classroom and for textbooks. Neither was providing some form of light, but fleeting, scientific entertainment the primary aim. The purpose of lecturing had been much more serious and profound. It had, as Huxley argued earlier, been fundamentally religious in its aims and effects, and beyond "mere science."[3] Science by its nature, after all, could not "stir the passions." The science lecturer, on the other hand, could and should. As a result, the public, made more sympathetic to science, would be inclined to act according to its wise counsel and resist the siren voices falsely claiming access to truth and public authority. Religious *feeling* would change and no longer be beholden to hoary dogma and superstitious beliefs. The lecturer was best placed to lead this revolution in moral and religious culture, as Huxley's career had shown.

This grand vision of what could be achieved by making use of the "living voice" may sound strange to twenty-first-century ears. Yet Huxley's appraisal of the power of speech motivated and directed the efforts of many prominent science communicators in the last decades of the nineteenth century. These efforts emerged in an era when public culture was saturated with speech of many kinds and in the service of many causes. Indeed, it was a period almost obsessively attentive to spoken address. Motivated by these kinds of observations, the argument of this book is that popular science lectures, understood in performative terms and as thoroughly embedded in a wider lecture culture, were a crucial means for shaping and extending the public authority, affective power and cultural meanings of science. In keeping with this, the book seeks to demonstrate that live lectures, and their afterlife in printed and platform responses, had a reach and impact that have not been sufficiently appreciated. Following five celebrity British scientists who toured the United States in the final three decades of the nineteenth century allows for a detailed account of spoken performances, their copious remediation into print and the energetic responses that frequently followed from them. It also provides an opportunity to explore the significance invested in lectures by those with a stake in securing and reinforcing the cultural authority and moral meaning of science in the late nineteenth century.

In following this line of thought, the book seeks to examine how popular science lectures were understood as a means to transform the moral ordering and religious aspirations of society. It does not pretend to offer a comprehensive overview. Rather, in selecting five popular lecturers (John Tyndall, Thomas Henry Huxley, Richard Proctor, Alfred Russel Wallace and Henry Drummond) and a particular context (the lecture circuit in Gilded Age America) it aims to recast and recalibrate the cultural significance of popular science lectures in this period. The five lecturers investigated in this book were all household names in their day. Their tours attracted profuse press coverage and commentary and this, along with a relative abundance of autobiographical and other archival material that records their own reflections on what it means to lecture well on science, helps to justify the selection. They were chosen, in other words, not because of their scientific stature (though hard to measure, there is clear variability) or because lecturing was a major part of their scientific career (this too varies significantly between speakers) or because they were necessarily talented lecturers (some arguably were, others perhaps not). Rather it is because they garnered more public and press attention than any other visiting lecturers who dealt with science on the American circuit. Even in the case of Wallace, who struggled to gain a popular hearing, the attention heaped on his tour was significantly greater than that enjoyed by the vast bulk of other visiting science lecturers. This fact allows the "voice of science," remediated through various sources, to be thoroughly reconstructed, contextualized and interpreted. The five lecturers, and their lecture tours, are especially useful for placing science lecturing within a wider culture of public speaking in a way that parallels and interacts with recent work on the coproduction of science and print culture in roughly the same period.[4] In addition all five, to a greater or lesser extent, have been the subject of close scholarly attention as individual scientists and lecturers.

Tracking the tours of the five figures that lie at the heart of this book makes possible a fine-grained account of their backgrounds, their American performances and the remediation of those communicative events into mass print. The latter is understood both as a transformative cultural process and as a vital part of the record and reception of the lectures themselves. This focus means that newspaper accounts are of central importance and are approached not only as a form of evidence to reconstruct the lecture tours but also as an integral dimension of lecture culture and, more generally, of the making and contesting of science's cultural meanings. Press reports are central here rather than incidental, and the book takes seriously the arguments made by others that newspapers—encouraged by technological, economic and political changes—had become a dominant and defining feature of public culture in both Britain and America.[5] Driven by commercialization and the quest for expanding readerships, newspapers helped fuel the growth of a mass media and transformed the dynamics and characteristic features of both political and more general society. This did not mean that oratorical culture was increasingly drowned out by a torrent of typeset words. On the contrary, the growth of a printed mass media facilitated the popularity, cultural prominence and commodification of public speech. At the same time, it is important to keep in mind that the "power of the living voice," insofar as it can be re-heard from an age before widespread voice recordings, was widely regarded as providing something more than could be easily captured in print or replaced by it. The popular lecture was, as Donald Scott noted, a "complex form of display," and its appeal was due in part to its "particular character . . . as a public ritual" and a "dramaturgical event."[6]

The performative dimensions of lectures are taken here to be of central importance, precisely because they were regarded as such by speakers, listeners and reporters at the time. The science lecture is approached, then, as a matter of vocal and embodied performance, as well as a part of associated cultural practices underpinned by assumptions about the power of public address. This is not to turn aside from the visual, but instead to explore how the visual and the vocal, or what was seen *and* heard, worked together to produce performances designed to provoke and direct cultural transformation in the name of science. Turning first to an account of the significance of public speech in the nineteenth century will help demonstrate why the popular science lecture deserves to be explored as something much more than a fleeting performance of little cultural moment.

A RISING VOLUME OF SPEECH

Throughout the nineteenth century, the power of speech to express and direct social and cultural change was widely affirmed. If anything, a fascination with public speech intensified in the closing decades of the century, not least because of the democratization of politics and public life. As Joseph Meisel has observed, by the 1880s the mass production and consumption of public speech had reached a zenith.[7] Lectures, sermons, political speech making, legal disputation and much else besides resounded widely and deeply, not only in their immediate context of delivery but also through the remediation of speech into print. Orality, far from fading in an age of mass print, was revivified and reinvented, creating a complex ecology of formalized talk. One contemporary commentator captured something of this in observing that the lecture had become a "thaumaturgic [wonder-working] agency," preternaturally driving cultural and political progress.[8]

The flooding of the public sphere with a cacophony of passionate voices was also true in the United States. Here, even more than in Britain, the lecture rapidly became a vital mode of spoken address. The lyceum movement, initially patterned on popular education initiatives in Britain (including mechanics' institutes), expanded rapidly in the early nineteenth century and became increasingly concerned with the organization of series of talks, enshrining the public lecture as a key agent of cultural expression and change.[9] From the 1840s onward, the lyceum movement became steadily indistinguishable from a more diffuse and diverse lecture culture that flourished first in the northeast before expanding to western towns and cities and, in the Reconstruction era, into the South. This growth was accompanied by diversification, commercialization and the rise of touring celebrity lecturers, often paid handsomely for their peripatetic lecturing. Lecture culture was increasingly scaled up, and the lecture tour, buoyed by an expanding and syndicated press, helped to sustain national and transnational discussions about topics of cultural and political significance. As Carolyn Eastman aptly describes it, by the final three decades of the nineteenth century, American platform culture "functioned as an explosive and innovative site for performance, criticism, deliberation, debate and the embodiment of ideas."[10] This was despite the fact that the American lecture system was widely regarded as being in decline from a previous golden age of public speech. Indeed, if the early years of the lyceum movement were looked upon as a period when instruction was properly prioritized over frivolous entertainment, in 1868 one commentator observed that too soon after the lyceum movement was founded the "scholar receded from sight and the impassioned orator took his place."[11] The emphasis on solid and edifying instruction was purportedly endangered by the rise of lecture bureaus, star celebrity speakers and high fees that became more common from the 1850s. Accounts alert to the possible distortive effects of this narrative of constant decline note the prominent role given to entertainment and emotion in the earlier period and later attempts to market lectures as a serious but stimulating alternative to less improving pastimes. Others place the point of decline in the 1880s, when lecture culture, by then apparently beholden to the demand for entertainment, could no longer compete with other staged forms of public excitement. Still, just as it was assumed to have degenerated or been superseded by print, formal education or other kinds of public entertainment, the lecture system was reinvented in new forms and according to new organizational arrangements. The 1880s witnessed new varieties of lecture culture emerge and achieve public prominence. It was at this point that the Chautauqua movement appeared and rapidly expanded, and lecture culture more generally, while perhaps fragmenting in certain respects, became yet more diverse and more vital in political and other terms. Arguably, the lecture tour by foreign visitors was only then reaching something of a high point, at least in terms of public visibility and national prominence.

The dramatic resurgence of orality in public life right across the nineteenth century, propelled and paralleled by the rapid growth in news media, made the lecture platform a potent and near omnipresent feature of American culture.[12] This was coupled with the ongoing development of a shift in the arts of public address that had begun much earlier. This change, sometimes referred to as the "elocutionary revolution," reconceived the effectiveness of public speech in terms of bodily performance and affective power. In her account of this trend, Sandra Gustafson notes how "gesture, facial expression and vocal tone" became "primary bearers of meaning and fundamental tools of persuasion."[13] With the rapid growth of print, the oral transmission of information was no longer the only way to communicate knowledge beyond a small circle of learned elites. This encouraged lecturers to reimagine their trade as an artful performance of individual creative "genius" rather than a learned recital of arcane knowledge.[14] An emphasis on embodied communication became critical for the emergence not only of the literary celebrity but also of the celebrity scientist. The lecture was reconfigured both as a form of speech that involved the public expression of emotions and as a vital technology of celebrity culture.

This trend has been noticed by other scholars interested in the political and cultural significance of oratory in the later decades of the nineteenth century. Jeremy Young, for example, has argued that by the 1870s American oratorical culture was thoroughly infused with the practices and assumptions that had marked the elocutionary revolution. In his study of religious and political orators in late nineteenth- and early twentieth-century America, Young shows that cultural leaders developed "a unique brand of emotional public speaking" that helped to dramatize political and cultural visions on a national stage.[15] It was a style of speechmaking that conveyed an "emotional availability" and deployed "sacralized language" to garner mass support for a variety of religious or political causes and to forge new identities for those struggling with the impact of social upheavals.[16] Young identifies the quest for emotionally persuasive speech as a search for a particular kind of "charisma" and argues that its roots lay in methods of address developed in the early nineteenth century and propagated through the subsequent decades via elocution manuals and private lessons in techniques of public speaking.

This same development influenced oratory in Britain in comparable as well as contrasting ways. Josephine Hoegarts, for example, has argued that vocal performance in parliament moved from being dominated by an "aristocratic theatricality" to a more constrained, conversational and demotic style of address. Emotion was not excluded, but it was more carefully managed and differently pitched. Hoegarts points out, however, that this had the paradoxical effect of reinforcing "the perceived connection between a . . . speaker's personality and the sound of his voice."[17] The embodied character of a speaker, even if more subtly expressed, did not diminish in importance. On the contrary, close attention to the physical aspects of speechmaking became more significant, not least because of increased press attention to the vocal performance of politicians at Westminster. The Liberal MP John Bright was a leading exemplar of this new style of political speech, studiously avoiding dramatic gestures and only modulating his voice slightly around a "lower G in the tenor clef."[18] Yet, Bright still managed to employ sufficient physical action and modulation to powerfully convey emotion, manliness and authenticity without risking charges of eccentricity or artificiality.

One consequence of these changes was increased attention to the embodied character of public speech performances. As Amanda Adams notes, the many mid- and late nineteenth-century accounts of lectures by celebrity speakers on both sides of the Atlantic "are at times preoccupied by *nondiscursive* elements of performance," including the actions and reactions of the lecturer's body.[19] Thomas Augst, in a careful analysis of the lectures of Ralph Waldo Emerson, confirms this point, observing that "literary discernment in the nineteenth-century lecture hall concerned itself with incantatory power, with being moved and with hearing what lay beyond 'mere words.'"[20] This fascination with the experience of listening and observing in the lecture hall was of central importance to assessments of the celebrity scientists who take central stage in the chapters to come, even if their bodies functioned in ways that were frequently contrasted to the meaning of corporeal actions deployed by other professional orators. A common refrain in journalistic depictions of speakers visiting from Britain was their relative lack of oratorical flair (it is not surprising that John Bright's style was taken as a model by at least some British lecturers). This was certainly the case for the lecturers on science who appear in this book. Even so, the nonverbal aspects of their performances were closely observed, dissected, criticized or, less commonly, commended and were often compared with those of accomplished homegrown platform performers. This broad trend toward the affective and performative aspects of public speech, however differently expressed in different places, has not been much noticed in the existing literature on popular science lectures.[21] As a result of this neglect, the related and deeply held conviction that lectures could dramatically reorient public "sympathies" toward science and its cultural import has often been obscured.[22]

A concern with deportment and vocal performance had, in more strictly delimited scientific circles, its own longer history. In the early and midnineteenth century, Michael Faraday was the most prominent example of a science lecturer who exhibited a conscious and determined commitment to the power of spoken address to create a deeper sympathy for science, and the moral practices and meanings that accompanied it, among a wider public. Faraday's celebrated career at London's Royal Institution involved, along with extensive experimental work, delivering popular lectures to a range of audiences over several decades. Faraday took the task of lecturing with utter seriousness. Early in his career, he attended classes on elocution and then took lessons from the teacher Benjamin Humphrey Smart. Smart was also invited to attend Faraday's lectures to provide feedback on his delivery. It was through Smart's influence, and by way of the example of his scientific mentor Humphry Davy, that Faraday develop his celebrated style of lecturing. Although Faraday studiously avoided the flamboyance of his mentor and avoided excessive displays of emotion, his lectures were designed to move his audiences. As one admirer observed, at the close of his lecture, "Faraday's enthusiasm sometimes carried him to the point of ecstasy. . . . His light, lithe body seemed to quiver with eager life. His audience took fire with him, and every face was flushed."[23] If that commentary tells us as much about the observer as it does about Faraday and his performance, it nevertheless captures a common conviction about the power of speech to deeply move both the speaker and the hearer. Faraday's career can be understood as a concerted effort to develop a prosody that, while suitably scientific, was carefully staged to resonate with, and develop, his audience's more noble affections.

By the time of his death in 1867, Faraday was widely regarded as the prince of science lecturers. Even if his scrupulous attention to the science of elocution was not followed by others, his recognized ability to command an audience's sympathy and absorb their attention was held forth as an ideal. This, along with his skill in the art of live experimental demonstration, was frequently commented upon. In keeping with wider trends within oratorical culture on both sides of the Atlantic, it was often less Faraday's capacity to impart knowledge or understanding that was remembered than the feelings that were stirred by his overall performance. Those asked to recall the power of his lectures routinely described his arresting presence, the emotional power of his speech and his ability to impart a lasting love of science. Ultimately, Faraday helped to make science resonate in a growing and intense market of passionate and appealing public speech. The importance of science to human flourishing was a deeply emotive subject. Successful science lecturers in the nineteenth century knew this and, often out of sheer necessity and in the face of considerable cultural pressure, regarded the lecture as the key mode of provoking and harnessing the power of public feeling.[24]

Science, of course, was only one topic among many that piqued the interest and stirred the passions of a lecture-loving public. As Martin Hewitt has suggested, while the science lecture dominated platform culture in early nineteenth-century Britain, its share of a growing lecture market shrank considerably as the century progressed. This paralleled a shift, recently explored by James Secord, in the relative cultural importance of scientific conversation.[25] These trends had consequences for the science lecturer seeking a successful career as a public speaker. If, in the earlier period, the spectacle of experimental demonstration was often sufficient to draw crowds, in the later period its appeal lessened. As lecture circuits became more extensive and marked by a greater diversity of approaches, voices and subjects, the science lecturer had to consider more carefully what Hewitt has aptly called the "spectacle of words" alongside the successful use of visual technology and live demonstrations.[26] Audiences were becoming increasingly familiar with, and often enraptured by, carefully crafted and delivered talks that relied much less than previously on visual effects to achieve the desired response. This is not to suggest, especially for the science lecturer, that visual aids or experimental demonstrations became unimportant. On the contrary, they retained for many a centrality that spoke of their continuing relevance. Nevertheless, word craft, vocal delivery and bodily comportment, if anything, increased in significance and certainly could not be ignored on either side of the Atlantic by those wanting success and fame on the lecture circuit.

Setting science lectures within the context of the norms and values invested in public speech draws attention to a competitive field that the lecturer on scientific subjects had to negotiate. But it is also important to acknowledge that each speech event, each lecture given, had its more local coordinates and proximate challenges. In addition to being understood as performances measured against, and dependent on, a set of expectations and aspirations around effective speech acts, science lectures in the nineteenth century can also be approached as events and spoken arguments shaped by more immediate considerations. The general importance of concerns with deportment and vocal performance certainly mattered, but so too did the relations between content, conduct and local context. Thinking in this vein, David Livingstone has drawn attention to the role of local "spaces of speech" in both constraining and enabling scientific argumentation and its reception.[27] Precisely where science was communicated in spoken form could profoundly shape what was said, how it was said and how it was heard. Lecturers, in attempting to harness or subvert particular protocols, controversies or expectations, often helped to reproduce and reinforce them. This constraining aspect of how lecture events generated certain kinds of meaning or responses cannot be followed through with the level of detail that might be possible if we were dealing with only a few performances or cultural locations. Even so, pausing longer in places where local disputes, reactions or concerns most evidently impinged on how a lecture was advertised, enacted and responded to points to the importance of geography, or cultural location, alongside a more general set of expectations and norms that informed the conduct of speakers and audiences.

Place, then, certainly mattered. But so too did the often-rapid dissemination of a speech event across a much larger swathe of cultural territory. To take a cue from Erving Goffman's reflections on formal verbal communication, there was something about the drama and ritual of the lectures that helped them move beyond a local platform to a much wider stage.[28] It is this, perhaps more than anything else, that helps explain why a number of events most associated with controversy over science and religion in the nineteenth century were first of all speech acts.[29] Thomas Henry Huxley's 1860 exchange with Bishop Samuel Wilberforce in Oxford and John Tyndall's Belfast address of 1874 quickly became lightning rods for generating and defining controversy about the authority invested in scientific and religious institutions and ideas. The power to provoke resided, at least in part, with the medium in which the provocations were made. The unstable character of speech, the difficulties in reporting exactly what was said and how it was said, all added to the allure, fascination and material for myth making. The appearance of spontaneity, the "once only character of the occasion," the exercise of institutional and social status, and the drama and emotional ambience could all facilitate a sense of the momentous.[30]

The power of speech events in general, and of lectures in particular, to gain cultural traction and public interest was not, of course, only about the magnetism of the speaker or the timing of an event. While this book singles out individual lecturers for close scrutiny, the purpose is not hagiographic or primarily biographical. Through focusing on individuals who achieved fame on the lecture circuit, it is possible to recover, at least to some degree, the lecture tour as a collective, as well as contested and multivalent, accomplishment made possible by the complex machinery necessary for consequential public speech. Just as book historians have drawn attention to the many agents, instruments, regulatory regimes and processes behind the production, promotion and consumption of published texts, so too studying the world of nineteenth-century lecturing requires attention to the complex infrastructure of lecture culture. As with print culture, platform culture required a "communications circuit" that could include speakers, hall managers, lecture agents, limelight operators and demonstrators, auditors, stenographers, civic leaders and members of learned societies or lyceums, among others.[31] More particularly, the complex circuitry of lecture culture was thoroughly enmeshed and invested in newspaper and print culture. As already noted, the power of the living voice to evoke the cultural and moral significance of science relied upon, and was in part constituted through, the operations and organizations behind an expanding and powerful fourth estate.

REMEDIATIONS INTO PRINT

Much of the vitality of the lecture tour in Gilded Age America, as elsewhere, relied on the flourishing and dynamic world of newspapers. There was, increasingly, a powerful symbiosis between platform and press, and newspaper editors saw the lecture as a powerful ally in sustaining and expanding sales. The immediacy of lectures, their trade in fresh talk, made them ideal material for transforming knowledge, and live speech, into news. The cult of personality, the rising culture of celebrity and the importance of publicity further reinforced lectures as worthy of close newspaper attention.[32] Lectures and speakers benefitted (as well as suffered) from this partnership, with newspapers acting as a crucial advertising agency both for talks and their authors.

Newspaper reports of the lectures themselves proved to be of great importance. This was captured well by a typically telling anecdote relayed by the American lecture agent James Pond. In his turn-of-the-century account of the American attorney and celebrated orator Wendell Phillips, Pond remembers how, during an address that could barely be heard because of a particularly raucous audience, the acclaimed advocate of abolitionism pointed to the row of reporters in front of him and instructed his hearers to "go on, gentlemen, go on. I do not need your ears. Through these pencils I speak to thirty millions [*sic*] of people."[33] Of course, the pencils of reporters and stenographers did not simply disseminate the content of what was said. One of the distinctive features of American newspaper reports of lectures by visiting celebrities was their attention not just to what was said but also to how it was said. This interest in the performative dimensions of platform appearances tracked a wider interest in lectures both as a vehicle for the expression of public emotion and as a vital component of civic culture. The science lecturer had to reckon with these realities. Even if lecturers on science wished to display a different kind of platform persona to the kind that most often attracted fevered attention, this was something that required a careful attention to deportment and delivery. There was a need to avoid a delivery so dry as to be written off as dull and dreary without being accused of a lack of restraint that betrayed the high ideals of scientific objectivity and self-effacement. This was a delicate task. But it was also critical. Often, there was a mutuality at work, with lecturers carefully combing newspaper accounts of their own performances and using them to adjust and improve their delivery.

Along with advertisements and copious records of lectures, the emergence of the interview—of journalists meeting with prominent public figures to record informal exchanges on topical issues of the day—also helped expand and cement the press-platform nexus.[34] It was a distinctively American invention and one that visiting speakers could either use to their own advantage or decry as an invasion of privacy or preparation time. Oscar Wilde, who as a young man toured the United States in 1882, provides the exemplar of a visiting speaker who quickly became a master of the art of the newspaper interview. During his tour, Wilde was interviewed by over a hundred journalists and quickly adapted the form to his own advantage. The interviews soon became carefully staged by Wilde and functioned as a platform arguably more important than his lectures.[35] In many respects, the American interview was the leading technology behind the minting of Wilde's celebrity. While this was not true to the same degree of other lecturers, the interview nevertheless became an important component of a lecture tour, often referring to things said, or not said, from the lecture platform and provoking controversies that later colored both how lectures were heard and the reactions they provoked.

The central role played by newspapers raises, of course, the question of using them as a source material for reconstructing the live talk and total performance of a lecture. On the one hand, the scrupulous (or sometimes unscrupulous) attention paid to lectures by journalists and stenographers makes newspaper reports a profoundly rich source for retrieving important aspects of live talk that otherwise leave little textual trace, including audience response both during and after the event. On the other hand, newspaper accounts could be deeply self-interested, turning lecture (and audience) performances to ends that routinely infuriated speakers and tour organizers. Inaccurate transcriptions abounded, either through sloppy recording or, less frequently, deliberate misquotation. The conduct as much as the content of a lecture event could be reported in dramatically contrasting and contradictory ways. Feuds broke out between newspapers over the cultural import of a particular talk. The recoding of lectures from performance to print was a highly political and precarious process. As such, newspaper accounts can be approached as a particularly significant kind of response to the lectures as much as descriptive reports.

It would be wrong, however, to dismiss the process of remediation, as was so often done by press-wary lecturers, as always and everywhere blatantly distortive. The reports, and reporters, were organically part of the performance; as such, the translation into print can be regarded (at least partly) as much constitutive of a given lecture as an after-the-event representation of it. There is a need to register the seriousness with which stenographers went about their task and to take the opportunity to triangulate between different accounts of the same event. This does not necessarily lead to a more "accurate" record, but it can help to determine the degree to which the message and meaning of a particular lecture was drastically disrupted or revised. Newspapers could be jealous of their reputation to provide exact reports of a particularly significant speech event. And stenographers also had reputations to guard. In some instances, the opportunity exists to calibrate the precision of a particular report against the written script used by a speaker. What emerges is the fact that the main features or arguments of a lecture were often reported with a significant degree of correctness, at least in terms of content. Moreover, it was not uncommon for newspaper editors to circumvent stenographers and simply print a preprepared script delivered to their offices by the speaker. There could also be a rough consensus among different journalists about the nonverbal features of an event—its atmosphere, audience response and the embodied character of a lecturer's performance. None of this is to deny what Tom Wright rightly calls the "productive flux" of meanings created by the newspaper renderings, reframings and textual placement of lecture reports.[36] But newspaper reports need not be thought of as *necessarily* any more accurate or distortive than the phonograph recordings that emerged toward the very end of the period considered here.

Whatever its varied effects, the central importance of newspaper involvement and interest in any lecture tour meant that visiting speakers often took pains to bring journalists onside. This, as we shall see, was certainly evident in the case of Thomas Henry Huxley who—in New York and in Nashville in particular—worked with leading newspapermen to ensure full and positive coverage of his lectures. This was a practice he had developed before he went to the United States (two years previously he had visited the offices of a Belfast newspaper late in the evening and spent two hours correcting reports of a lecture he had delivered), and he put it to good use when he arrived.[37] The cultivation of good relations with newspaper editors and newspaper readers was, if anything, yet more significant to Richard Proctor, who, as Joshua Nall has recently shown in his study of the astronomer's alliance with the new journalism of William Stead and others, quite consciously incorporated a journalistic and demotic style into his writing and speaking.[38]

In cases where newspapers editors could not be easily brought onside, a more negative approach could also be adopted. On occasion, Henry Drummond, for example, was known to request midflow that no record of his lecture be taken. For Drummond, the recording of the kinds of lecture-cum-sermons that made up part of his repertoire could change their character and detract from their positive influence over the spiritual well-being of his audience. He was also shrewd enough to know this tactic often raised rather than reduced curiosity about what he had to say. But as well as this, he was particularly sensitive to the fact that a record of his words could allow publishers to take advantage of still rather inadequate copyright laws and produce extremely rapidly full transcripts of lectures he had intended to revise and print in book form. On this issue, Drummond would have his day in court.

Drummond's approach points to another vital way in which the transfer of vocal performance into the printed word was an integral part of the lecture tour business. Many visiting speakers had their eye not only on increasing sales of existing works they had authored but also on exposing listeners to thoughts that would later be gathered into popular books. The potential financial benefits of lecture tours, particularly in the new and expansive markets for books found in the United States, went well beyond speaking fees. It is well to remember this when interpreting reports of profits made or donated by visiting speakers. There is no easy way to measure how much speakers made through increased book sales or through the production of works based on a successful series of lectures delivered in several US cities and widely reported in the press. That Drummond went to court so that tens of thousands of pirated copies of his US lectures would be pulped (and hurriedly prepared his lectures for official versions) points directly to the importance of this dimension of lecture tours. Indeed, benefitting in this way was often justified precisely because books written by visiting speakers had so often been pirated in the United States. The lecture tour might be regarded, among other things, as a form of pecuniary revenge.

In sum, then, the machinery of print and platform culture, and the interlinked trends in speaking, hearing and reporting—or what Tom F. Wright calls the "complicated dialectic of reciprocity and resistance between lecturers and reporters"—provide a vital context for understanding the science lecturers who make up the empirical mainstay of this book.[39] At the same time, of course, the tours had a longer history and built on trends and personalities that had come before. The longer history of British men of science speaking in America provides another, more concrete historical context in which to place the later tours of Tyndall, Huxley, Proctor, Wallace and Drummond.

BRITISH SCIENCE LECTURERS IN ANTEBELLUM AMERICA

One place to begin an account of the travels, travails and triumphs of British men of science who embarked on tours in Gilded Age America is with the institution that became, for a number of key figures, a platform not just for a lecture series but also for publicizing and funding further speaking engagements. By many measures, the Lowell Institute in Boston led the field in financing the traffic of visiting speakers and attracting celebrity authors. The institute, funded by $250,000 left for that purpose by the Boston merchant John Lowell Jr., supported the first set of lecture series in 1839–1840 and continued throughout the nineteenth century to use the endowment to sponsor series by speakers of national and, increasingly, international reputation. Although the subject matter of the lectures ranged widely, in the early years at least there was a definite lean toward science. Another more unusual (but not unique) feature was that the lectures, though ticketed, were free of charge.[40]

For the first thirty years or so, the institute did not draw large numbers of foreign lecturers, though a number of those who did accept invitations, such as Louis Agassiz and Arnold Guyot, later became leading figures in American science. Between the commencement of the Lowell lectures in 1839 and 1867, only three British or Irish lecturers crossed the Atlantic to give them. Most famously, the geologist and man of letters Charles Lyell visited and lectured at the Lowell Institute on three separate occasions. The Irish botanist William Henry Harvey and the Scottish chemist James F. W. Johnston both spoke at the institute during the 1849–1850 season. It was not until 1867, when the classical scholar D'Arcy Wentworth Thompson gave a series of lectures on the philosophy of education, that another speaker from the United Kingdom stood before a Lowell Institute audience. Thompson, however, was the first of twenty-four lecturers from Great Britain and Ireland to speak at the Lowell Institute in the period from 1867 to 1898. Of these, the majority spoke on scientific topics, but only a smaller portion embarked on a subsequent lecture tour. Even fewer were able to generate significant levels of press attention or make their tour profitable.

Though relatively uncommon, lecture tours by prominent British men of science nevertheless made a significant impact on the American circuit. Charles Lyell's three speaking tours in 1841, 1845 and 1852 set a precedent and pattern for others that followed. In his first tour, the Lowell Institute provided the capital and the springboard to launch a successful tour of East Coast cities. If Boston newspapers did not cover Lyell's lectures in detail, his words appeared verbatim in the pages of the *New York Tribune* after he repeated some of them in New York. The general reception of his first tour was positive, though Lyell's poor speaking abilities were often singled out. As Robert Dott Jr. has argued, it was only the content (including the striking visual aids) of his lectures and his high reputation that overcame his hesitant speaking, weak voice and lack of "imposing . . . action."[41]

Lyell's first tour, did, however, generate a controversy with a lasting and distasteful legacy. As Robert Silliman has shown, a number of American geologists became increasingly concerned that Lyell was engaging in a form of intellectual piracy.[42] Gleaning geological information and ideas from local practitioners, Lyell could then, without due acknowledgment of sources, revise his leading works, produce new editions and make money from their sale. For Lyell, generally scrupulous about acknowledging his sources, the problem was precisely the opposite. It was the loosely regulated American print industry that was profiting from pirated editions of his works. His lectures, too, could be recorded by stenographers and published with alacrity. Intellectual property rights remained a problem for visiting speakers throughout the nineteenth century, one that came to a head when Henry Drummond took an American publisher to court for selling an unauthorized and fraudulent account of his Lowell lectures.

Other science lecturers from Britain, though not invited to give Lowell lecturers, nevertheless toured America with considerable success in the 1840s. The natural philosopher and astronomer Dionysius Lardner was among the most noteworthy. Beginning in November 1841, Lardner spent four years lecturing in towns and cities across the United States, reputedly earning an astonishing $200,000. Lardner started his tour in New York with a series of twelve lectures delivered in Clinton Hall. By the time he reached the midway point, the *New York Tribune* was printing his lectures verbatim on the front page. The paper later gathered them together with other lectures in the form of a book that went through several editions.[43]

As Jo N. Hays has argued, Lardner's lectures were not designed to provide fully comprehensive accounts of scientific developments.[44] Instead, Lardner emphasized character formation and social change. His stress on the practical utility and accessibility of science made his lectures popular beyond a learned elite. According to Hays, these features of Lardner's lecturing success were among the causes of his declining reputation after his American tours. Yet the centrality of character and social relevance to the purpose and power of science lectures was revived to great applause later in the century, making the disintegration of Lardner's always fragile social standing rather than his philosophy of lecturing the more important reason for his failing reputation.

The style of Lardner's lectures was also, by his own account, a vital ingredient in their success. Lardner quite deliberately used extempore address and paid close attention to deportment and vocal delivery. His lectures had a dramatic flair that he had developed earlier in his career through involvement with theater companies in London. Lardner, like many after him, aimed to move his audiences as much as instruct them. The aesthetic and affective appeal of his lectures was further enhanced through novel use of the magic lantern and other arresting visual aids.[45] This, of course, presaged the thinking and techniques of the British science lecturers who visited later in the century, even if their style was generally more understated and restrained.

The Scottish astronomer and author John Pringle Nichol followed in Lardner's footsteps when he traveled to the United States in November 1847. His lectures in Boston, New York, Cambridge and elsewhere drew large audiences and newspaper attention. It was, once again, in New York that his lectures received the most detailed treatment. The *New York Tribune*, employing the celebrated stenographer Oliver Dyer, recorded Nichol's lectures on astronomy in full and republished them as a pamphlet in the same way they had done for Lardner. The *Tribune*'s tribute to Nichol as a speaker noted his style was "at once chaste and impassioned, as eloquent and finished as it is rich and glowing." Like Lardner before him, Nichol had "elevat[ed] the hearts, and kindl[ed] the imaginations and the religious sensibilities of thousands."[46] This description would not have been out of place in reports of lectures by scientific figures who toured America over the next several decades, demonstrating the prevalence and persistence of the conviction that successful science lectures spoke as much to the heart as to the intellect and were often understood as a means to strengthen moral culture and stimulate self-improvement. What changed later in the century was the insistence that science itself provided the most reliable guide to a moral life.

In speaking in the United States, Lyell, Lardner and Nichol joined American science lecturers in benefitting from, and contributing significantly to, the rapid development and commercialization of American lecture culture. As Angela Ray has shown, by the 1840s the lyceum movement had shifted from a set of institutions dedicated to local learning through small-scale lectures, libraries and reading rooms to a more diffuse set of lecture societies that provided the infrastructure for an ever-expanding American circuit. The opportunities to make money through lecturing similarly multiplied, with the expansion of transport links, particularly the railroad, acting as a further accelerant. The three British men of science joined the ranks of American speakers in becoming star performers on that circuit. The American astronomer Ormsby MacKnight Mitchel, for example, drew similar numbers to equivalent applause and approbation, on occasion provoking audiences to "leap to their feet and cheer as at a sporting event."[47] Others, such as geologist William Barton Rogers and the chemist Benjamin Silliman Jr., were highly regarded platform performers.

Silliman in particular was instrumental in promoting science as a central topic for popular lectures in the 1830s and played a substantial role in the establishment and early success of the Lowell lectures.[48] Silliman's efforts to garner philanthropic and public support for science and science communication were not, however, unique. As Marlana Portolano has shown, the celebrated rhetorical skills of John Quincy Adams were not only put to use to cultivate public interest and support for astronomy. Adams's oratorical abilities proved critical in persuading Congress to use James Smithson's generous bequest to fund an institute dedicated to "pure" scientific research and the dissemination of science through a dedicated lecture program. It was fitting, then, that Joseph Henry, first secretary of the Smithsonian Institute, helped to design not just a program of talks by eminent men of science but also a lecture room that was hailed as a "triumph of acoustical science applied to public buildings." Shaped in the form of "an immense trumpet," it provided a model space for science lectures until it was destroyed by a fire in 1865.[49] This disastrous event was not a portent of the end of well-supported science lectures in America. Instead, it occurred on the cusp of dramatic growth in the popularity of science lectures. In the same year, for example, the Wagner Free Institute for Science opened its lecture theater, the design of which was inspired by the one destroyed at the Smithsonian.[50] Another dramatic development in the years that followed was the rise in the number and popularity of science lectures delivered by foreign, celebrity visitors. This boom time provides the more immediate historical setting for the five lecturers explored in the subsequent chapters of this book.

SCIENCE AND LECTURE CULTURE IN THE GILDED AGE

During the 1850s and 1860s, there was something of an understandable hiatus in terms of science lectures delivered by British speakers in more elite and established American venues and institutions. John Tyndall's tour of 1871–1872 was the most dramatic reentry of a scientific figure after the Civil War. From that point, British lecturers became a common feature on lyceum programs, particularly on the East Coast. Tyndall's visit came in the wake of Charles Dickens's dramatic reading tour in 1867, which marked the start of a new wave of celebrated speakers visiting America. Dickens's success has been described as "a catalyst behind the boom in such Transatlantic tours," in part because the celebrity author found it to be "golden campaigning ground."[51] Dickens was not, however, alone. The radical lecturer Henry Vincent was another star attraction on the American lecture circuit in the late 1860s. With a long-established reputation for stirring oratory and a public record of support for the North in the Civil War, Vincent captivated American audiences in northern cities in the early years of the Reconstruction era.

From the late 1860s, speakers crossed the Atlantic in greater numbers, trading on their reputation as celebrity authors or orators, or both. As Amanda Adams has argued, both British and American authors anxious to expand the market for their books and create a persona recognized by the masses capitalized on the opportunities afforded by lecture circuits on both sides of the Atlantic. Through public performances of novels by their authors and talks by literary celebrities, the lecture platform became, in Adams's description, "a central part of the international literary world" and one regulated by an "ideal meritocracy of personality."[52] Of course, it was not only novelists like Charles Dickens and Wilkie Collins who benefitted from this trade in spoken words.[53] Authors (and, perhaps more significantly, publishers) of all kinds quickly recognized the marketing opportunities opened up by an expanding transatlantic lecture culture.

The traffic of British speakers crossing the Atlantic was recorded in early fall each year through newspaper announcements of coming attractions. John Tyndall, for example, was placed among a group of British lecturers who included the dramatist Edmund Yates, historian and novelist Anthony Froude, novelist George Macdonald and women's rights activist and celebrated orator Emily Faithfull.[54] His lectures on the physics of light had to compete with presentations by leading literary and political subjects and figures. The presence of Emily Faithfull among the tranche of visiting lecturers demonstrated the predominance of male lecturers but also indicated a growing appetite for female voices on the American lecture circuit. This appearance of visiting female speakers was not a new development, but the relative novelty still attracted interest and often carried a strong political charge. Faithfull was part of a larger countermovement that challenged the ideology of the public sphere as a space for the exercise of exclusively male authority. A female voice, while still marginal in both British and American

lecture culture, had the capacity, not least in terms of its "vocality" or physical characteristics, to unsettle dominant norms, and, as a result, garner
considerable public and press interest.[55]

The visiting science lecturer had to be prepared, then, for comparisons
with other speakers on the circuit and competition for column space and
critical acclaim. This was not only secured through performance on the
platform. The general fuss and elaborate ritual accompanying a lecture tour
were also of vital importance. This was most vividly apparent in the tour
of the young Irish aesthete and playwright Oscar Wilde in 1882. Unlike
nearly all other visiting lecturers, Wilde was virtually unknown and unpublished. His remarkable rise to fame over the course of a year in the United
States was driven less by his lectures and more by the captured poses, press
interviews, impromptu (but highly staged) public appearances and carica-
tures that turned so many heads and produced such sensation. This was
no more evident than in the twenty-seven portraits of Wilde produced by
the New York photographer Napoleon Sarony. The only photographs taken
of Wilde during his one-year sojourn, they circulated widely, helping to
secure his public image and, as David Friedman has shown, dramatically
increased the desire to not only hear but also see him.[56]

The extent of the control Wilde had over his persona, image and message is a matter of contention among Wilde scholars. On Friedman's account, Wilde's "genius" lay in his ability to create and control his identity as
a celebrity. Others, most recently Michèle Mendelssohn, lean more toward
Wilde as the object rather than master of forces that conspired to make him
(in)famous.[57] Mendelssohn's revisionist account of Wilde's American tour
also stresses the loss of control of Wilde's image and reputation, and their
changing and conflicting meanings, as they circulated through different
cultural constituencies. His image, never a singular thing, splintered and
proliferated and was quickly pilfered for causes that distorted and derailed
the intentions of its original creators. The overlooked backdrop for Wilde's lecture tour was racial politics. Depictions of Wilde as a simian-like or
black "Paddy" did more than cramp the aesthete's famed style; it threatened
the central objective of his campaign. But wherever the emphasis should
lie—on Wilde as pawn or Wilde as self-made celebrity—his tour points to
the vital roles played by the accompanying apparatus of the speaking tour.
Staging a lecture tour meant much more than hiring halls and advertising
and delivering talks from the podium. Controlling and contesting the message and the reputation of the speaker on tour was critical.

The five scientists selected in this book for special attention mostly fared
well in the game of comparison that was so often played, at least in terms
of the level of attention given to them. This is one reason why they and
not others have been chosen for detailed investigation. Indeed, this accom-

TABLE I.1 BRITISH AND IRISH LOWELL INSTITUTE LECTURERS, 1841–1898

Lecturer	Year(s)	Subject
Charles Lyell	1841; 1845; 1852	Geology
William Henry Harvey	1849–1850	Cryptogamia
James Johnston	1849–1850	Agriculture
D'Arcy Wentworth Thompson	1868	Education
John Tyndall	1872	Light and heat
Richard Proctor	1873; 1875	Astronomy
Bonamy Price	1874	Currency and finance
John Turtle Wood	1875	The Great Temple of Diana
Archibald Geikie	1879	Geographical evolution
Lyon Playfair	1879	Arts and sciences/public health
W. Boyd Dawkins	1880	Primeval man
Edward A. Freeman	1881	The English people
James Bryce	1881	The Greek and Turkish East
William B. Carpenter	1882	Physical geography of the deep sea
Rev. J. G. Wood	1883	Structure of animal life
Robert S. Ball	1884	Modern astronomy
Edmund W. Gosse	1884	From Shakespeare to Pope
Rev. Hugh Reginald Haweis	1885–1886	Music and morals
Alfred Russel Wallace	1886–1887	Darwinism
James Geikie	1890–1891	The Ice Age
John Murray	1891–1892	Oceanography
Henry Drummond	1892–1893	The evolution of man
Edward Poulton	1893–1894	The colors of animals
Thomas William Rhys Davids	1894–1895	Buddhism
Lloyd Morgan	1895–1896	Habit and instinct
George H. Darwin	1897–1898	Tides
Michael Foster	1897–1898	Brain work

plishment should not be taken for granted, for it was hard won. It was quite possible to fail spectacularly on the US lecture circuit even if one arrived with a reputation as an attractive speaker or with great fame as a scientific author. One example of this was the successive tours of the science popularizer and Anglican clergyman John G. Wood. Although Wood's works on natural history had sold spectacularly well in America and the sketches he executed while he lectured were widely praised, his tours damaged both his bank account and his reputation. This was a view passed on privately to Alfred Russel Wallace by a critic of Wood. Wallace then blamed his own difficulties in securing lecture invitations on his having been billed, like Wood, as a great English naturalist.[58] Once Wallace reached America, there was apparently little appetite to hear another great English naturalist speak. Other visiting British scientists did not exactly fail, but either attracted much less attention or simply gave a handful of lectures in one or two places and thus did not really participate in anything like a speaking tour. For example, only a handful of the sixteen British scientists who gave Lowell lectures between 1867 and 1898 (for details see Table I.1) achieved more than regional coverage of their lectures and significantly capitalized on their otherwise high-profile Boston appearances.

The dramatis personae explored here have, of course, been selected over and against many other possibilities. Keeping within a British lecturing world (including, in this period, Ireland), one place to start would be to look at the invitees to the Lowell institute from its foundation to the end of the century. Table I.1 provides a list of British figures who gave Lowell lectures. About three quarters of these can aptly be described as scientists (in the contemporary American sense of the term). A number of them lectured only for the Lowell Institute, making their talks of less interest to this particular study. Others did use their Lowell lectures as a springboard for a wider speaking tour but attracted much less attention than the five examined in detail in this book. A few did make a go of it, including the Irish astronomer Robert Stawell Ball. Ball first lectured in America in 1884, after an invitation from the Lowell Institute and, employing the lecture agent James B. Pond, toured more extensively in 1901. Ball appears to have been successful in both financial and reputational terms.[59] In many respects, Ball entered a field left open by Richard Proctor's decision to quit the lecturing scene after his tour of 1879–1880. This followed a pattern set on the other side of the Atlantic. Ball had replaced Proctor as a Gilchrist Educational Trust lecturer in 1880.[60] The similarities between the two lecturers are not exhausted by their shared expertise in astronomy. Ball, like Proctor, was strategically circumspect about his own religious views, which tended toward agnosticism. But Ball's American lecture tours, however successful financially, did not provoke anything like the public commentary that Proctor had garnered in the 1870s. Ball may have attracted sizeable audiences, but the newspapers paid him only limited attention. In contrast, all five lecturers investigated here attracted significantly high levels of press interest and left a detailed record of all dimensions of lecturing in the period. Indeed, it is difficult to find an equivalent level of newspaper reporting for public science lectures in the nineteenth century.

The all-male cast is symptomatic of the marginality of female voices speaking explicitly about scientific topics on the elite and profit-making lecture circuit that traded on the celebrity status of foreign speakers. As suggested earlier, success on the American circuit was not foreclosed to visiting female speakers, but the science lecture, understood in contemporary terms, remained a strongly male preserve. Female voices had long had a place in American educational culture, though one strongly defined by gender norms related to audience, topic and style of address. In the early national period, female educationalists such as Almira Hart Lincoln Phelps offered oral science instruction, later transmitted in printed form, garnering large readerships.[61] The American astronomer Maria Mitchell, too, had occasionally lectured outside her classroom at Vassar College.[62] The commercialization of lecture culture and its rapid expansion in the postbellum period opened up many more opportunities for marginalized groups to participate in, and repurpose, public speaking. African American lecturers achieved prominence alongside female speakers and used the amplification that lecturing facilitated to champion various urgent social causes.[63] At the same time, the emphasis on "star lecturers," reputation and celebrity meant that to succeed as a science lecturer was in large part dependent on a well-established scientific or public reputation. It was extremely difficult for anyone other than a credentialed male scientist to receive invitations to deliver the kinds of high profile, and well-paid, lectures performed by John Tyndall and those who followed in his wake.

In part because it remained difficult for women to establish themselves as authorities on scientific matters, science lecturers continued to convey a strong commitment to rational discourse naturalized as male. In other spheres, there was a perception that an emphasis on rational instruction had been eclipsed by the importance of personality and embodied presence as the primary driver of celebrity status and attention.[64] In science, even though personality and embodied performance did count for much, the rhetoric of science as the (male) voice of reason rather than of emotion meant that female lecturers, negotiating with and attempting to subvert gender prescriptions, faced a struggle. It is not that female lecturers neglected scientific topics, but more often than not their public appearances were embedded in talks and events whose focus was on subjects—such as suffrage, temperance and public health—for which women were (increasingly) given a platform.[65] As late as 1895, when Richard Proctor's daughter Mary lectured to an audience in Worcester, Massachusetts, on planets, suns and infinite space, a journalist noted that "it was rare to hear a woman speak authoritatively on science, much less on such an abstruse and exacting science as astronomy."[66] It is noteworthy, too, that late in her career Mary advised women anxious to perform some "life's work" to "refus[e] to be drawn aside" by the diversions that the female sex were particularly prone to. Instead, like her, they should give themselves to it "absolutely" so that there was "nothing else to absorb [their] energy or take any of [their] thought or time."[67] Mary Proctor had to work extraordinarily hard to elicit the kind of praise and attention lavished on those male celebrity scientists who, in the previous two decades, had, to greater or lesser degrees, triumphed on the American circuit.

As well as being an all-male cast, the five scientists whose tours are followed here are also wholly British. In this sense, they are unrepresentative of the diversity of visiting science lecturers who were heard on the US lecture circuit during the Gilded Age. But it is difficult in that period to find lecturers from other countries who garnered anything like the same degree of public attention or sparked the same level of public discussion and controversy. One exception toward the end of the period considered here was the German physicist Hermann von Helmholtz. Helmholtz, as David Cahan has shown, was catapulted into stardom when he visited the United States in 1893. Like Huxley and Tyndall before him, Helmholtz became the darling of American scientists and educational elite optimistic about their country's scientific future. Here was a leading figure and "grandee" of German science, a product and champion of German education, who could help America to consolidate its position in international science.[68] The timing was auspicious, and the tour a triumph. It is worth noting, however, that unlike Tyndall, for example, Helmholtz's performances as a speaker were generally regarded as lackluster. This barely mattered, it seems. By the time of his visit, support for American science, and especially the kind modeled by Helmholtz, was well consolidated. The lecture tours of Tyndall and Huxley in particular, twenty years previously, had helped lay the necessary groundwork. Given his reputation, it was difficult for Helmholtz to do wrong in the eyes, and ears, of his supporters. In the end, however, his tour was more circumscribed and of less moment than those of his British counterparts in the years before.

Another key characteristic of the five lecturers at the empirical heart of this book is that they were all routinely described as scientists. This label, coined by the British philosopher and scientific statesman William Whewell in 1834, was rarely if ever used in Britain even by the 1890s. But in America the situation was different. There, it was already common currency when Tyndall arrived in 1872.[69] In attracting the level of newspaper attention they did, the tours of Tyndall and Huxley in the 1870s might even be taken as instrumental in securing the word's long-term use on both sides of the Atlantic. In speaking in the United States and attracting mass attention, Tyndall, Huxley and the others who followed helped to forge the public identity of the scientist and shape the cultural meanings and deportment of science as a form of knowledge and a way of life.

There was one final uniting and, I would argue, crucial feature of the five lecturers at the heart of this book that deserves fuller comment. All five dedicated themselves to using their lectures to give science, and a scientific age, a transcendental meaning and referent. If this is most obvious in the case of Wallace and Drummond, recent work on the religious tenor and intent of scientific naturalists like Tyndall and Huxley (a group that Proctor, somewhat loosely, eventually aligned himself to) points to its relevance in their case as well.[70] The final section of this introduction explores some of the reasons why this aspect of the lectures was such a critical ingredient for success.

SCIENCE FOR THE PRESENT

When the American historian Moses Coit Tyler surveyed Britain's "lecturing system" in 1869, he diagnosed a fatal flaw. In his view, British lecture culture suffered from an overbearing imposition of "neutrality" on lecture programs. This had led to "discourse upon all subjects, except those which men are most interested in, those vast fascinating problems of political and religious thought." Whether or not Tyler's diagnosis was correct, he had put his finger on a reason for popular support of lecture culture in the United States. Instead of lectures on "sapless shavings pared off from the dead trunk of the past," the American circuit offered discourses enlivened by the "throbbing political and religious interests of the Present." Without this essential aspect, lectures would become merely "instructive," generating an appetite not for noble thoughts, whether from philosophy, art or science, but for "musicians and buffoons."[71] This view was shared by one of the key actors in Gilded Age lecture culture, James Redpath. As director of the leading lyceum bureau, established in 1868 to organize tours for notable American and visiting speakers, Redpath's view was that any local committee or lyceum "afraid of political or other living questions are sooner or later consigned to bankruptcy."[72]

Tyler's and Redpath's appraisal of American lecture culture sits uneasily with the regulative ideal of neutrality that was commonly appealed to on both sides of the Atlantic. Unlike parliament or a church, the lecture hall was designed to be nonpartisan, a liberal space for free expression of opinion on subjects that united rather than divided a community of listeners. Science increasingly could be, and was, formulated in such a way as to make it the exemplar of liberal speech, neutral on matters political or religious and free from prior metaphysical or ideological commitments. By the 1870s this powerful rhetorical construct—science as free and liberal speech— could take different forms. Some refashioned a natural theology that had, in early decades, been claimed as a common discourse to unite otherwise opposing religious groups. Others were drawn to a conception of religion as private feeling and science as the only valid route to definite, accountable knowledge. But however conceived, the idea of science as free and liberal inquiry, and therefore as a fitting topic for a public lecture, faced a challenge hinted at by Tyler and Redpath. How could it avoid degenerating into instruction about the material world that carried an indefinite or nonexistent transcendental meaning and lacked the urgency (and excitement) of moral imperatives or political debate?

One way to overcome this was to pursue a high ambition to unite religious meaning and sentiment with an elevated view of science but in ways ostensibly free from partisan ambition or intent. For the lecturers considered here, evolution (biological or cosmic) supplied a common grand theme that provided a way to reconceive the place of religion in a world of science and render the latter meaningful in broadly religious terms. The exact outworking of this ambition differed from speaker to speaker. But all five harnessed the power of speech to convey a lively religious vision for a scientific society. None of them saw the lecture as a particularly effective way of conveying ideas or information. All subscribed to Huxley's view that the lecture was a potent means for provoking and transforming affections and sympathies. Among other things, their lectures operated as ritual forms for the passionate promotion of a new mythos to replace more traditional articulations of religious belief. They were each, in their own way, following Emerson in using the lecture to create what Augst describes as a "visceral immediacy" and a "secular conversion experience."[73]

For Tyndall and Huxley, this took the form of an appeal to the ineradicable but nebulous world of feeling, which for them was beyond the reaches of science even while necessary for its very possibility. For Proctor, the power and allure of science lay in its termination in the mystery of the unknown that he, more than Tyndall and Huxley, was content to call God. For Wallace, evolutionary science reached its material limits with the emergence of spirit, a conviction that thoroughly colored his entire tour. And for Drummond, science in general and evolution in particular offered patterns and trends that revealed the truths of a gospel of love in ways that revivified, even as they transformed, traditional Christian belief.

To some degree, the lecture tours also carried certain political and social messages and, as a result, provoked a range of energetic responses. John Tyndall, for example, was happy to gently chide his audiences about the state of American science while also charming them with counterbalancing praise. Tyndall thus articulated a form of Anglo-American relations that played well in postbellum United States (in the north at any rate). Richard Proctor, on the other hand, was much more effusive in his direct praise of American science, particularly American astronomy, in ways that reflected and reinforced his antagonism toward institutional science in Britain. The political opportunities afforded by the presence of leading British scientists in American auditoria were also used to advantage by local supporters. The fact that more than one of the visiting speakers attracted the incumbent

president of the United States and many other elite statesmen further signals the political complexion of the tours.

Though these political notes undoubtedly helped to generate public
interest, the religious bearing of the tours provoked the most vocal and
voluminous public reaction. Lectures responding to lectures, reactionary
letters to the newspapers and pulpit condemnations and commendations
were only some of the ways in which the lecture tours fueled the engines
of public debate. Each tour generated these and other types of responses
from different constituencies and on different grounds. Whether provoking
these reactions was intentional on the part of the lecturers, or whether it
represented a primary aim of their tours, the religious implications and
inflections of their spoken arguments played a dominant role in commentary on the meaning and wider consequences of their lectures. It was this,
perhaps more than anything else, that gave these speakers cultural traction and helped them give their audiences performances that pulsated with
metaphysical and religious interest.

The performative dimensions of speaking took on particular importance
when metaphysical and religious matters came into play. If, as was generally
the case, science as such—in an ostensibly pure form—was thought to be
best delivered in a style of speaking confined to a narrow emotional range,
any religious or moral implications of scientific ideas or a scientific attitude
tended to invite a more exuberant performance. Audiences, or at the very
least, journalists had a particular ear and eye for that element of speech
that was, as Erving Goffman phrases it, "about more than textual transmission." When a speaker moved to matters ostensibly beyond science this
was regularly noticed and commented upon. The perceived religious stance
of science lecturers tended to produce vigorous and locally conditioned responses. David Livingstone's close study of engagement with Darwinism in
several Calvinist conclaves provides one compelling demonstration of this.[74]
It is not surprising, then, that the clotted and complex religious culture of
Gilded Age America meant that metaphysical claims provoked contrasting
and hotly contested reactions as a lecturer moved from place to place.

The chapters that follow do not pretend to provide a comprehensive or fully
representative account of science lectures in Britain and the United States
during the late nineteenth century. They also intentionally leave largely in
the background aspects of science lecturing in this period that have been
well studied by others, not least the place of demonstration, live experimen-
tation and new visual technologies.[75] Instead, tracking the formation and
fortunes of five British science lecturers who were, or became, household
names on both sides of the Atlantic allows for an analysis that brings verbal
and embodied performance, remediation and response in the context of

oratorical culture into closer view. It permits, in other words, an elaboration and detailed exploration of the book's argument that science lectures were widely understood to be vital agents of cultural change and as embodied performances designed to inform and restructure the beliefs and feelings as much as the visual intelligence or intellects of their hearers and readers. The emphasis given to style as much as substance is not to deny or to ignore the importance of the content, rhetorical or substantive, of the lectures as textual products. But it arises from an approach that seeks to uncover a hitherto neglected but, as I want to argue, crucial aspect of speaking for and about science in an age of popular oratory, namely the evident value placed on the conduct as much as the content of science lectures with mass appeal.

Each chapter begins with a detailed account of the origins and development of the lecturing careers of the speaker, foregrounding their style, attitudes to lectures and experience of delivering them. Attention then turns to the performances, places and cultural correlates that strongly informed the main aims and outcomes of their American lecture tours. This means paying close and sustained attention to how each lecturer cultivated their approach and attitude to public speaking and to how, where and to what ends the five speakers delivered their lectures. It also provides a focused sense of the wider platform culture in which their lecturing careers began. With that individualized context in place, the chapters then reconstruct the speaking tours in the United States. The huge public interest they generated has left a rich and diverse set of descriptions that allow a detailed account to emerge of vocal and embodied performances, of cultural location and local circumstances and of the intricate, dialectical relations between platform and print (especially newsprint) culture. The stories that follow will, I hope, offer dramatic and colorful reconstructions of the dynamic landscapes of talk and text that, in the ears and eyes of many, helped make science lectures events of deep, if always contested, cultural significance.

CHAPTER 1

SCIENCE, SPEECH AND CHARACTER

John Tyndall's Lectures on Light

It was John Tyndall's firmly held conviction that character was the wellspring of scientific endeavor. Without a moral charge, and a call to act, his career in science would never have begun. The inspiration came from "three unscientific men," Thomas Carlyle, Ralph Waldo Emerson and Johann Gottlieb Fichte. They pointed Tyndall to "what I ought to do in a way that caused me to do it."[1] There was something inscrutable about this. The origins of moral impulse terminated in mystery. But even if shrouded in darkness, the decision to act could be released by "the proper word spoken."[2] Tyndall was careful to stress that once activated, the individual will and the feelings that directed it must submit to the scientific method and the physical reality it alone was competent to describe and comprehend. As Tyndall warned in an essay on Goethe's theory of colors, not doing so risked overextending subjectivity and mingling self and science, to the detriment of both.[3] Despite this risk, it was essential to affirm the two cardinal components of the human constitution—the affections and reason. Character was the matrix that bound them together.[4]

Tyndall's conviction that good character was the bedrock of culture was widely shared.[5] The language of character provided, in Stefan Collini's words, the "chief structuring vocabulary of political reflection" in Victorian Britain. However, as Collini further observes, character discourse, while pervasive, was fundamentally unstable and ambiguous. Character determined the will and yet was formed by consciously chosen habitual action. It was never clear whether habit or character had primacy and the specter of determinism lurked in the background. This was true of Tyndall's reflections on the subject. Character, he argued, was inherited, a product "of all the ages that preceded us."[6] We are, he suggested, "indubitably bound by our organisation."[7] He echoed Ralph Waldo Emerson's definition of character as an "undemonstrated force," a "natural power" and a preexisting "moral element."[8] At the same time, the unknown "potentialities" of character could be released through the work of clearing away obstacles.[9] This work was secular in more than one sense. Like changes in the physical world, it was slow work. But it was also radically independent of religion, which could not alter character.[10]

Whether philosophically coherent or not, making character fundamental to cultural life raised crucial questions. How does individual character manifest itself? And what was its measure? Emerson's view, given its influence on Tyndall, is instructive. In his essay on character, Emerson argued that it could be discerned in "a sense of mass." A person of strong and good character "shall stand stoutly in [their] place." They will not conform to convention but will present "resistance" and "a new and positive quality." Evidence could also be found in a person's bearing. In a man of genius, his character could be looked for, "agitating and embarrassing his demeanour."[11] Tyndall seems to have concurred and was convinced that good character could be emotionally discerned. In an essay published in 1871, he avers that the "affections or sympathies" offer the "best guide [to] . . . moral goodness." There was, he argued, a "moral congruity between outward goodness and inner life."[12]

When it came to lecturing, Tyndall believed that the character of a speaker, or their ability to be true to themselves and speak truthfully about the world, was key to success or failure. As he noted in remarks on his philosophy of education written in 1874, "instruction is only half the battle." Borrowing directly from Emerson, Tyndall argued that the other half consisted in "provocation," or the "power of the teacher, in the force of his character."[13] This conviction was articulated again in a lecture delivered at the Birkbeck Institution in October 1884. There Tyndall noted that "knowledge is not all. There may be knowledge without power . . . a power of character must underlie and enforce the work of the intellect." This was a lesson he had learned as a lecturer at Queenwood College in the late 1840s. Without resolve and a "strong and earnest character," intellectual "expertness" was but "the bright foam of the wave without its rock shaking momentum." With character, the lecturer could operate a "lever to lift . . . growing minds."[14] The centrality Tyndall gave to character made attention to self-presentation crucial in the lecture theater. It was vital that a speaker signal his earnestness, sincerity and honest manliness. The ideal teacher should, as Tyndall expressed it, "merge" themselves with their subjects. In doing so, they would energize and animate scientific knowledge and connect emotionally with their audience.[15] Lecturing, even on science (or, to follow Tyndall's logic, especially on science), ought to be a manifestation of character.

In making this argument, Tyndall was, once again, articulating a common view. The importance of oratory in the mid- to late nineteenth century was often tied to the belief that it could be a powerful expression of character. Taking the celebrated orator and Liberal MP John Bright as an example, Patrick Joyce notes how his admirers thought of him as an icon or talisman of the kind of moral self necessary for progress in a democratic age.[16] Bright's reliance on the spoken more than the printed word facilitated this presumption. Oratory was assumed to permit access to inner character, and the orator's body and words were received as the incarnation of deeply held moral convictions. This was true of lecture culture in the United States in the same period. Emerson, among the most celebrated speakers on the American lecture circuit, provides a noteworthy example. Tom Wright, for instance, has noted that Emerson's body presented a "beguiling social text" that was scrutinized by his audiences.[17] Bonnie O'Neill, too, has noted that "audiences consistently looked to Emerson's physical presence in the lecture hall for evidence of his moral character."[18]

The relations between science, speech and character surfaced in a variety of ways in Tyndall's lecturing. One of these was his command of his nerves. In Emerson's influential view, self-restraint was a preeminent trait of good character. This kind of mastery was often associated with the practice of science itself. The strength of will required to patiently investigate the operations of nature without succumbing to personal predilections or passions was relevant to the sphere of moral action.[19] That same resolution of will could be applied to the threat of extreme nervousness. Writing shortly after Tyndall's death, Thomas Henry Huxley noted that one of the secrets of Tyndall's lecturing career was his abiding fear of public speaking. Huxley confessed that he "had never met with anyone to whom an impending discourse was the occasion of so much mental and physical disturbance."[20] Tyndall himself had described this turmoil when he relayed the feelings he experienced just before the delivery of his first lecture at the Royal Institution in 1853. He felt like "a prisoner going to be hanged."[21] These preperformance nerves were described in a language reminiscent of the terror Tyndall experienced during his Alpine climbs. Controlling such powerful feelings of nervousness represented a remarkable strength of character. When Huxley spoke of Tyndall's "lecture fever," he was not seeking to undermine his compatriot. Rather, he was reminding readers of his friend's marvellous capacity to sublimate his emotions.

The kind of resolve that Tyndall exhibited in the lecture theater did not mean that his manner of speaking was dispassionate in style, tone or substance. On the contrary, his lectures, as one admirer put it, were a "work of enthusiasm," able to stimulate an audience without risking a loss of control. In his own reflections on effective lecturing, Tyndall confirmed this view. Commenting in 1855 on a discourse delivered by the Italian scholar James Laciata on Dante's *Divine Comedy*, Tyndall noted that it gave him "great satisfaction" even while he struggled to determine its worth. The "secret," he decided, lay in the lecturer's "assured and animated" delivery and in his ability to "keep . . . sympathy from running into extravagance."[22]

A balance between restraint and the free flow of enthusiasm was also perceived in the kinesics of Tyndall's lecture room performances. As numerous observers noted, in keeping with his suspicion of artifice, Tyndall avoided choreographed bodily action. There was no conscious use of formal gestures to drive home a point or to stir his audience's passions. He lectured, one report noted, with his hands behind his back, folded across his chest or with his elbows leaning on a lectern.[23] A parallel can be drawn here with Emerson, also known for his infrequent use of oratorical gestures. The stillness of his body concentrated the attention of his audience on the power of the words uttered. Emerson's bodily stillness was not, however, absolute. According to one observer, his left hand was often in motion "as if the intensity of his thought were escaping, like the electricity from a battery."[24] Another auditor noticed a continual rocking motion, with Emerson standing on his tiptoes for emphasis. These subtle movements were frequently interpreted as the natural outflow of his thought rather than as the conscious acts of a polished orator.

Like Emerson's, Tyndall's body was rarely motionless. His apparently involuntary movements suggested to some observers a man possessed by his subject. The American editor and popular science lecturer Edward Youmans noted that, while lecturing, Tyndall "was not still a moment, but bending in all possible shapes, as if he had the St. Vitus's dance twisting his legs together . . . and working and jerking himself in all directions."[25] This outflow of physical energy was read as a sign of Tyndall's felt commitment to his science. It was a side to Tyndall's lecturing that was perhaps less often noticed than his perspicacity, cogency and dramatic verbal imagery. But the kinesics of Tyndall's platform performances was not simply an unconscious physiological reaction unrelated to his lecture. His edgy and fidgety movements created an impression of a performer utterly absorbed by, and devoted to, his subject.

Among the other material indicators of the character of a lecturer was their voice. During John Tyndall's celebrated career as a lecturer, this conviction was commonplace.[26] Speech revealed degrees of civility, authority and authenticity. "Savages" spoke coarsely, the civilized with polish and refinement. A weak, faltering voice suggested timidity or vanity. A clear and confident voice was a mark of intelligence and benevolence. Or as Ralph Waldo Emerson put it in a lecture on eloquence first delivered in 1867, "The voice, like the face, betrays the nature and disposition, and soon indicates what is the range of the speaker's mind."[27] Such assumptions invested layers of social and cultural meaning in vocal performances and were tied to a pervasive culture of oratory.

Whether or not Tyndall fully subscribed to such views, he was certainly convinced of the need to attend to the effective use of his voice and took steps to ensure he spoke with sufficient volume and verve to be heard and admired. A revealing letter written by the botanist Joseph Hooker to Charles Darwin in 1866 offers some confirmation of this point. Hooker described an encounter with Tyndall shortly before Hooker was due to give an evening discourse as part of the program of public lectures organized for that year's meeting of the British Association:

> He [Tyndall] came up to me in the forenoon, evidently most anxious for my success, & questioned me about it. When I told him it was a written discourse, & that I intended to *read* it, his countenance fell & I saw he was cut.—he turned away first, but came back & with great delicacy & loving kindness gave me some hints; to learn passages by heart &c—(I had done this copiously already) & to put myself en rapport with the audience &c &c. I saw in short that he prognosticated a dead failure, & I spared no pains that afternoon in preparing myself to succeed in his eyes. I hope I did.[28]

Tyndall was clearly agitated, caught between a desire not to offend or patronize and an aspiration to offer advice about what made for an effective vocal performance. It was the latter that won out. By this time, Tyndall was acknowledged by people as prominent within science as Hooker and Darwin to be a consummate public lecturer. Still, according to the remembrance of at least one commentator, even Tyndall was not always successful. In an obituary, the zoologist Chalmers Mitchell noted that Tyndall's voice had been "not notably sonorous" and that "on several famous occasions his efforts to be heard were unsuccessful."[29] Criticisms of Tyndall's voice were, however, rare. Like everything else about his lectures, he made careful preparations to ensure effective vocal delivery. To take just one example, before delivering a lecture on matter and force to working men during the meeting of the British Association in Dundee in 1867, Tyndall positioned himself "at the most distant point from the platform, to get some idea of the strength of voice necessary to fill it."[30] His expertise in acoustics placed Tyndall in a good position to ensure his voice was sufficiently strong to avoid charges that he lacked vocal strength and, by extension, strength of conviction.

As well as addressing concerns about volume, Tyndall was conscious of judgements that might be made based on his accent. Writing in 1862, Edward Youmans observed that while Tyndall had "Irish blood and temperament," there was no indication of an Irish brogue when he spoke.[31] During his American lecture tour ten years later, his "strong English accent" attracted comment, something that amused close friends like Thomas Henry Huxley, either because his accent was so obviously Irish not English or because Tyndall's efforts to efface his origins by altering his accent made them chuckle.[32] In 1886 Tyndall's Irish accent was detected in the last lecture he delivered at the Royal Institution, but only "now and then."[33] In at least some people's hearing, Tyndall's transformation from the son of an Irish shoemaker to England's most celebrated science lecturer could be detected in his voice.

In addition to accent, the tone of Tyndall's voice was subject to scrutiny. On at least one occasion, a sharp rebuke directed at one of his technical assistants, William Barrett, not only sparked a breakdown in relations between the two men (and one that ultimately led to Barrett's resignation) but also, according at least to Barrett, "caused a 'jar' in the audience." This incident threatened Tyndall's standing as a speaker of charm and grace and he acknowledged that his habit of complaining "in a low tone" when something was not right with an experiment was a bad one that "must be got over."[34] The more significant threat, however, was to Barrett, whose reputation as a competent experimentalist in his own right was still in the making. A public admonishment from Tyndall could do serious damage and demonstrated the influence of the inflections of his voice, even those apparently incidental to the lecture being delivered.

The paralinguistic indicators of character ran alongside, and at times reinforced, signals found in the cognitive content of Tyndall's lectures. One of the important considerations in composing them was the fine line between manly and offensive speech. The difficulties of this task were most acutely felt when Tyndall stepped beyond what he took to be strictly scientific concerns to comment on subjects of a more controversial nature. It might be argued, however, that Tyndall's reputation as a brilliant and brave lecturer was not so much endangered by his forays into public controversies as, in part, built upon it. There are reasons to argue that this was true at least until he delivered his notorious Belfast address in August 1874.[35] If his views on matters such as prayer, miracles, materialism and the limits of human knowledge—all broadcast in lectures and articles through the 1860s—set him in opposition to orthodox Christian belief, they also provided a vehicle for expressing his manly character. Being outspoken about such matters demonstrated his fearless commitment to the search for truth. Tyndall was drawing on what Stefan Collini has termed the political aesthetics of mid-Victorian "muscular liberalism." This emerged from the "mingling of physiological and ethical properties" with "bodily and moral vigour . . . cultivated by the same means." A lack of resolve, the opposite of manliness, "could be walked or climbed out of the system," and the truth faced without fear whatever the cost to cherished convictions.[36] Tyndall's skilful combination of mountaineering and daring metaphysics was a classic example of this dominant aesthetic.[37] Tyndall's version of muscular liberalism was a philosophy that could be articulated with particular force and clarity in the lecture theater.

A more fine-grained analysis of Tyndall's expressions of "manly" devotion to truth points to what might be termed a geography of provocation. Tyndall was committed to obeying the speech protocols in place at the Royal Institution, avoiding controversial political or religious concerns and resolutely sticking to scientific topics, uncontentiously construed. Once he strayed further afield, whether in print or on a different lecture platform, he "claimed intellectual freedom" and was known to utter more provocative remarks.[38] Yet, depending on the mode or location of his communications, he did not claim that liberty in quite the same way. In articles written for publication or in letters to newspapers, Tyndall on occasion pushed harder at the self-appointed task of separating religious feelings from mistaken beliefs about the physical world. In lectures he was more often circumspect, not wishing to offend his immediate audience or to be heard to contravene the presumed neutrality of public speech. However, that did not prevent him from interlacing his speech with hints of a more radical agenda even before his outspoken attack on religious cosmogonies in Belfast.

Tyndall's concern with carefully managing the ways in which his lectures instructed and stimulated his audience without compromising ideals of character and conduct informed other aspects of his lectures, including the experiments that were so often a central component. The meticulously prepared demonstrations that habitually formed the backbone of his lectures were a vital part of his strategy for taking science to the wider world. But as well as helping to inform and inspire his audiences, they became a mark of his expertise and personal commitment to the painstaking and disciplined investigation of nature. Their live performance was another way to display not just nature's hidden processes but also Tyndall's own inner character and the ideals that informed it.

As several commentators have pointed out, through his dramatic demonstrations Tyndall added a sense of theater to his platform performances.[39] This placed them within a wider competitive field shaped by an appetite for spectacle. Doing so risked inviting criticism from those concerned about the corrosive effects of such popularizing strategies on moral culture. One of the tactics Tyndall used to deflect this criticism was to elevate the importance of exposition without downplaying the power of experimental demonstration. As he noted of his mentor Michael Faraday, he "did not confine himself to experiments;" rather, he labored hard to identify "the most effective methods of scientific exposition." While Tyndall summarily dismissed the elocution lessons that Faraday had taken early in his career as a lecturer ("happily" Tyndall opined, they did not "damage his natural force or grace of delivery"), he expressed admiration for Faraday's efforts at cultivating vigorous, dramatic and manly speech.[40] Tyndall was also writing about himself. Balancing exposition and experiment offset accusations that undermined Tyndall's reputation as someone committed not to adulation and applause but to the clear communication of scientific truth.

For all that, controlled and clear exposition that aimed to explain the science behind the experiments did not offer a straightforward solution to the complex task of combining science and character without compromise. The spoken parts of the lectures also had to be stimulating and contend with what Martin Hewitt has described as the "spectacle of words" that characterized mid-Victorian lecture culture. This had become increasingly important as the latter shifted toward a broader range of more literary and political topics that relied less on visual technologies and more on the "pictorialism of the word."[41] Tyndall was well placed to meet this expectation. His love of poetry, and his belief that it played an essential role in cultivating the aesthetic faculty, aided his efforts to give scientific exposition an emotional charge through vivid description. Poetic fancy could aid as much as obstruct the unprejudiced search for scientific truth by reminding audiences that science invariably terminated in an unknown power. It was vital to Tyndall that his audiences felt "the mystery of the universe without giving it rigid form."[42] The ability to do so was a fundamental mark of good character and high eloquence. In this Emersonian frame, the theatricality of experiments and exposition did not necessarily threaten the character ideals that Tyndall subscribed to. Properly employed, it could aid the task of standing firm on scientific fact while stirring, purifying and elevating the emotional lives of his hearers. In Emerson's words, Tyndall came to his audiences "armed with reason and love."[43]

Given the importance Tyndall attached to "levering" an audience or "pulling the trigger" that would prompt them to moral action, the nature of their response to his lectures was also of supreme importance.[44] Tyndall's concern with pleasing and stimulating his audience is evident in his own reflections on his lecture performances.[45] Writing in his journal after a talk delivered at the Royal Institution on January 17, 1854, he noted that "my lecture being so theoretic I almost feared it was a failure . . . a hearty applause however was granted me and I believe on the whole people were pleased. This, of course, is to me the principal matter."[46] This might be interpreted as Tyndall's desire for applause. But it is better read as underscoring his belief that if an audience responded coolly to his lecture, there would be no evidence that he had done anything more than merely instruct them.

Others have rightly noted that Tyndall's audiences shaped his performances. Against the idea of science popularization as the dissemination of knowledge from expert practitioner to a passive public, Jill Howard has argued that Tyndall's auditors were active participants in the overall speech event.[47] This might be taken to suggest a significant disruption of any rigid or permanent hierarchy between speaker and listener, or expert and novice. Whether or not that was the case, it was not how Tyndall understood the relationship between speaker and hearer. In his lectures, Tyndall presented himself as the expert and his listeners as witnesses of his superior grasp of science.[48] Tyndall maintained a view of science as the purview of a select number of original investigators endowed with exceptional cognitive and creative powers. This view, in part, flowed from his understanding of character and the related concept of genius as something given in advance by nature to a few. But it was also informed by an Emersonian conviction about what constituted a successful lecture. Character again was key. As Emerson put it in his first lecture on eloquence, a piece Tyndall almost certainly read, "There are good speakers who perfectly receive and express the will of the audience, and the commonest populace is flattered by hearing its low mind returned to it with every ornament which happy talent can add. But if there be personality in the orator, the face of things changes. The audience is thrown into the attitude of pupil, follows like a child its preceptor, and hears what he has to say." Though more often than not positioned physically beneath his audience, Tyndall's strength of character was displayed by his speaking from an intellectual and moral height. He was expressing his "strong personality" and occupied "the highest platform of eloquence . . . moral sentiment."[49]

The importance of character to Tyndall's lecture performances can also be detected in some of the criticisms leveled by his detractors.[50] The parodic verse penned by James Clerk Maxwell during the Edinburgh meeting of the British Association for the Advancement of Science in 1871 is perhaps the most illuminating example. Four stanzas of the poem, entitled "A Tyndallic Ode," were published in *Nature* shortly after the meeting. Another three, more explicitly critical of Tyndall's scientific epistemology, remained unpublished until after Maxwell's death.[51] In the first four verses, Tyndall is made to speak of various theories and experiments familiar to those who had heard or read his lectures. The contested and speculative nature of Tyndall's explanations of molecular motion, clearly registered in the poem, provided grounds to suggest there was too much of the lecturer in the science presented from the platform.[52] As Gowan Dawson has suggested, Tyndall's reputation for flirting with the female members of his fashionable audiences is hinted at in the second verse, in which molecules "with fierce desires, shiver in warm embraces."[53] As well as implying that exciting admiration was more important than attending to scientific accuracy, the allusion was also a slur on Tyndall's moral character. Perhaps more significantly, this description of molecules drew attention to Tyndall's well-known efforts to blur the boundaries between the organic and inorganic. This point was made again later in the poem through a reference to the fishlike form of a cloud conjured by Tyndall in an experiment he had reported in the *Proceedings of the Royal Society.*[54] All of this made Tyndall's performances more about his own "fancies" than about the hard work required to properly appreciate scientific facts and principles.

The final published stanza of Maxwell's ode went further and fully merged the lecturer and his subject matter:

> I light this sympathetic flame,
> My slightest wish to answer,
> I sing, it sweetly sings the same,
> It dances with the dancer:
> I whistle, shout, and clap my hands,
> I hammer on the platform,
> The flame bows down to my commands
> In this form and in that form.[55]

In its description of one of Tyndall's popular platform experiments, the verse collapsed the distinction between the lecturer and his representation of natural phenomena. The lecturer, through theatrical sounds and movements, made nature mimic his own flickering form, which itself became like a flame, mesmerizing his audience.

This interpretation is given weight by the final three unpublished stanzas. These mounted a more direct attack on Tyndall's "fancy scientific" and presented him as building a veritable tower of Babel out of "mental bricks," which "conquer[ed] gravitation." In doing so, they expressed in verse what Maxwell had earlier articulated in a response to Tyndall's widely discussed address on the scientific use of the imagination. The implication of Maxwell's response was clear. Tyndall had encouraged the undisciplined use of the imagination in scientific research and communication to the detriment of the more disciplined step-by-step construction of mathematical analogies between different natural phenomena. In Maxwell's view, the only way to make science popular without forgoing its scientific character was through a "profound study and copious application of those principles of truly scientific illustration which . . . depend upon the mathematical classification of quantities."[56] There was no room in Maxwell's view for "desiring" molecules mysteriously creating beautiful and orderly forms in ways uncannily like sexual reproduction. With this hermeneutic in place, Tyndall's unscientific character, most powerfully apparent on the platform, became an obstruction to the progress of science and a threat to moral culture.

Tyndall's critics notwithstanding, in the late 1860s and early 1870s his reputation as someone worth listening to was riding high. This was true not only in Britain but also in the United Sates. The New York publishers D. Appleton and Company had issued several of Tyndall's books. Included among the Appleton list were Tyndall's Royal Institution lectures on heat and sound. In addition to books based on science lectures, Appleton published two collections authored by Tyndall the year before his visit. The first, *Fragments of Science*, differed from the other publications in being essays and talks originating outside the Royal Institution. Some of them dealt directly with more controversial metaphysical concerns and included essays on prayer, miracles and materialism. Notably, the book was dedicated to his "friends in the United States" who had encouraged its publication.[57] The second, *Hours of Exercise in the Alps*, recounted his mountaineering exploits, often in a poetic style that underlined his worship of nature's wondrous works. Both books confirmed his reputation as a risk taker in the world of metaphysics and in the high mountains.

In 1869 a biographical sketch of John Tyndall published in *Appleton's Journal*, a publication edited by Edward Youmans, provided a particularly telling portrait. The first section was mostly lifted from a sketch that Tyndall had "concocted" with his close friend Thomas Hirst several years earlier.[58] In it, Tyndall's ancestry is traced back to members of "an old English family of Tyndales," who migrated in the seventeenth century to Ireland's "eastern Saxon fringe."[59] This playing up of his English and Saxon roots was deliberate. Tyndall elsewhere complained about a sketch that had him born among the "Irish peasantry."[60] His father's character matched his venerable ancestry and was marked by "intellectual power and personal courage, combined with delicacy of mind and feeling." Tyndall's career as a surveyor evinced the same kinds of virtues and was characterized by "extreme caution and accuracy, together with dauntless perseverance under difficulties." As a teacher at Queenwood College he eschewed corporal punishment and overcame "insubordination" by sheer "force of character." This emphasis on character continues through the remainder of the sketch, which extends beyond Hirst and Tyndall's original. His "rare gift" as a writer and lecturer was to "give us the poetry of science without impairing the quality of science itself." In the lecture hall, "his movements are rapid and decisive," not conforming to oratorical ideals but expressing his "highly vitalized and restless temperament." As a man of "enlarged and independent views," his opinions "carried weight and force with the public."[61] It was this Tyndall who received a letter of invitation to lecture in America signed by nineteen scientists, engineers and university administrators and by Tyndall's American lodestar, Ralph Waldo Emerson.[62] This was part of the slow but steady build up to Tyndall's lecture tour of the eastern United States.

FRAMING THE TOUR

Before Tyndall arrived in New York on October 9, 1872, his tour promised to be a sellout success. From Boston to Brooklyn, his course of six lectures on light was expected to attract crowds well beyond the capacity of the auditoriums in which the famous English lecturer was to perform. Tyndall had been offered the largest honorarium—$500 per night—ever extended to a lecturer on the American circuit.[63] Although his Lowell lectures in Boston—the first he would deliver—would be less lucrative, it was likely that he would make a very healthy profit. Like Charles Dickens before him, Tyndall could expect to earn a significant amount of money.[64] This prospect did raise some suspicions. Not long after the delivery of his Boston lectures, a report in the New York newspaper the *Golden Age* described one disgruntled woman who refused to attend Tyndall's lectures. Her reason? They were thought to spring from a conspiracy among English visitors to "storm the American lyceum in order to recapture the fifteen millions [*sic*] of dollars awarded us for the Alabama Claims."[65] This story was only half in jest. English visitors going home with large sums earned from the American lecture circuit damaged their reputation as gentlemen of honor who valued duty over monetary reward.

There were reasons why such discussions were taking place. Tyndall was not alone in traveling from Britain to the United States to lecture. The winter lecture season in the northeast was to include the historian James Anthony Froude, the novelist George Macdonald, the journalist Edmund Yates and, in the New Year, the women's rights activist Emily Faithfull. Froude's visit was undoubtedly the most controversial, arranged to counteract the growth of anti-English feeling and Fenianism. But political intrigue aside, the unprecedented number of notable British lecturers arriving in New York raised expectations and interest. It also provided material to make comparisons between their abilities in the lecture hall and their character as public performers.

Tyndall was in a good position to top the field. It was widely thought that his lectures would encourage the growth of a scientific spirit that would contribute to the health of the nation. His books were celebrated as models of lucid and lively description and explanation. But the distribution of his books was one thing. His presence was another. It was only in encountering the man himself that his scientific message, and his enthusiasm for scientific research, would take full root in American soil. As an editorial in the *Popular Science Monthly* put it, a visit from someone of Tyndall's stature demonstrated "not only what manner of books they write but also what manner of men they are." Audiences had an advantage over readers in being brought immediately under the vital magnetic influence of the lecturer's personality.[66]

FIGURE 1.1. Albumen carte de visite of Tyndall by the London photographers Elliott & Fry (c. 1872). NPG Ax18206 © National Portrait Gallery, London. A sketch based on this image was published in the *Popular Science Monthly* in advance of Tyndall's visit.

That personality would be fully displayed in the lectures, not least because of Tyndall's commitment to avoiding display for display's sake. A writer for the *Galaxy*, a popular monthly published in New York, announced that "as an original and skilful experimenter Prof. Tyndall is unrivalled. Fertile and ingenious in contrivances for bringing out his points, the effects are always telling and impressive. Yet the experiments are never the main things; they are always subordinate to the idea with which he is dealing—helps to its presentation. He is never eclipsed by his own pyrotechny, but holds the attention of his listeners closely to the question under examination."[67] It was the exposition as much as the experiments that would transform the teaching of science. This chimed with Tyndall's own commitment to subordinating demonstrations to character-building speech. Writing in advance of his visit he asked his American cousin Hector Tyndale to bear in mind that he could not compete with the "brilliant" and "fiery" experiments of American lecturers (he perhaps had Henry Morton, president of the Stevens Institute, particularly in mind).[68] It is notable that John Henry Pepper, former director of the Royal Polytechnic Institution in London and a popular performer of sensational physical demonstrations, was visiting America at the same time and demonstrating his spectacular experiments in New York's Steinway Hall.[69] Tyndall, against the run of scientific showmen whether American or British, was committed to making his lectures "sound science, aiming to address the intellect more than the eye and only addressing the eye to make it the avenue to the intellect." While he worried that his lectures would, as a consequence, appear "cold," he made a virtue of his steely commitment to solid exposition.[70]

For all that, Tyndall would have to walk the line between stimulating the emotions while making them subordinate to the dictates of reason and good character. In advance of his visit, it was important that his demonstrations were widely praised as esthetically as well as scientifically satisfying. It was also crucial that the appeal of the vocal and corporal aspects of his platform performance was given due attention. The popular press did not disappoint. In the same article in the *Galaxy*, Tyndall's unusual platform performance was carefully described, whetting the appetite of Americans who had the chance to see and hear him in the lecture hall:

> Prof. Tyndall manner as a lecturer is individual and unique. He never reads, but holds his audience by the power of lucid and forcible extemporaneous statement. He is not what would be called a fluent or even speaker. . . . He is not an elegant disclaimer, whose measured cadences are accompanied by graceful and appropriate gestures. He is irregular and sometimes hesitating in speech, and unstudied in gestures and movements. His habit of speaking has been formed in connection with his habit of experimenting. . . . clearness, force, vividness of description, felicity of illustration, and the eloquence inspired by grand conceptions are the striking features of his style. . . . He is intensely in earnest. . . . [and] is a remarkable example of self-forgetfulness upon the platform, always being absorbed in his subject.[71]

Others replicated this portrait. A sketch in the *Christian Union*, a weekly religious periodical with sales over one hundred thousand, commented that Tyndall's "language indeed was better than his action, which is too angular and quick to be called graceful."[72] He was nevertheless "endowed by a vigorous physique and unbounded enthusiasm; nervous, sympathetic and ardent." He was "no orator," but his "extreme lucidity of style never fails to carry the audience completely with him." With the "imagination of a poet," he was able, despite the lack of polish and grace, "to throw round the driest details of scientific fact . . . a rich glow of feeling and fancy." His lectures would do much "to stimulate and foster" science in America.[73]

In addition to being a visual indicator of the poetic and emotional appeal of his lectures, Tyndall's body was used as a proxy for his national and racial identity. When *Harper's Weekly* described him as "the very type of a burly Englishman," it earned a rebuke in *Appletons' Journal*. Rather than being like the heavyset "Saxon," Tyndall, "an Irishman," was "of a wiry, elastic physique, and nervous temperament, resembling far more the typical American."[74] Ironically, although this description was ostensibly more accurate than the one given in *Harper's Weekly*, it pushed against the grain of Tyndall's desire to represent himself as being of Saxon stock and an Irishman by birth rather than ethnic origin. In part, as Tyndall himself was aware, political expediency was at play. While relations between Britain and America had improved by the time Tyndall was to visit (the Alabama claims had been settled and the Treaty of Washington signed), tensions remained. Picturing Tyndall as an Irishman closer in physique and temperament to the typical American may have helped to ease the strain in Anglo-American relations.

Near the beginning of Tyndall's lecture tour, the issue of Anglo-American relations was raised again, this time on the other side of the Atlantic. Writing in the *Pall Mall Gazette*, a commentator offered a warning to "the Englishman traversing the States on a lecturing tour" who "seeing courteous clique after clique" returns home with a completely distorted sense of the strength of the bond between England and America. A warm welcome was a mark of American hospitality but did not necessarily signal agreement with the message brought by the visitor. Indeed, it no more indicated that than a Russian embrace of a visiting German opera singer signaled "the approach of a political alliance between Russia and Germany." The lecturer on an American tour was, equally, "only a performer."[75] These remarks were largely aimed at Anthony Froude, who traveled with Tyndall to the United States and lectured there on the Irish question. The highly political nature of his lectures meant that he faced considerable opposition, even while being extravagantly feted by his hosts. But the lesson also applied to Tyndall. Applause was not necessarily a sign of outright approval. Tyndall would have to work hard to prove himself more than simply a performer.

As well as navigating the politics of being an "Englishman" in America, Tyndall had to manage the religious controversies that he had already helped to spark before his arrival in the United States. In the lead up to his tour, Tyndall had issued his most forthright statement to date on the issue of the efficacy of prayer. In the July issue of the *Contemporary Review*, Tyndall introduced a proposal made by the London surgeon Sir Henry Thompson for testing the power of prayer to heal the sick.[76] The test would involve "believers" praying for patients in a general hospital ward over a three-to-five-year period and studying the mortality rates to see if they deviated in a numerically significant way from known percentages for the illnesses involved. This would either provide experimental confirmation of the power of prayer to heal physical illness or supply evidence that the human body was not subject to "providential action." That the latter was Tyndall's view was already well known. To Tyndall, nature was a closed system governed by permanent laws. Prayer was thus "impotent in external nature."[77]

Tyndall's promotion of Thompson's proposed experiment rekindled debate in the United States as much as it did in Britain. It was impossible for Tyndall to lecture in America in a way free from his association with the debate over prayer. Tyndall's intentions in this regard, to avoid any direct confrontation with those who opposed his view on the issue, seem clear. In a letter to John Clausius written after his tour had ended, Tyndall remarked that he had resolved to "let no foreign element, or motive, mingle with what I regarded as a purely friendly visit."[78] He was there to promote cordial relations between England and America and stand up for "pure" science. To engage in sustained theological controversy would not have served that end. Yet despite this declaration, it is not clear that Tyndall wished to disentangle his lectures from his wider philosophical arguments. In November, Edward Youmans published in the *Popular Science Monthly* a short piece by Tyndall that had appeared a month earlier in the *Contemporary Review*. In it, Tyndall restated his position and clarified his argument. He was not, he insisted, denying anyone the right to hold to a theory about ultimate reality that differed from his. What he did insist upon was the need to subject conceptions of reality to empirical testing. The example used was telling. Newton held the view that light traveled faster in water than in air. This supposition could not rest on Newton's authority and was overturned through the Fizeau-Foucault determination of the speed of light. Tyndall thus placed a theory of light (Newton's) in the same category as a metaphysical account of nature or arguments about the possibility of miracles and implied that all required experimental demonstration to secure their veracity. Without these, they remained "a mere figment of the intellect."[79]

Tyndall's six lectures, to be performed in six East Coast cities, thus had a potent metaphysical charge. The discourses would communicate recent work on the nature of light. But they would also demonstrate in fine and arresting detail the power of experiment to validate or discredit theoretical conceptions generated through the legitimate but limited operations of the human imagination. Prayer would get no explicit mention in the substance of his lectures, but its efficacy remained on trial even as Tyndall expounded on the wave theory of light. In his careful stage-managing of such a

TABLE 1.1 DATES, VENUES AND LOCATIONS OF TYNDALL'S LECTURES ON LIGHT. (TWO FINAL LECTURES WERE DELIVERED AT YALE COLLEGE, NEW HAVEN.)

Dates	Venue	City
October 15; 17; 19; 21; 23; 25	Lowell Institute	Boston
November 11; 13; 15; 18; 20; 22	Horticultural Hall	Philadelphia
November 26; 28; 29 [three lectures only]	Peabody Institute	Baltimore
December 3; 5; 7; 9; 10; 12	Lincoln Hall	Washington
December 17; 19; 21; 26; 28; 31	Cooper Institute	New York
January 4; 7; 10; 14; 17; 20	Academy of Music	Brooklyn

controversial subject, Tyndall continued to claim intellectual freedom and to demonstrate his manly resolve to speak the truth without the risk of offending more than a small portion of his hearers. This anticipated mix of tenderness and tough-mindedness was a preview of the character audiences in America could expect to encounter in the coming months.

TYNDALL'S CIRCUIT FROM BOSTON TO BROOKLYN

In his first essay on eloquence, Emerson called attention to the competition between "the demands of the hour and the prepossession of the individual." How the rivalry between orator and occasion played out could be used as a measure of the eminence of the speaker. A lack of manliness becomes apparent when the circumstances into which a lecturer speaks overwhelm their own convictions. A more "substantial personality" creates an occasion rather than takes the lead from it. To "instruct an audience in what they already know" is a sure way of losing an audience.[80]

Such considerations may well have played on Tyndall's mind as he traveled across the Atlantic to begin his tour. He did not bring a set of six finished scripts to New York. His lectures were written aboard ship and were added to and refined as his tour progressed. As Tyndall later described it, his lectures in America were marked by "a kind of growth which reached its most impressive development in New York and Brooklyn."[81] Even so, the subject of light and heat was one Tyndall had lectured on and written about many times before. The London-based publisher Longmans, Green & Co. had published Tyndall's notes from a course of nine lectures on light delivered at the Royal Institution in 1869.[82] The experiments he was to display were tried and tested, and, as Jeremy Brooker has shown, Tyndall had worked hard at scaling up his optical projections to meet anticipated American demand for innovative and visually striking illustrations.[83] Bringing two assistants with him and some portable equipment, Tyndall's lectures were well supported. On arrival in New York he visited the hall he was to speak at two months later and made the short trip to Brooklyn to familiarize himself with the venue for his lectures there. When he stood up to speak at the Lowell Institute, inaugurating his tour, he was well prepared.

The Lowell lectures had run since 1839 and had a reputation for being a cut above run-of-the-mill lyceum courses.[84] Under the conditions of Lowell's bequest, tickets were issued free of charge. For Tyndall's lectures, all available tickets were distributed within twenty minutes of being released.[85] At the close of the series, the *Boston Daily Advertiser* reported that "no course of lectures on any subject has ever been given in Boston upon which the favored ticket-holders have given their attendance with more eagerness."[86] With the illustrious history of the Lowell lectures in mind, this was a striking claim for a Boston newspaper to make.

That Tyndall gave the first airing of his lectures in Boston sparked civic rivalry. Newspapers in Boston and New York competed in demonstrating the warmth of audience reactions to Tyndall and in using the unprecedented popularity of his lectures as a measure of the health of local intellectual culture. Tyndall's lectures were also the occasion for trading insults. One New York newspaper, for example, noted that Tyndall "made his debut to an American audience in the great dingy and ugly hall of the Lowell Institute."[87] Civic rivalry did not entirely play into Tyndall's hands. There were suggestions, for example, that his Boston audiences found his lectures too straightforward for their advanced tastes and level of scientific education. Tyndall caught wind of this. Nearly four weeks after the end of his Boston lectures, Bence Jones wrote to Tyndall to report that "Emerson is here & gave Sir H. Holland a flourishing account of you but he thought you did not realize how much the Yankees know of science."[88] This critique was also used to advantage by New York newspapers. One report noted that "it is said that Boston was disgusted with the distinguished lecturer because he made everything he touched so plain that all his hearers could understand him."[89] That which apparently appalled Bostonians would be welcomed in New York. The *New York Tribune* added to the insults by claiming that Tyndall "exerted himself more here than elsewhere" to make his science accessible. In reply, the *Boston Evening Journal* carped that "probably he *had* to exert himself more in New York . . . to make himself clear to his listeners."[90]

Tyndall was sensitive to precisely this issue. Writing to Thomas Hirst after delivering three lectures in Boston, he complained that his "main difficulty is want of acquaintance with the intellectual level of my audience."[91] When he later heard that Emerson of all people felt he had pitched his lectures too low, the criticism must have stung. For all the warmth of his reception in Boston and for all the praise his lectures were given in local newspapers, there were suggestions that he had failed to impress on this crucial score. His character, the substance of his personality, had not made its presence felt to the degree desired.

From Boston Tyndall traveled to Philadelphia, to repeat his lectures at the town's Horticultural Hall. There, as elsewhere, Tyndall was sensitive to his immediate reception, looking out for signs of approval from his audiences. In a letter to Hirst, Tyndall commented that "the Quaker element is strong in this city and at first the audience to a man accustomed to applause might have felt clammy and cold." By the end of the course, however, "they behaved like other Christians—and a heartier burst I never heard than that which greeted my farewell last night." Press attention, however, seemed rather lackluster. In Tyndall's self-confessedly brief glance at Philadelphia newspapers, he found them "as far as science is concerned, . . . dull, commonplace, and ill-informed."[92] Tyndall's experience of Philadelphia was also clouded by the illness of one of his two assistants, George Miller. Miller had contracted typhoid fever in Philadelphia before Tyndall arrived and died after he left, on November 30. A personal blow, this also presented a severe practical challenge—Tyndall had to rely on only one assistant to ensure his demonstrations and optical projections were successfully conducted. When things went wrong like this, the wider infrastructure of a successful lecture tour could come more materially into view.

After Philadelphia Tyndall stopped off at Baltimore to give three lectures at the Peabody Institute. In his correspondence there are hints that all was not well with this part of the tour. The day after completing his series in Boston, Tyndall wrote to Joseph Henry to ask if it was possible to drop Baltimore from his itinerary. The local organizer was "much exercised regarding money."[93] As became clear by the end of the tour, the financial dimensions of his lectures were of serious concern to Tyndall. This was not because he was looking to profit directly from his lectures in America. On the contrary, it was vital to Tyndall that his audiences knew that he had no such intention and was committed to investing any profits he made in promoting the cause of science in America. His lectures had to be the embodiment of that ethic. Tyndall's trip to Baltimore nevertheless went ahead, even if the number of lectures given there was curtailed. Whatever the initial misgivings, Tyndall reported that his lectures in the city were warmly received.[94] A few years later it was reported that his audience was the largest ever witnessed at the institute.[95]

From Baltimore, Tyndall took the short trip to Washington to deliver his full course of six lectures at the Lincoln Hall. This location opened up new opportunities for reaching a rather different audience. Tyndall was fully aware that his listeners would include a large number of congressmen and other members of the political class. Writing to Tyndall several weeks before the Washington lectures, Joseph Henry observed that "your lectures will induce the most remarkable effect in Washington where you will be heard by men of the most active minds in the country. There is a growing disposition in Congress to make liberal appropriation for scientific objects and a few words from you on this point will I am sure produce a marked effect."[96] Certainly, "active minds" were present. The newly elected president, Ulysses Grant, attended Tyndall's second lecture (and was apparently delighted with it) and, according to one report, up to thirty congressmen stayed the course.[97] As elsewhere, Tyndall underlined his message about the importance of valuing the original investigator at the conclusion of his sixth lecture. But in Washington he had an audience composed in part by political movers and shakers adjusting to Grant's second term. He exercised there a degree of "pertinence"—for Emerson, a mark of an orator's ability to override the occasion and command an audience—in reminding his audience of Alexis de Tocqueville's doubts about whether a republic like the United States could cultivate basic scientific research that had no clear practical utility. Tyndall's challenge to his Washington audience was to prove Tocqueville wrong and to invest in the "original investigator in natural truth."[98]

While Tyndall's lectures received significant press attention throughout his tour, when he traveled from Washington to New York there was a noticeable increase in the extent of newspaper coverage. His lectures at the Cooper Institute in New York had originally been placed second after Boston but were postponed on account of the presidential election (it took place on November 5). The venue, the Cooper Institute, was a second choice, secured when the larger and more attractive Academy of Music could not be booked for the whole series. By the time Tyndall lectured there, the institute had become a sign of a more general decline in American lecture culture. Tyndall, at least, had been told that the "magnates of New York" had deserted it.[99] Still, partly driven by the flurry of press attention, and despite severe winter weather, Tyndall defied the odds and attracted unprecedented numbers of hearers.

The significant coverage given to Tyndall was due in no small part to the *New York Tribune*. Tyndall facilitated this by submitting a full transcript of his lectures to journalists (done because, to that point, reporters had made a "hash" of describing his experiments).[100] The *Tribune* not only provided a verbatim record of the lectures but also included illustrations of many of the demonstrations. The resulting reports were later gathered as an "extra sheet" and sold for three cents. Sales reached two hundred thousand within a few weeks and the *Tribune*'s extra was serialized by several other newspapers and popular periodicals. The astonishing sale figures were presented by the *Tribune* as evidence of Tyndall's exceptional popularity and as an indicator that Americans, whatever their reputation, were fully disposed to support scientific education and study.[101]

The *Tribune*'s role in raising Tyndall's profile to new heights was not accidental. The paper had strong associations with transcendentalism, reform and liberal Republicanism and Tyndall was a useful ally to recruit in support of at least two of those causes. For Tyndall, receiving assistance from the *Tribune* was low risk in political terms. By the time he arrived in New York, the newspaper's long-serving editor and defeated presidential candidate Horace Greeley had died and was being eulogized even by his political foes. That may have aided Tyndall's efforts to remain entirely nonpartisan with respect to American party politics and to receive warm backing from the *Tribune* without risking association with a recently defeated political bloc that included the Democrats.

The final airing of the six lectures was given at the sumptuous hall of the Academy of Music, Brooklyn. In a letter to Bence Jones, Tyndall described his audience as "most impressive" and the venue as "exceedingly fine." The sea of faces to his left and right, and the "storm of applause" at the end of his lectures, confirmed his view that he was more loved in America than he could ever have believed before he arrived.[102] Reports in Brooklyn's *Daily Eagle* confirmed the warm reception. Here again, a certain amount of civic rivalry was in play. Unlike Louis Agassiz, who some years previously had "trifled" Brooklyn out of money spent on tickets by performing weakly there but strongly in New York, Tyndall had given his Brooklyn audience "his best thoughts" and most excellent experiments.[103] Such a state of affairs reflected well on Tyndall. So too, did the attention given to his words as much as his demonstrations. Writing to Thomas Hirst halfway through his Brooklyn course, Tyndall noted that his audience had remained "as loyal to the spoken words, to the expositions, as to the experiments."[104]

Apart from two lectures given at Yale College, New Haven, Tyndall's formal lecturing concluded in New York. Whatever the differences between the venues and occasions, throughout his tour Tyndall presented himself as a teacher of science translating and amplifying the language of nature, heard through experiments but usually in a way "too feeble for the public ear." Tyndall's task was to become for his audience the very voice of nature. Taking one "department of natural philosophy," Tyndall aimed to show "the growth of scientific knowledge under experiment." [105] At the same time, his character—what Emerson called "nature in its highest form"—was put on display and reports of his lectures scrutinized the performative dimensions of his speaking as much as the substance of his message.[106] His features, voice, style and gestures all attracted commentary. On rare occasions, they provided material to find fault. The Washington correspondent for the *Cincinnati Commercial*, for example, offered a less than complimentary sketch of Tyndall's motions: "His style on stage is peculiar. He talks right off, as a schoolboy would a well-learned piece, and without attempt at gesture. He leans over on his table with elbows supporting his weight and his hands either clasped or his fingers drumming against one another. To this add a gentle sway back and forth and you have the Professor as he appears in his lectures. His posture reminds one of a country merchant who has sold a bill of goods and is leaning over his counter talking to his customer."[107] Overall, however, this kind of critical commentary was drowned out by the fulsome praise repeatedly heaped on Tyndall for the attractiveness of his lectures in all respects, not least in terms of visible evidence of his good character.

Tyndall himself was aware that his voice and manner communicated more than the meaning of the words he spoke. When quoting Alexis de Tocqueville's conclusion that "there are few [civilized peoples] in which the highest sciences have made so little progress as the United States," he was sensitive to the dangers of being misunderstood. To avoid this, he first drew attention to his tone and posture. Using an appropriate metaphor, he urged that "an invisible radiation from my words and manner will enable you to find me out and will guide your treatment of me here tonight." His conclusion confirmed his intentions. America could support the scientific geniuses "scattered here and there," by freeing them from other responsibilities and allowing them to pursue "science for the sake of truth."[108]

The "invisible radiation" that emitted from Tyndall announced his character to an audience scanning his lectures for evidence. Tyndall was happy to confirm the significance of this. In his final lecture he announced that "there are things better than science. There are matters of the character." One of the heroes of his lectures, the French natural philosopher Augustin-Jean Fresnel, happily combined "upright character" and "high intellect," demonstrating integrity in claiming no more for his contributions than was just and placing public "suffrage" beneath the pleasure of discovery.[109] And it was not only Tyndall who thought this of the highest value. An editorial in the *New York Tribune,* published immediately after Tyndall's lectures in the city, noted that the "whole secret of his success" was that he had been "thoroughly in earnest." In the editorialist's view, audiences appreciated this more than any other aspect of his lectures. Tyndall's own public performances were designed to display the same kind of devotion to scientific truth, without thought of financial gain or public approval. If his voice and actions did not match those of accomplished orators, to many who heard and witnessed his platform performances they expressed something of primary importance—a sincere and unstinting devotion to truth.

TRANSCENDENTAL NOTES AND RELIGIOUS REACTIONS

One of the most severe tests of character that Tyndall faced in America played out in the arena of religious controversies. If his science, and his brief forays into political concerns, demonstrated his earnestness of character, his dealings with theological questions threatened to undo that accomplishment. Engaging with religious and moral questions nevertheless provided opportunities for Tyndall to lend his lectures a significance that would deepen their appeal. More than this, provoking some opposition, if carefully managed, could provide evidence of a manly resolve and a commitment to truth in the face of disapproval. As Emerson put it, introducing moral considerations into lectures could secure "the advantage of the assembly," by hinting at fresh sentiments that overturned "fossil conservatism."[110]

Tyndall's lectures in America did not spark the kind of controversy that his British Association address would ignite less than two years later in Belfast. Religious or moral issues were not addressed in any sustained or direct way from the platform. As one observer put it, "occasionally in his lectures we have a hint of [his scepticism] but on the whole they are as acceptable as they are interesting."[111] It was straightforward enough for anyone wishing to celebrate his skilful presentation of current science to do so without having to endorse Tyndall's views on religion. Nevertheless, hints of more controversial matters transpierced his otherwise safe lectures on light and attracted considerable attention.

A number of these intimations appeared in his opening lecture, particularly in the introductory sections concerned with the intellectual background to the modern science of light. Early on, for example, Tyndall noted that the notion of causality in nature arose among "savages" who attributed to transcendent beings the "moods" discerned in "the fury and serenity of natural phenomena." Here lay the origins of a belief in supernatural acts of any kind. This may have been innocuous to some, but others took exception. One commentator writing in the *Christian Intelligencer*, for example, accused Tyndall of using a lecture on light to "wipe out . . . the whole biblical faith in a first holy man, in primitive monotheism, and in a primitive revelation from a personal God."[112] Tyndall was being accused of impropriety and of misleading his audience about his true intentions.

Another concrete claim made "in passing" near the start of the first lecture also caught the attention of religious critics. Commenting on the "defects" of the human eye, Tyndall set out to discredit the "common delusion" of perfect sight by demonstrating the effects of spherical aberration and listing other flaws such as asymmetry, chromatic aberration and opacity. One response came in the form of an editorial in the *Methodist*, a weekly newspaper published in New York. Expressing a concern with "how patronizing the new school of sceptical science is to the Almighty," the editorialist criticized Tyndall for claiming to know the "Creator's mind." No one could "see all of the Creator's designs" and nature's operations should not be assessed on the basis of assuming knowledge of what perfection might entail.[113]

Tyndall's incidental remark about the eye, however, was not simply an attack on naïve natural theology. It was an argument made to reinforce his more general point. The science of light could overcome such defects of vision to reveal what was invisible. Experiment and demonstration pushed beyond human limitations and demonstrated the incalculable and incomparable power of science to secure knowledge of the unseen. In other ways, too, Tyndall was more nuanced than his critics allowed. He pointed out that while the eye may not be "theoretically perfect," measured against more instrumental criteria, and bearing in mind "adjustments" that counteracted aberrations, it "remained a marvel to the reflecting mind." At the close of his lecture, Tyndall developed this point to affirm his own philosophy of nature. The eye, along with the brain, allowed humans to register the "exceeding complexity" of light. His peroration on the astonishing capacity of humans to plumb nature's depths ended on a high philosophical note: "Would it not seem that nature harboured the intention of educating us for other enjoyments than those derivable from meat and drink? At all events, whatever Nature meant, and it would be mere presumption to dogmatise as to what she meant, we find ourselves here as the issue and upshot of her operation, endowed with the capacities to enjoy not only the materially useful, but endowed with others of indefinite scope and application, which deal alone with the beautiful and true." This passage presented in brief compass Tyndall's romantic, or "transcendental," materialism. It functioned as a defense against those who might accuse him of a "rank materialism" that undermined morality and civilization. His peroration echoed and deflected the biblical verse often flung at those thought to be flirting with materialism: "If the dead be not raised . . . let us eat, drink, for tomorrow we die" (1 Corinthians 15: 32). But it also provided a philosophical rationale to support the original investigator who was not driven by pressures to meet the material needs of humanity but instead was using the capacities that, whatever their origins, conferred deeper meaning on human existence.

Tyndall briefly alluded to the issues raised by his reflections on the eye in later lectures. In his second discourse, he relayed to his audience a recent conversation between himself and the Scottish natural philosopher David Brewster. In Tyndall's telling, Brewster expressed his reluctance to accept the wave theory of light because it required the "clumsy contrivance" of ether, something God would never have produced. Tyndall begged to differ. Science had a legitimate "quarrel" with those who presumed to "know too much about the mind of the Creator." Here Tyndall anticipated the argument made against him in the *Methodist* and used it to his own advantage.

Tyndall also used the limitations of the human eye to introduce his fifth lecture, stressing that the invisible parts of the spectrum were much more extensive than the part that could be detected by the eye. This fact not only allowed him to demonstrate the power of the science of light but also gave him opportunity to introduce his audience to the future scope of evolution. As reported in the *New York Tribune*, Tyndall observed that "if the theory of evolution be true, then there are in store for man visual inspections far grander than he has yet experienced. These eyes of the future should find beyond the violet and actinic rays which produce chemical effects, rays which produce the decomposition of carbonic acid in plants and which have so great a power in all the processes of animal and vegetable life."[114] Most people in his audience knew, of course, that Tyndall was convinced that the theory of evolution was true. It was a theory that extended beyond the origin of species and accounted for the evolution of the cosmos, of life and human society. In his third lecture on the refracting effects of crystalline forms, Tyndall introduced his audience to the hidden architecture of nature, built by "the play of polar forces." In exploring the molecular processes that produced exquisitely beautiful mineral structures, Tyndall invited his listeners to ask whether vegetable and animal forms might not also have been constructed by the same molecular action. The fernlike arrangements produced by electrolysis demonstrated the tendency of inorganic nature to produce structures that strikingly resembled patterns found in the living world. Tyndall was a critic of theories of spontaneous generation he deemed unsupported by experimental demonstration, but in his American lectures he hinted at his belief that the puzzle of life's origins would be found in the self-organizing nature of matter.

In the very same passages where he offered such metaphysical teasers, Tyndall also compared the productive forces of nature to the workings of human society. In a vivid metaphor, he compared overly rapid crystallization to a nation that resists "healthy and natural change." The anarchy of revolution was the result, an outcome that might have been avoided had society not become "oversaturated with the desire for change." In using this analogy, Tyndall gently pushed in the direction of the naturalization of social change and left his audience to wonder whether he was referring to a society that refused to accept the implications of scientific developments for cherished but exploded beliefs.

In these and other ways, Tyndall's romantic naturalism occasionally but significantly bubbled to the surface and imparted a frisson to his otherwise safe lectures on light. As Tyndall had done in previous lectures, he inserted ideas that, in fuller, clearer form, would so offend many of those who would hear his Belfast address some eighteen months later. In America, suitably enough, Tyndall adopted the Emersonian technique of interspersing comfortable content with radical asides, giving his lectures a transcendental tincture.[115] Long an admirer of Emerson, Tyndall met him in Boston during his lectures there and quoted him liberally in his American addresses. Even before this, Tyndall's transcendentalist credentials had already been confirmed. In October 1869, for example, he had been the subject of high praise and detailed description in a letter published in the *New*

York Tribune by another prominent transcendentalist, George Ripley. In his letter, Ripley adopted Tyndall as an honorary participant in what was by then a diffuse but still influential force in American intellectual culture.[116]

The passing references to his metaphysical proclivities rarely registered in the newspaper reports of Tyndall's lectures. They periodically flickered amid dazzling projections and scientific exposition but did not result in a significant newspaper controversy. What did attract considerable and sustained commentary were Tyndall's recent interventions in the debate about prayer. Tyndall himself was fully aware of this. In a letter written two months into his tour to the London-based chemist Heinrich Debus, Tyndall exclaimed, "you cannot imagine the stir this prayer question has excited in America."[117] Throughout Tyndall's tour, accounts of his lectures were punctuated with sermons, letters and opinion pieces attacking his proposal to apply an experimental test to prayers for the sick.[118] Even before his arrival in New York, the Fulton Street Prayer Meeting in New York called for churches across the United States to "petition for Mr. Tyndall's conversion."[119] This was not a small concern. The Fulton Street Prayer Meeting, founded in 1857 and associated with a nation-wide religious awakening, attracted thousands and had the ear of large swathes of American evangelicals.

Unlike the preachers who used their pulpits to defend the efficacy of prayer, Tyndall did not directly address the issue in the substance of his lectures. As we have seen, the message about experimental demonstration and epistemic authority as it related to matters such as prayer, providence and the miraculous was subliminal and subtle. On at least two occasions, however, Tyndall did refer to the prayer debate in ways that his audience immediately recognized. The first came at the close of his opening lecture delivered in Washington's Lincoln Hall on December 3. As noted, Tyndall was fully conscious of where he was delivering his lecture and who was likely to be in the audience. It was an opportune moment to emphasize the friendly relations between Britain and America, no longer clouded by tensions generated during the Civil War. It also provided an occasion to remind his audience that he was in no sense against "prayer," if it was understood as the expression of the deepest longings of the human heart. In what was later referred to as his "Washington prayer," Tyndall expressed "the profound gratification with which I read this morning the words of the President of the United States, that not a shadow now rests on the friendly relations of the governments of America and England. . . . I am not, as the proprietors of this beautiful hall [the Young Men's Christian Association] well know, in the ordinary sense of the term, a praying man. But if prayer be the heart's sincere desire then I may claim to pray, that the hopes and wishes of General Grant may receive ratification and fulfilment in the coming time."[120] Tyndall's peroration associated a sentiment that his audience could not but welcome with a candid but carefully worded confession about his views on prayer. It was a skilfully constructed piece of oratory designed to elicit warm praise and generate tolerance for his heterodox opinions.

It was a sermon on prayer in New York by John Hall, a prominent Presbyterian minister, that prompted Tyndall to make his second public remark about the prayer debate.[121] On this occasion he referred to it indirectly, though obviously. At the close of his fourth lecture delivered at New York's Cooper Institute, Tyndall noted his regret that "in the course of some observations which he had put forth, and when endeavouring to judge matters by purely scientific rules, he had aroused some bad feeling on the part of certain clerical gentlemen, Presbyterians, principally, he believed." Those clergymen had "been tempted to say bad things of him." According to the *New York Tribune*, Tyndall "could only reply in the language of a book he deeply venerated from his childhood and obey the order to 'pray for them that despitefully use you and persecute you.'"[122] Other reports differed. The *New York Times* (more accurately as it turned out) had Tyndall say that he was "resolved to act according to an old book which he greatly revered 'not to render railing for railing, but contrariwise blessing.'"[123] Whatever the exact form of words and scriptural quotation, all reports agreed that the audience warmly received Tyndall's comments. The account published by the *New York Times* observed that the applause lasted for ten minutes, a fact that "seemed to thrill the lecturer, for his face flushed proudly and his eye gleamed and . . . his face smiled all over as if the inner man was exceedingly comforted."[124] Tyndall had skilfully presented the view reinforced throughout his lectures that verification by experiment was the only trustworthy route to knowledge while expressing reverence for a book many took to be the only reliable source of knowledge about supersensible realities.

As subsequent correspondence published in the *New York Times* made clear, Tyndall believed his clerical critics had accused him, among other things, of coming to America simply to make money. He noted this in a letter to the *New York Times*, but then apologized for being misinformed, stating that no such slur had been uttered.[125] In response, the Reverend John Hall wrote (anonymously) to accept the "friendly hand offered to him" but used the occasion to criticize Tyndall for employing a rhetorical trick associated with "third-rate politicians." At the close of his lecture, Tyndall had used "the readiest and cheapest weapon against a clergyman" by claiming greater respect for "practical religion" than those charged with its proper profession.[126] Tyndall had also accused not just a single clergyman of attacking his honor, but Presbyterians in general. It was to them that

Tyndall owed a full and public apology. Hall's letter amounted to a direct attack on Tyndall's moral character as displayed on the lecture platform of the Cooper Institute.

There is no evidence after the New York incident that Tyndall spoke publicly about the prayer debate. This was despite the fact that he received a request to directly address the issue in his lectures in Brooklyn. In the letter to Tyndall calling for a response, the correspondent, Albert Jackson, noted that another popular Presbyterian minister, Reverend Thomas De Witt Talmage, had preached against Tyndall's views on prayer and "enlarged on the infidelity of science" more generally.[127] What made this worse in the eyes of Tyndall's correspondent was the fact that Talmage—who attracted a congregation of thousands—was now preaching every Sunday in the Academy of Music (his tabernacle had burned down a few weeks before Tyndall's scheduled lectures). Despite this immediate juxtaposition, Tyndall did not respond and there are no reports of him addressing the prayer debate in his Brooklyn lectures.

In his letters written during and shortly after his tour, Tyndall barely mentions these skirmishes. His encounter with Hall, "an ill-tempered Irish Presbyterian," is dismissed as "a slight passage of arms" not even worth mentioning.[128] Others, however, were more troubled by the steady stream of criticism leveled at Tyndall by the religious press. His supporters were certainly anxious to protect his reputation as someone who could be trusted not to abuse the lecture platform by using it to promote political or religious aims. It was vital, then, that the banquet to be held in his honor before his departure to London be carefully choreographed and controlled. The reality turned out to be a rather different story.

"A WAKE OVER THE REMAINS OF A DEAD RELIGION"

Before his departure home, a banquet for Tyndall was organized at New York's celebrated Delmonico's restaurant. At that time, the restaurant was a powerful cultural institution, which in the words of one chronicler "influenced the manner, tastes and customs not only of the city but of the nation."[129] Five years before, Charles Dickens had been lavished with French cuisine and fulsome praise at the end of his reading tour in America. Anthony Froude's publishers had organized a banquet for the start of his lecture tour, forestalling D. Appleton and Co., who had hoped to organize a similar event to welcome Tyndall to America. Tyndall's time came, and the banquet was set for the evening of February 4.

Tyndall had, in fact, been anxious to cross the Atlantic after he concluded his lectures at Yale. As he expressed it in a letter to Joseph Henry, though eager to get home and inclined against public dining, "many excellent men here think that in the present state of the public mind; if men of weight in science could be collected together, and would but express their views, something of permanent value to the science of the country might accrue."[130] To make it a success in these terms, identifying someone to chair the after-dinner speeches was of some importance. Neither Joseph Henry nor Louis Agassiz was able to attend, so the lawyer and statesman William Evarts was approached and appointed.

On the evening of the banquet, nearly two hundred "scientific and prominent men" were seated around tables at Delmonico's. Listing many of the diners, the *New York Times* noted that "seldom have so many distinguished gentlemen assembled together from all parts of the Union to do honour to a single individual."[131] As was usual for banquets celebrating an English visitor, the dining hall was decorated with the flags and arms of England and America. Once the "cloth" had been removed, the speeches began with Evarts's "wise and witty and fluent" opening remarks. Tyndall's speech followed and provoked an extended comment in the *New York Tribune* describing his performance:

> Mr Tyndall, while he never transcended the limits of an after dinner speech, alternated his play of humor and fancy with a serious and inspiring appeal to the young men of America on behalf of pure science. There was something indefinably dignified and austere about his voice and his manner, speaking as he did with the 'authority of a student of science who had never attempted to make the world a dollar richer by it.' Americans have all learned to honor Prof. Tyndall as a scholar and to admire him as a lecturer, but all who had the pleasure of hearing him . . . will have gained a new regard and respect for him as a Man.[132]

This portrait might readily have been applied to his entire tour. But it was fitting that his last public performance had revealed him "as a Man" and put on display the character-building potential of his scientific and metaphysical vision of nature and morality.

The coupling of science and character was certainly a recurring motif in several of the speeches delivered at Delmonico's. Tyndall gave it a first airing in stressing the importance of duty and self-sacrifice in pursuing a scientific education. He had sought to master science not for "material reward" or for the "glory" of success but out of a "sense of duty" that lent "tone" to his character. This theme was picked up again by Andrew Dickson White, president of Cornell University, who pointed to the relevance of science for political progress by picking out the character traits of scientific workers. Such workers demonstrated a "zeal for truth as truth," along with "bravery," "a sense of duty" and "faith that there is a Power in the universe that forbids any honest truth seeking to lead to lasting evil." These virtues were vital if the country were to witness "political regeneration" and overcome a skepticism and materialism that valued "mere comfort and wealth" and rejected the idea that "the true and the good were one."

During these proceedings, Tyndall further underlined his own rejection of this kind of money-grubbing materialism by presenting himself not as "a servant of mammon" but rather as someone committed to strengthening and enlightening minds. It was this, he urged, that motivated his appeal to America's "men of wealth" to support the cause of "pure" rather than "practical" science. In this, Tyndall led the way. The profits of his lectures would be donated to a fund designed to support the progress of science in America. A few days letter, Tyndall signed a deed of trust that placed the $13,033 earned from his lectures to an American board of trustees responsible for using it to aid the "advancement of theoretic science."[133]

Tyndall's resolute commitment to truth no matter what the cost to personal wealth or beliefs supplied the theme of the speech by Henry Ward Beecher, a celebrated minister of Plymouth Street Congregational Church in Brooklyn and by then notorious for being the target of accusations of adultery.[134] Beecher sought to remind his audience that clergymen had assisted scientific progress in America by following truth "wherever it led them." He commended those who, like Tyndall, taught scientific truth "without regard to the social or civil or the theological results." Religion, after all, was not to be defined according to doctrinal commitments but rather as "a personal thing, another name for manhood." Science, which aimed to produce better men, was "in the direction of true religion." As a result, Beecher could hail Tyndall as "a priest, ordained and in the same diocese."[135]

For all the plaudits, two speeches sounded a dissenting note and raised a question mark over the positive impact of Tyndall's visit and his widely broadcast views. The first, by Professor Roswell Hitchcock of the Union Theological Seminary, was short but pointed. Noting that he was not going to apologize for being "one of those Presbyterians whose pulse had been quickened a little" by Tyndall's visit, he denied that religion and science were locked in a struggle to the death. If such were the case, religion, which unlike science was essential to humankind, would win out. It was fortunate, then, that "the path was broad enough for both science and religion."

Parke Godwin, editor of New York's *Evening Post*, dilated upon such concerns at much greater length. Although responding to a toast for the press, Godwin spent his time reminding those present that "real science" was clearly confined to "its own sphere." It dealt with facts and their coordination and resisted "wandering away into the void inane." The examples Godwin gave of those who transgressed this clear demarcation were telling: a "German doctor" who reduced everything in the universe to matter and force, another "outside teacher of science" who subscribed to a "cosmic evolution" that traced an Iliad or a Shakespeare back to a nebulous gas, and a "French litterateur" who reduced mind to the motion of molecules. While it remained unspoken, many of Godwin's listeners would have been aware that Tyndall had already associated himself with all of these "gratuitous" propositions. Godwin made it clear that anyone making such claims were like Grub Street hacks defaming science by extrapolating well beyond what facts and their coordination allowed.[136]

The efforts to prevent such speeches spoiling the occasion were not helped by the fact that one response to Tyndall's banquet lauded the retreat of "faith" in the face of scientific advance. Written by the physician and man of letters Oliver Wendell Holmes Sr., the response—printed in several newspaper reports of the banquet proceedings—was a manifesto for the "renovation" of the intellect and dispelling the "thick mists" of supernaturalism. According to Holmes, Tyndall was a representative of those who "give their lives to the sincere study of Nature," men who were "more and more our leaders in this morning twilight of knowledge." They were "oracles," "prophets" and "apostles" sweeping away the "oxidated beliefs of the past" and declaring faith "bankrupt." Such "attacks" were for the wider good. Science would teach how to "look for truth at first hand and not be afraid of it." In closing, Holmes announced that what would remain, after all "opinions disproved by facts" had expired, was "the law above all laws which shows our reason what we ought to be and commands our will in the authoritative accents of duty." [137] Such sentiments were judged by some to be in bad taste, offending the beliefs of many who were at the banquet and undermining the moral fabric of civil society. As an editorial in the *Cincinnati Daily Gazette* put it, Holmes's letter had made the banquet "virtually a wake over the remains of a dead religion."[138]

In its aftermath, the banquet for Tyndall generated something of a food fight between those who sympathized with his vision of science and those who found it overextended and damaging to a Christian conception of nature and morality. The *Popular Science Monthly* became one forum in which these tensions played out. As editor, Edward Youmans sparked several exchanges when he offered a sustained critique of Godwin's speech. In turn, Godwin's response found support from John Christopher Draper (the son of John William Draper) in letters Youmans published and then critiqued. The nub of the matter concerned the limits of scientific knowledge and the materialist tendencies of Tyndall-style science. The tense set of exchanges culminated in a letter from Tyndall rebutting the charge that he had set out to attack religious faith in his lectures. Specifically, he denied that he had targeted Presbyterians as a group; he further accused the Reverend John Hall of "wilfully twisting" his words to give that impression. In an important statement, Tyndall asserted that he had "steadily refused to quit the neutral ground of the intellect during my visit to the United States." The witnesses were his audiences in the urban centers where he lectured. They could "testify whether a single word relating to religion was heard in any lecture of mine delivered in those cities." His "heroic" New York audience would confirm that his "attack" on Presbyterians was in fact a reference "in mild language, to the reported words—*reported, I would add, by a Presbyterian*—of the intemperate occupant of a single Presbyterian pulpit." Tyndall insisted he was telling the truth (and his opponents on this question, "unless they be in a state of invincible ignorance," were doing the opposite). Pushing his message home, he declared that "he would fearlessly trust to the manhood of any Young Men's Christian Association in the Union for a verdict in this matter."[139]

Protecting Tyndall's reputation in America as a speaker who avoided giving unnecessary offense was also of concern to some of his key supporters. A few weeks after the New York banquet, Professor Alfred Mayer wrote to Tyndall to assure him that the "demonizing" speeches of Hitchcock and Godwin had taken the organizers by surprise. They had been allowed a hearing by the inept chair and had regrettably taken the place of speeches by those "eager to make the parsons feel that science could take care of herself without clerical patronage." In a bid to limit the damage, Mayer had "taken a decided stand against the publication of these homilies." He further reported that "Youmans . . . will control the Appleton in the matter." In place of the rogue responses, Mayer proposed including letters of support from "prominent scientific men." Publishing those "will give proper balance and character to the celebration."[140] Still, against these intentions, while these letters of support were indeed included in the official proceedings published by D. Appleton, Godwin and Hitchcock's speeches also appeared in full.[141]

It was not only Tyndall's more vocal religious critics who worried that he was driving a wedge between science and conventional religion. Joseph Henry, who was so instrumental in organizing Tyndall's entire tour but who was unable to attend the banquet, wrote to the geologist Benjamin Silliman Jr. to express concern that Tyndall's participation in theological controversy had given the impression that "cultivators of science" were antagonistic to Christianity.[142] His tour had become a lightning rod for debates about the future of American religion and American character as much as American science.

Writing in the New York *Galaxy* magazine nearly three years after Tyndall's visit, the American literary critic William C. Brownell offered a survey of "English lecturers" who had visited the United States over the previous several years. Of all the lecturers, only two were judged to have been a success: John Tyndall and the astronomer Richard Proctor. The others that Brownell considered, including Anthony Froude and George Macdonald, had failed not because they came with nothing of interest to say but because they lacked ability in the lecture hall. Froude had read his lectures "very rapidly and with no oratorical eloquence" and Macdonald had "none of the merits of elocution" that might otherwise have covered up the mistake of sermonizing in lyceums. Edmund Yates's error was perhaps yet more fatal. He had come to make money, to sell second-rate "foreign wares" from the lecture platform. Tyndall, on the other hand, had displayed "all the electricity of a natural orator" and had "moved" his audience as much with his scientific exposition as with his experiments.[143]

An earlier assessment in the same magazine had provided a different, but complementary, account. Tyndall's remarkable popularity was due to "the character of the man" and the application of his "genius" for exposition of a subject that otherwise lacked the interest provoked by more controversial topics. Lecturing on light lacked the excitement of "novelty," scientific controversy or "heresy." Without those features, the fact that Tyndall's course drew unprecedented audiences, "both in number and intelligence," could only be explained through an appeal to the "personality of the lecturer."[144] His earnest and sincere manner—exhibited in his platform performances and confirmed by his donation of the profits of his tour to further American science—had made his tour an outstanding success.

This explanation parallels Graeme Gooday's argument that Tyndall's authority as an expert on electrotechnical science relied upon compelling lecture performances in and beyond the Royal Institution. Tyndall's reputation as an authoritative expert was, Gooday suggests, "a matter of trustworthy performance."[145] When verbal performance was judged a failure, Tyndall's status as expert came under strain. Underlying this emphasis on the performative nature of Tyndall's reputation was a concern with the character of the scientific man. This was not only Tyndall's conviction, born of sustained engagement with the prophets of character and personality, Emerson, Carlyle and Fichte. It was also a conviction that helped to shape political and intellectual culture and informed the judgements made of "prominent men," whose public visibility often rested on evidence from platform performances that they had "master[ed] the science of persuasion."[146]

The debate over Tyndall's conduct in his dealings with religious matters were couched in the language and politics of character. The struggle to present himself as a man of character and feeling was animated by an undercurrent of debate over the true sources of a moral self. Tyndall's own religious vision had the notion of character at its heart. Given the inscrutable mystery of the "power" that lay beyond human knowing, the human character was invested with considerable metaphysical significance. This not only echoed an Emersonian elevation of character as a "supreme power."[147] It also resonated with a Carlylean celebration of great men—the heroes deserving of worship—as the most striking revelations of the divinity found in nature. Tyndall later confessed that in his conversations with Carlyle he had earnestly tried to persuade the Scottish sage that there was none more fitting of the epithet hero than the "man of science."[148] It was no wonder, then, that Tyndall's lectures projected an image of the man of science as a man of character and stressed the ultimate importance of such recognition. He embodied the Emersonian understanding of the public lecture as "being concerned with being moved, hearing what was beyond mere words" and as a "performance of philosophical disposition."[149]

Tyndall's reputation in America as an attractive speaker appears to have persisted even after the fallout from his Belfast address in 1874. Some three years after Tyndall's tour, a newspaper commentator compared his performances with those of another visiting English scientist of high repute, Thomas Henry Huxley. In "mere manner" Huxley was not as attractive as Tyndall. The more athletic of the two, Tyndall had an "easy erectness of bearing" and had a "stronger and clearer voice." He also, apparently unlike Huxley, "spoke entirely without notes and had the art of putting himself in more complete sympathy with his audience." Huxley's lectures also lacked the "brilliant and surprising experiments" that had enlivened Tyndall's discourse and captivated his New York audience.[150] These judgements were not made in complete independence from the subject matter dealt with by the two men. Where Tyndall lectured on light with minimal reference to religion, Huxley dealt with evolution, with significant attention to the consequences for religious belief. There were, in the lectures themselves, more reasons to provoke the ire of those offended by Huxley's arguments. That this expressed itself in the form of attacks on his lecturing capabilities is not surprising given the significance attached to the man speaking. As we will see in the next chapter, such attacks targeted aspects of Huxley's lecturing style that he had deliberately cultivated to effectively communicate the truths of evolution in the public sphere.

CHAPTER 2

REASON'S RHETOR

The Scientific Oratory of Thomas Henry Huxley

In his *Descent of Man* (1871), Charles Darwin includes, among a mass of evidence for human evolution, a description of the natural history of oratory. The public speaker, Darwin observes, varies pitch, rhythm and tempo to express and stimulate vivid emotions in a way strikingly similar to the use of tonal diversity in communication among monkeys. For Darwin, the "cadences of impassioned oratory appear . . . like mental reversions to the emotions and thoughts of a long-past age." Elaborating on this hypothesis, Darwin notes that in stirring the feelings of his auditors, the orator "little suspects that he uses the same means by which, at an extremely remote period, his half-human ancestors aroused each other's ardent passions, during their mutual courtship and rivalry."[1] Such speculations, however intended by Darwin, chimed with a widely held view of oratory as inimical to scientific reason because it inflamed rather than restrained the emotions.

It was often supposed that Thomas Henry Huxley, self-appointed spokesman for Darwin's theory of evolution, embodied this negative appraisal of the oratorical arts. By 1871 media outlets regularly portrayed Huxley as a speaker who studiously avoided the "tricks of oratory" and labored hard to speak plainly.[2] This style of plain and passionless address did not just stem from word choice and logical argumentation. It was also a matter of pace, diction and posture. Huxley achieved his reputation for restrained and precise exposition as much by a steady tempo and clear pronunciation as he did by skillful composition and transparent terminology. His bodily stillness also contributed to the overall effect. Carlo Pellegrini's much-reproduced caricature of Huxley, published in *Vanity Fair* in 1871, captured the convergence of substance and style. Arms folded, eyes set and feet apart—Huxley's pose was "wonderfully matter of fact."[3]

Huxley affirmed this image as he looked back on his long career as a lecturer and public controversialist. In brief autobiographical reflections

FIGURE 2.1. Thomas Henry Huxley (Men of the Day no. 19) by Carlo Pellegrini. Chromolithograph published in *Vanity Fair*, January 28, 1871, 306. By permission of Special Collections, Queen's University, Belfast.

written near the end of his life, Huxley recounted how he missed out on "that mellifluous eloquence which, in this country, leads far more surely than worth, capacity, or honest work, to the highest places in Church and State." Instead he had restricted himself to "saying what I mean in the plainest of plain language."[4] As he had put it some years before, this disavowal of elaborate or contrived speech was accompanied by an aversion to the "superstition" that anyone who wants to "write and speak English well should mold his style after the models furnished by classical antiquity." For Huxley, the only rule for producing an effective style of address, written or spoken, was to "strive after the clear and forcible expressions of definite conceptions." This alone would lead to "true eloquence" and steer any novice writer or orator away from the "slipshod copiousness" of so many apparently "distinguished contemporary speakers."[5]

In making these remarks, the elderly Huxley was living in the past. Even at the start of his career as a public speaker, parliamentarians were moving away from a style modeled on classical oratory. As Josephine Hoegaerts has argued, an era of political reform and increased franchise required a more "democratic" voice that spoke in a language and manner that could be clearly heard by reporters and appreciated by middle-class voters. A more controlled and less dramatic performance marked by a lack of verbal and nonverbal artifice secured perceptions of authenticity.[6] The liberal MP John Bright exemplified this new form of political oratory. Contemporaries judged his disciplined delivery, giving "his words force without theatrical gesture," coupled with minimal modulation, as better adapted for a democratic age.[7] Huxley expressly admired John Bright's speeches and adopted a similar style.[8] Despite Huxley's retrospective remarks, his platform performances were in tune with those increasingly favored by practitioners and critics of parliamentary speech.

The shift from rousing speech that made elaborate use of bodily movement to plain speaking delivered with minimal use of gesture also resonated with a growing mistrust of the body as a vehicle for communicating truth. The body's natural motions, and the passions that they provoked, were regarded by those skeptical of nature's role as a repository of divine meaning. If Michael Faraday could regard "natural" (if carefully stage-managed) gesture as a handmaiden to the revelation of scientific truth, Huxley could only regard chironomy as a distraction from the task of imparting true knowledge of nature.[9] There was for Huxley no natural theology of the human body. The latter was not designed to communicate truth and required careful management to avoid deception or undue distraction.

Taken together, these influences fed a suspicion of the classical arts of oratory and pressed in the direction of an ideal that imagined a disembodied voice free from the deceptive play of emotion. Not everyone, of course, shared this new ideal. Writing for *Fraser's Magazine* in 1874, the journalist Thomas Escott diagnosed a general decline in the quality of parliamentary speeches and singled out Huxley for blame, arguing that literature and not science generated noble political oratory.[10] Public speaking that derived inspiration from science was characterized by a lack of charm and transformed the statesman into a specialist. To someone like Escott, scientific oratory alienated listeners and devalued the public import of political speech.

Still, despite appearances, Huxley's scientific oratory should not be too readily aligned with the caricature presented by Pellegrini and attacked by Escott. That image was indeed a parody and a half-truth. Huxley's own account of science lectures was more complicated and conflicted. In a revealing defense of the genre written late in life, Huxley admitted that the "purely intellectual value" of lectures was modest. Perhaps more important was that "the living voice has an influence over human action altogether independent of the intellectual worth of that which it utters." Among other consequences, "the most zealous of popular lecturers can aim at nothing more than the awakening of a sympathy for abstract truth, in those who do not really follow his arguments; and of a desire to know more and better in the few who do."[11]

These seasoned thoughts summarized a view Huxley had held for much of his career as a science lecturer. In 1853 he advised his friend John Tyndall to work above all else on "clear powers of exposition—so clear that people may think they understand even if they don't." After all, he continued, "not a tithe" of the audience that listened to Michael Faraday, the acknowledged prince of science lecturers, understood him.[12] Huxley's remarks described here, made decades apart, suggest a different account of public speaking to the one commonly associated with him. Speech about science was effective when it provoked a feeling that one had been stirred by truth not yet comprehended. In his lectures, Huxley aimed to illustrate what a scientific account of the world feels and looks like without expecting the bulk of his audiences to grasp or remember much of the cognitive content of his oral discourse.

The tension between speech that constrained emotions in the service of communicating and comprehending scientific truth, and speech that stimulated a certain kind of emotion, a sympathy for science, found expression in Huxley's ambivalence toward training in the oratorical arts. Huxley, like John Bright, made it clear that he had benefitted from a *lack* of such training. The arts of public speaking, as Darwin's natural history of them had made clear, had little to do with ratiocination. They were best avoided in the interests of developing a style of speaking that facilitated dispassionate scientific reasoning. At the same time, Huxley conceded the need for a strategic embrace of those otherwise disreputable arts. After his much-discussed clash with Bishop Samuel Wilberforce in 1860, Huxley declared that he had "changed my opinion as to the practical value of the art of public speaking, and that from that time forth I should carefully cultivate it and try to leave off hating it."[13] Later in life he described how, in his early attempts at lecturing, he had paid careful attention to the rudiments of effective speech. His first efforts, he confessed, were marked by "every fault a speaker could possess (except talking at random and indulging in rhetoric)." Those same efforts provided vital opportunities for the "oratorical experiments" necessary for forging a career as a public speaker.[14]

Among those aspects of effective speech that Huxley worked at was an ability to deliver an address extemporaneously. At least from the late 1860s, commentators frequently observed his ability to speak without notes, giving a powerful impression of spontaneous invention. Huxley's own thoughts on this skill, again expressed toward the end of his life, are revealing. On the one hand, he emphasized the importance of writing out a lecture in full (especially the parts likely to cause the speaker to "give way" to their feelings) and then sticking resolutely to an outline. In this way, the lecturer would be more likely to produce a logically cogent address and also resist the "strange intoxication" produced by "the breathless stillness of a host of absorbed listeners."[15] At the same time, Huxley confessed to an admirer of his lecturing style that the "stimulus" of an audience often produced "better words and phrases than I have thought of at my desk."[16] It was only as an inexperienced lecturer that Huxley relied on a full script. As his career matured, he generally brought only a skeleton outline to the lectern and increasingly ignored his notes altogether. With the necessary controls in place—painstaking preparation and a condensed sketch of a logical sequence of thoughts on hand—Huxley could discard notes and, to all appearances, play to the crowd.

In this, again, Huxley was adapting to rather than resisting wider trends in oratorical culture. Here the pulpit provided a particularly prominent and pervasive model. By the 1860s the trend, particularly outside Anglican circles, was toward an extempore style, even if some celebrated preachers continued to read their sermons. As Robert Ellison has argued, by the middle of the nineteenth century most guides to preaching advised against reading and argued for the superiority of extemporaneous delivery, especially if the majority of a congregation was made up of less-well educated people.[17] The reasons for this advice largely turned on the presumed vivacity of extempore address over and against the bookish coldness of reading. None of this meant, however, that preachers were encouraged to enter the pulpit unprepared. If the words used were invented during delivery, the thoughts they expressed must be settled well in advance. As Emily Cope has shown, that same advice exerted a powerful influence among American preachers.[18] It is certainly significant that one of the most successful of evangelical preachers during this period, Charles Haddon Spurgeon, echoed this line.[19] Astonishingly popular on both sides of the Atlantic—as well as addressing thousands each week in person, his published sermons sold in the millions—Spurgeon's pulpit style helped to promote extemporaneous speech as authentic and authoritative.

Given this, Huxley's favored form of address—particularly when communicating with audiences outside of elite institutions—might reasonably be characterized as evangelical in style. The evangelical and Nonconformist character of Huxley's efforts to promote a scientific "gospel" has, of course, been noted before.[20] But what is most relevant here is that this was true in a very specific sense—his platform performances adopted and adapted a speaking style very familiar to evangelical preachers and their hearers. It was not quite, as Adrian Desmond put it, that Huxley adopted the style of the "tub-thumper . . . preaching old-time hellfire" when speaking to arti- sans.[21] Rather, he cultivated a form of address that treaded a path between a classical style that emphasized literary quality and was often fully scripted and the spontaneous and volatile speech associated with revivalist or radical evangelicalism.

As well as working at an extemporaneous delivery, Huxley also labored at other aspects of speech performance considered crucial by contemporary critics of public address. One of the most important, and challenging, was vocal volume. However successful Huxley was perceived to be in adopting a widely admired speaking style, his struggle to project his voice in large venues was regularly commented upon and criticized. One of the earliest instances comes from a private report of his exchange with Samuel Wilber- force at the Oxford meeting of the British Association for the Advancement of Science in 1860. His friend Joseph Hooker observed in a letter to Charles Darwin that Huxley, despite admirably answering the bishop's attack on Darwinism, "could not throw his voice over so large assembly or command the audience."[22] The same vocal weakness was noted toward the end of his career. In a report of Huxley's Romanes lecture, delivered at the University of Oxford's Sheldonian Theatre, the volume of his voice was judged "too low." Most in the audience "heard with difficulty" and "many hardly heard at all."[23] On that occasion, he had been recovering from a throat infection that had exacerbated the already considerable challenge he faced in speak- ing loud enough.[24] But the two struggles to be heard at Oxford were book- ends to a career-long battle to speak with adequate volume while lecturing in large auditoriums. It was likely this reputation for vocal weakness that led the *Popular Science Monthly* to note in its column advertising Huxley's visit to the United States, perhaps in anticipation of criticism, that "Hux- ley's manner as a speaker is very quiet . . . but his discourse is clear, finished, deliberate and strong."[25]

Huxley's low tones were more than just a practical inconvenience. Vocal strength was a quality that elocutionists often highlighted as a key compo- nent of effective public speech. John Broadus and Charles Haddon Spur- geon—two of the most influential commentators on effective pulpit oratory on both sides of the Atlantic—both insisted on the vital importance of a well-projected voice. Spurgeon was typically direct: "What is the use of a preacher whom men cannot hear? Modesty should lead a voiceless man to give place to others who are more fitted to the work of proclaiming the messages of the King."[26] Broadus was equally forthright. Quoting Cice- ro, he argued that "for an effective and admirable delivery . . . the voice, beyond doubt, holds the highest place." An effective speaker, Broadus ar- gued, should have full command of the voice's four "powers": its "compass," "volume," "penetrating power" and "melody."[27] A weak voice, lacking in resonance and reach, was the result of an unmanly body. These judgements pointed to the connections made between a strong and "manly" voice and public credibility and reputation. It also meant that when critics turned their attention to a speaker's voice, they could draw on a large repertoire of terms to critique not just a speaker's vocal power but also the credibility of their spoken words. It was perhaps for this reason that the American journalist George Smalley—one of Huxley's most ardent admirers—recalled in 1895 that Huxley's voice was "rather deep, low, but quite audible, at times sonorous and always full." For Smalley, Huxley's platform performances gave "an impression of sincerity, of solid force, of immovability. . . . He was masculine in everything. Look, gesture, speech."[28] As we will see, this was not the only view of Huxley's platform presence and vocal power when he spoke in America.

Huxley's strenuous efforts to communicate effectively, and to move an audience in the right emotional direction, were also shaped by the reporting practices of newspapers. Controlling what appeared in the press was, for all public speakers, a huge, even impossible, challenge. This applied to representations not only of what was said but also of how it was delivered. In a period that saw an explosion in the recording of public speech in print, reporters played an increasingly important role both in increasing the audience of speech events and in creatively reworking the content and connotations of spoken discourse.[29] While this was particularly apparent in reports of parliamentary deliberation, it applied to all manner of oral performances. Consequently, public speech was increasingly written and performed with more than an eye on what might appear in the columns of the national and local press. Huxley was himself acutely aware of the dangers of misrepresentations of oral discourses in newspapers. This was one reason why he stuck resolutely to a prepared outline, refusing to be seduced into saying something he would regret reading the next day in the local newspaper. He also labored hard to ensure that reporters provided adequate summaries of his lectures, particularly when these were on controversial subjects. A slow tempo—about one hundred words per minute—and clear diction were crucial if stenographers were to record his lectures with any degree of accuracy.[30] If possible, checking the resulting transcription was also important. In 1874, after his lecture on animals as automata delivered during the Belfast meeting of the British Association for the Advancement of Science, Huxley visited the offices of the *Northern Whig* late at night and spent two hours reading and correcting the report that would appear in its pages the next day.[31]

The problems Huxley faced in controlling how his lectures were represented in the press point to another facet of his efforts to strategically embrace the arts of oratory. It was crucial to Huxley's success as a scientific orator to develop a flexibility that allowed him to adjust his speaking style for different audiences and occasions. In that sense, the cultural geography of his science lectures—or the more local influences that conditioned his speech acts—needs to be carefully considered.[32] Huxley's adaptability was vital given the sheer variety of institutional and civic spaces in which he spoke. Yet such versatility can also be viewed from the opposite angle. The spaces in which Huxley spoke—understood as material, institutional and cultural sites mobilized and modified through speech acts—shaped representations of his platform performances in ways that were well outside his control. Huxley's efforts to vary his address to suit or stir particular audiences can be approached not simply as a product of his own agency but rather as a result of negotiations with sometimes dramatically different local norms and material settings.

Throughout the 1860s Huxley spoke in venues that ranged from elite spaces such as the Royal Institution, the Royal Academy and the Royal Society to provincial auditoriums and drill halls. He also had to manage the varying expectations associated with different types of public address: after-dinner speeches, lectures to students at the School of Mines, more formal addresses to learned institutions and his self-described lay sermons, to give just some examples. Huxley undoubtedly struggled to control his message in the face of the powerful cultural associations of these kinds of speech events. His high-profile involvement in Sunday lecture societies, for example, helped reinforce his reputation, despite his own protests to the contrary, as a preacher of materialism who was undermining public morality.[33] His lectures to "working men," inaugurated in 1855 at the School of Mines and considered by Huxley and others to be among his most successful oral discourses, were frequently heard and read in ways that Huxley could not approve. The social cross-pressures, stirred up by the public and dramatic nature of speech events, not only shaped the substance and style of Huxley's lectures "for the people" but also destabilized and reworked their meaning to suit class-based political causes at odds with Huxley's own agenda.[34]

An analysis of the specific variables at play in particular venues helps to further confirm the importance of attending to the local forces that influenced the content, character and consequences of Huxley's lectures. Take, for example, Huxley's controversial lectures on the relations between man and other animals to the Edinburgh Philosophical Institution in January 1862. Part of Huxley's aim in delivering the lectures was to provoke a backlash from Edinburgh's evangelicals, a move designed not to court consent but to whip up rivalry. This, he suggested, would circulate his arguments "through regions which they otherwise would not reach."[35] But he also looked for approval from his socially elevated audience of men and women from Edinburgh's large and diverse middle classes. While Huxley's own version of events suggests he was successful on both counts, others demurred. Joseph Hooker felt Huxley "had no idea in how small a circle he makes a noise," and that his efforts to influence religious politics in Edinburgh were futile.[36] While some reports of the lectures, including Huxley's own, suggested that the audience had warmly welcomed his efforts to chart the ape ancestry of humans, others observed that his auditors had not "expressed the usual forms of approval."[37] These alternative viewpoints were, arguably, inevitable. Huxley had attempted to both offer a polite polemic suited to the audience who heard him and stoke up controversy beyond the lecture room. The unstable form of his lecture was, at least in part, a product of the place in which he spoke and the city that he hoped to reach.[38]

Whatever challenges Huxley faced in controlling the meaning and effects of the message he communicated verbally in different places, by the 1870s his reputation as England's leading scientific orator was firmly established. Arguably, when he visited the United States in August 1876, he did not just bring arresting scientific ideas. Nor did he simply promote a scientific epistemology or theory. He also imported a mode of speech, one that shared a common ancestry with other familiar and favored forms of address. His American itinerary, lectures, speeches, scientific investigations and interactions with leading thinkers have been recounted elsewhere.[39] Here, Huxley's tour presents an opportunity to track a series of speech events in a country that knew him from his writings but had never encountered him in person or heard his living voice breathe passion into his controversial scientific and metaphysical proposals. In America, close attention was paid to the "man talking" as much as to his message. It remains to be seen what kind of hearing Huxley, thinking on his feet, elicited among audiences anticipating a speech sensation.

PLATFORM PREPARATIONS

Huxley's visit to America was long in the making. His reputation as a scientific celebrity was already well established. His books had been published and pirated, his articles reprinted and extensively circulated, and his views appreciated, debated and vilified across different political and religious constituencies. It was in the interests of his American publisher, D. Appleton, to arrange and carefully stage-manage a lecture tour. As both William Peirce Randel and J. V. Jensen point out in their accounts of the tour, substantially increasing the sale of Huxley's books was very much on the minds of his American publisher.[40] Huxley was amenable to the idea of a tour. His sister, Eliza, lived in Montgomery, Alabama, and he had not seen her for thirty years. He was fully aware of the success, measured more in reputational than financial terms, of John Tyndall's tour. Other American supporters, such as the philosopher John Fiske and Daniel Coit Gilman, president of Johns Hopkins University, joined in a growing chorus of appeals for Huxley to make the trip. The appeals resonated, and in August 1876 Huxley and his wife boarded the *Germanic* and crossed the Atlantic to New York.

Edward Livingstone Youmans—one of Huxley's most effective American advocates and editor of the International Scientific Series published by Appleton—used his own journal, the *Popular Science Monthly*, to present Huxley as "next to John Bright, the best orator in England." To reinforce the point, Youmans reported that he had witnessed Huxley give "one of the most successful efforts ever witnessed on the platform," a lecture to working men on physical geography delivered in the School of Mines. Youmans's editorial commendation ended with a warning. If readers did not secure tickets soon for Huxley's three lectures to be held at the end of his tour in New York, they would face bitter disappointment. The lectures were already almost sold out.[41]

As well as having the support of leading American intellectuals, Huxley worked hard to gain the backing of one of the giants of America's fourth estate—the *New York Tribune*. The paper's London correspondent, George Smalley, was on board the *Germanic* and Huxley forged a life-long friendship with him during the crossing. He also courted the favor of the *Tribune*'s editors. As the *Germanic* approached New York, Huxley noted the twin towers of the *Tribune* and the Western Union Telegraph and expressed his pleasure that the skyline was dominated by "centers of intelligence" rather than church steeples.[42] A few days later, Huxley and his wife visited the *Tribune* building, toured its editorial offices and climbed to the top of the tower to survey the city.[43] His tried and trusted strategy of bringing on board an influential newspaper he knew would be sympathetic to his platform pronouncements was put firmly into action in New York.

Another crucial facet of platform preparations was the week Huxley spent with Othniel C. Marsh, professor of Vertebrate Paleontology at Yale University and curator at the Peabody Museum of Natural History in New Haven, Connecticut. This visit provided Huxley with at least two topics that he could tackle to connect with American audiences. The museum's exceptional collections of fossils, secured and curated by Marsh and others, spoke of a continent containing untold riches that would provide direct evidence of evolution. Huxley's fulsome praise of the collections redounded well on Marsh, a leading American scientist in the making. The second, more specific discovery, was the existence of fossil remains that allowed a detailed reconstruction of the evolution of the horse. In examining this collection, Huxley could produce an ideal object lesson that suited his Pestalozzian approach to instruction. It provided him with an American equivalent of his piece of chalk, an object made famous by his by then celebrated lecture on that topic delivered to working men in 1868 during the British Association's visit to Norwich.

Together with Youmans and Appleton, the *New York Tribune* advertised the three lectures on evolution, scheduled to take place in the final days of his visit, as a much-anticipated climax of Huxley's tour. Everything else—his time with Marsh and other leading American intellectuals, his participation in the American Association for the Advancement of Science meeting in Buffalo, his trip south to Nashville and his lecture there and even his inaugural lecture at Johns Hopkins University—was, from this point of view, as much scene setting as anything else. It was the arrival of England's leading scientific orator in New York and the promise of lectures on the theory of evolution that generated the most interest. Before considering the New York performances, however, Huxley's other public appearances deserve attention.

OPENING GAMBITS

Huxley's maiden address in America came in the form of an impromptu speech given during the 1876 meeting of the American Association for the Advancement of Science. Held in Buffalo, the convention welcomed Huxley as an honored foreign delegate. As a past president of the British sister organization, Huxley was lionized and invited to participate in the scientific business. By his own account, Huxley had no plans to speak during the meeting. Even so, his arrival in Buffalo on August 24 was announced in the newspapers, along with an expectation that he would address the convention. Huxley obliged on the fourth day of the meeting. During the morning session the association's president, the geologist William B. Rogers, introduced Huxley who then offered some informal remarks. These were prefaced with apologies. He had "no scientific matter to communicate" and was unprepared. He further confessed that the "emotion" of the occasion made speaking a "difficult task." He could not find the words to "express how much I feel." He was "not by nature a man of many words." It was fortunate then, that the "highest eloquence" was marked by concision. What followed was a series of compliments to his hosts—the good character of the American people (womenfolk in particular), their hospitality and generosity in privately funding science education and their nation's uniquely rich store of prehistoric antiquities. Huxley closed his brief address by singling out Marsh's collection. It was a dramatic peroration. Before Marsh gathered his fossils, "evolution was a matter of speculative reasoning." Afterward, "it is a matter of fact and history." This startling realization was ultimately to blame for Huxley's inability to "find thoughts appropriate to this occasion."[44] Huxley appeared overwhelmed by the scale and significance of this "direct evidence" of evolution (this gave him the title of his final New York lecture). On Huxley's account of public speech, the whole oral performance, delivered without preparation and without notes, could not have been anything other than an inadequate expression of emotion.

The remarks Huxley made in Buffalo were widely reported. The *New York Tribune*, among other papers, printed it in full and included it in its extra, a supplement printed on the day Huxley left for England that brought all his American addresses together. Significant variation in the reports of the speech in different newspapers suggests it was hastily recorded by shorthand. Some days later, it also appeared in full under "American Notes" in several British newspapers. It was the only speech that Huxley made during his tour that was extensively reported in Britain. In America, it was a taster, stirring expectations for more. Reports of it also lingered longer in American papers than might otherwise have been the case, thanks to a short paragraph inserted in a religious periodical by the Episcopalian bishop of the diocese of Western New York. Writing in the *Orbit*, after Huxley's appearance in Buffalo, the Reverend Cleveland Coxe described Huxley as a "gratuitous assailant of revelation" and someone who "demands our pity" for his rejection of an "intelligent creator."[45] Huxley's display of emotion over compelling evidence for evolution had provoked the kind of religious reaction that bought him more press attention. It was a tried and tested method.

After his speech in Buffalo, Huxley joined members of the host association in making a visit to the Niagara Falls before traveling south to Nashville. Despite numerous invitations to lecture on evolution, or on anything at all, Huxley had no plans to do so until his previously arranged address to be given in Baltimore at Johns Hopkins University on September 12. While on a four-day visit to Nashville, however, he conceded to give a single lecture, an agreement given on the understanding that his audience would "put up with an extemporaneous address."[46] While unprepared, in Nashville Huxley had the opportunity to control the meaning of his lecture in a way that he could not in other venues. Huxley's host there was editor of one of the town's leading newspapers, the *Daily American*. Albert Roberts, married to Huxley's niece Edith, had recently launched the newspaper and was making it a leading voice for progress in the American South. Roberts used his paper to oppose more conservative religious views and Huxley presented an ideal opportunity to push against those anxious to resist the modernizing impulses of science.[47]

The invitation to lecture was made by a committee of prominent citizens that had met the day after Huxley arrived to determine how to mark the visit of such a celebrated scientist. Though Huxley welcomed it publicly, it was not entirely in tune with his own aims. The chair of the committee, George Stodart Blackie, was anxious to exercise control over the meaning of any lecture. Blackie, a surgeon and botanist trained in Edinburgh, had been professor of botany at Nashville University before the Civil War and was on record as a defender of the harmony between science and the sacred record. He was joined, among others, by the comptroller of the state of Tennessee, Colonel John Burch, and Thomas O. Summers Jr., a professor of comparative anatomy and histology at Nashville University and son of a rigorously orthodox Methodist theologian resident at Vanderbilt University. An exchange between Huxley and Thomas Summers Senior at Vanderbilt the following day provided the *Daily American* with the opportunity to alter the script and emphasize Huxley's reputation as a controversialist and opponent of dogmatic theology of any kind. Billed as a "jolly, good-natured representation of the 'Warfare of Science and Theology,'" the altercation between the theologian and scientist put on show Huxley's reputation for goading the self-appointed guardians of religious orthodoxy.

Advertising the lecture provided a further opportunity to create an impression of Huxley that would attract large crowds and create a public spectacle. Tickets were sent privately to local dignitaries before being distributed more widely at no cost. The *Daily American* announced Huxley as the great advocate of popular education who, for that cause, "has made a reputation, very rare indeed, in England, for ease, grace and fluency in off-hand speaking. He is the most distinguished extempore speaker in Britain."[48] The venue for the lecture, the Masonic Hall, was considered the best theater in the South outside New Orleans. Renovated in 1875, the décor was sumptuous. With a seating capacity of 750 and space for 400 to stand, the theater could accommodate a large crowd of Nashville's citizens. A band was employed to set the atmosphere and Thomas O. Summers Jr. asked to introduce the speaker.[49]

In the event, Summers introduced Huxley as the "great apostle of modern science," whose "magnetic influence . . . would awaken impulses which have already been kindled by the electric fire of his ready pen." The lecture lasted for close to one hour and was an apologia for science education cast as a "sermon in stone." Huxley opened his address by stressing again that he was speaking without a "set speech" and would tackle instead "such observations and such reflections as have suggested themselves to me during my unfortunately brief residence here." In the first section, Huxley expressed admiration for the natural wealth and education provision that he had witnessed during his brief visit. He urged his listeners to make every effort to further improve training in physical science. This was, he argued, the only way to exploit the natural riches that surrounded them. But, more importantly, it was the only way to teach the proper method of discerning truth from falsehood. This was Huxley's real quarry. The rest of the lecture illustrated a distinction between what he termed "historical evidence" and "archaeological evidence." The first relied on human testimony. The second rested on material remains unaltered by the "stupidity or malice of man." The geology of Tennessee and the "cataract of Niagara" were used to provide material or "archaeological" evidence of the vast duration of the earth's history. It was an account of that history that was "very different from the one that is commonly taken."[50] Huxley's task was not only to show that the account supported by material evidence was true but, more importantly, to convince his listeners that knowing how to judge that evidence was crucial to cultivating citizens who could discern the difference between truth and falsehood.

In making this argument, Huxley was rehearsing what would be a major theme in his opening lecture in New York two weeks later: the superiority of material evidence over human testimony. This was, of course, a rhetorical construction and one that Huxley had to work hard to make persuasive. His message had to be clear without immediately offending his audience. The Bible, as usually read, lacked the credibility of the results produced by the careful scientific study of nature. In making this argument, Huxley had to convince his audience to believe his testimony. He aimed at holding forth a moral vision, a scientific ethos, that his auditors would feel compelled to accept. Through his extempore lecture, Huxley was wooing his audience.

It is not at all clear whether he produced the desired effect on those present. One newspaper report suggested that, half an hour in, some of the audience started to "buzz, fan, and to leave." According to the same report, the hyped expectations of a great address were dashed as soon as Huxley opened his mouth. Instead of a rush of high eloquence, "not a sound was heard expect a sort of mumbling from the speaker, but whether it was English, French or Chinese, not a mortal could tell. There was no articulation, no emphasis, no action, no grace of position . . . nothing but a confused jumble of inaudible, inarticulate and indistinct sounds." So bad was the speaker, that it was doubted whether he was indeed the great Thomas H. Huxley. He had shown no knowledge of the "principles of speaking" and was judged "bogus" by members of the audience who had heard "Spurgeon, Cumming, Gladstone, Disraeli, Manning and others of equal renown."[51]

The careful recording of Huxley's lecture by the stenographer Duncan Dorris, employee of the *Daily American*, helped rescue what might otherwise have been a complete failure. Huxley praised the final transcription, which filled three and half columns of the newspaper the next day, as the best ever done.[52] Partly as a consequence, extracts from his hastily prepared Nashville lecture found its way into newspapers across the country. The most frequent extract was his description of Niagara, with his argument about the importance of discovering and affirming the truth of a scientific account of earth history almost invariably appended. As a result, a careful record of his spoken words likely reached larger audiences than the output from his "ready pen" had ever done. And it did much to confine the damage of a poor oral performance to brief reports of it in a rival local newspaper.

CEREMONIAL POLITICS IN BALTIMORE

The impromptu nature of Huxley's Nashville address presented both opportunities and unwelcome setbacks. Despite all the resources Huxley had at his disposal—not least the support from Roberts—the lecture was judged a dismal failure according to the standards for eloquent speech by at least some of his audience. When Huxley traveled to Baltimore to deliver his lecture at Johns Hopkins University on September 12, there was greater harmony between what he hoped to achieve and the expectations and preparations of the organizers. The event had been carefully planned. Daniel Coit Gilman, president of Johns Hopkins University, had mentioned the possibility of a lecture at the university several months beforehand and had extended a formal invitation to Huxley in June, seeing him as an ideal candidate to inaugurate a series of lectures to run during the first year of the new institution. With that advanced notice, Huxley could carefully plan what he was going to say. The two men had in fact met the year before in London and Gilman had listened to Huxley deliver lectures. Both knew that they could serve a shared agenda—the promotion of an extra-Christian education free from denominational rivalry and religious opposition to modern learning. Both handled the place of religion with care. Gilman stressed the importance of religious instruction and religious engagement beyond the formal confines of academic study. Religion remained in Gilman's representations a central component in the cultivation of good character and civil society. Huxley likewise attacked theological dogma but preserved untouched the civic virtue of religion. Accepting Gilman's invitation thus made sense. It presented a stage on which to declare a vision of a scientific future untrammelled by theological restrictions of any kind.

One of the most crucial aspects of Huxley's address at Johns Hopkins was what Erving Goffman has termed the "environing social fuss" that surrounded it. As Goffman has argued, certain types of public lectures function as powerful "celebrative occasions" that serve as "fundamental organizational forms of public life."[53] Such lectures have a ritual density that marks them out as significant social events. They also generate particular pressures that threaten to overwhelm the lecture itself. The politics of ceremony and the tactical alliance between speaker and organizer can operate in ways that direct attention away from content and the textual self of the lecturer to the ambient meanings generated by the occasion taken as a larger whole. In Baltimore, however, Huxley does not appear to have feared an overshadowing of what he said by the ways in which his address was framed and represented. The reasons for this require some further exploration.

By the time of his appointment as president of Johns Hopkins University, Daniel Coit Gilman was already highly adept at negotiating between different parties with a stake in higher education. The new university, funded by a bequest from the Quaker philanthropist Johns Hopkins, offered an unprecedented opportunity to develop a model institution that drew on American traditions of education and European influences to create something at once familiar and new. Science education and research would be at the heart of the new institution, something that Gilman had pushed for throughout his career as an academic administrator. The justification was not primarily utilitarian. It was, instead, moral and religious. Gilman, as a liberal Congregationalist, was convinced that science was an effective means of cultivating moral culture in keeping with a nondogmatic and extraecclesial form of Christian belief. At the center of his moral imagination was a concern with character formation, understood in loosely Christian terms. Gilman succinctly expressed this outlook in his inaugural address as president: "the object of the university is to develop character."[54] Scientific learning involved a set of practices that disciplined the mind and body and thus fulfilled one of the fundamental purposes of an institution dedicated to higher learning.

Unlike Huxley, Gilman was comfortable with traditional ideas about science as the search for God's laws and as a method for discerning divine purpose in nature, without giving much specific content to those notions. For Gilman, one of the ruling ideas of university education was the lack of conflict between the "immutable truths" revealed in God's word and His works.[55] Nevertheless, there was ample scope for concord between Huxley and Gilman. As Paul White has shown, Huxley did not reject religion or all religious language but rather redefined their boundaries, scope and character. Indeed, in Huxley's hands, science became a form of religion, and scientific practice "an embodiment of the Christian virtues of humility, earnestness, and devotion."[56] Huxley was also highly skilled at cooperating with a liberal religious elite. By 1876 he had spent years working closely with liberal Anglicans to renegotiate the relations between science and religion as they played out in the school classroom and in public culture more generally. Whatever disagreements there may have been about the evidential basis or credibility of particular religious claims, there was substantial agreement about the need for reformation, for overcoming hidebound traditionalism, sectarianism and dogmatism without rejecting the ethical impulses and moral feelings that were felt to mark the development of Christian beliefs. Huxley's aim was not to promote an unfeeling religion of science but to present a new religion built upon scientific method yet directed by patience, love and wholehearted commitment to the pursuit and promotion of truth.[57] This vision was manifest through Huxley's involvement in the London School Board in the early 1870s. In that context, Huxley promoted nonsectarian religious instruction in schools, with the Bible as the core text. Arguing against secularists who opposed this arrangement, he made a key distinction between laws of conduct determined by science and "the engagement of the affections" in support of those laws. People's desire to act morally was "quite beyond mere science" and formed "all that has any unchangeable reality in religion." Such "religious feeling" was the "essential basis for conduct," and Huxley was convinced that, at least for the time being and among the less educated classes, the predilection for ethical action could not be nurtured or sustained without the Bible.[58]

Gilman was fully aware of Huxley's educational philosophy, in which there was much with which he could agree. Even if Gilman wanted to define religion—like so many of Huxley's clerical collaborators—as more than simply a felt commitment to ethical ideals, he quite clearly prioritized that basic feature over more substantive religious claims.[59] Gilman, like Huxley, was also convinced that science "possessed the emotional characteristics of religion," particularly moral earnestness about truth and a propensity for self-discipline.[60] This understanding was useful for marketing a new university free from the guardianship of church and state. Having Huxley address the Johns Hopkins and Baltimore community a day before the start of the first academic year made a great deal of sense.

In opening his address at 11 a.m. on September 12 at the Baltimore Academy of Music, Huxley made the most of the meeting of minds and the celebratory character of the occasion. He reminded his audience that "the University founded by the munificence of Johns Hopkins commences tomorrow." This made the occasion of his address a "unique event." It had become, in effect, a commencement address offering guidance about what a new university ought to be. This may not have been how it was officially understood—that was a matter of dispute after the event. But it was how Huxley framed it in rhetorical practice.

The fact that it was held at the Academy of Music was also significant. Opened on North Howard Street in January 1875, the academy's auditorium was praised as "the finest theater in the United States." Used for musical performances and other forms of entertainment, it also functioned as the main hall for Johns Hopkins University. It had a seating capacity of 1625 (including the orchestra), with the stage "distinctly visible" from every position.[61] Famous for the enormous chandelier that hung from the central crystal dome, it was a venue that added luster to Huxley's address and to the whole event.

The task that Huxley had set himself was to reflect on the extent to which the plans for the new university agreed with the educational principles to which he subscribed. Toward the end of his address he confirmed the alliance between speaker and host. Of the arrangements made by the trustees, Huxley declared, "I have little to do but to applaud them." Most of his reflections in between were taken up with recommendations for elementary, university and medical education. Conspicuous by its absence was any reference to religious instruction. Yet his closing peroration engaged the affections of his audience and appealed to them to make the university a "fortress for the higher life of the nation." It was a nation that faced unique challenges as its population rapidly expanded. The United States, celebrating the centenary of its founding, was undertaking the "greatest political experiment that has ever been performed" and success was in the "highest interests" of all nations. America, if guided "in wisdom and righteousness," would be "great in true glory." The only condition for accomplishing this great task was "the moral worth and intellectual clearness of the individual citizen." Johns Hopkins University, on the cusp of its first year, should look to fulfill its high purpose in helping to meet that condition. In his final sentence Huxley expressed hope that his address would long be remembered as delivered by one permitted to "feel as if your cause were his cause, your success his joy."[62]

Despite his reputation and against his usual practice, Huxley's address was not delivered extemporaneously. Instead, he labored to recall it word for word from a script recorded by a "reporter friend" the day before.[63] This had been done to allow the *New York Tribune* to print a full and accurate transcription. Huxley had planned to use the report as notes, but the "flimsy" paper on which it had been written rendered the words illegible to his reading eye. So, like William Morley Punshon, an evangelical orator who had come to Huxley's notice several years before, he spoke from memory.[64] Unlike Punshon, a past master at memorizing a script, Huxley faltered. He spoke for over an hour under extreme pressure, fearing that what he said would deviate significantly from what was to be printed.[65]

Responses to the content and style of Huxley's address were mixed. One report bluntly described Huxley's address as "rather dull."[66] The reporter in the Baltimore *Sun* may have been damning the address with faint praise in commenting that Huxley's voice was "low, clear and distinct and could be heard with attention in nearly all parts of the house."[67] Gilman, recalling the lecture many years later, remembered it as lacking "glow."[68] Huxley's closing peroration, on the other hand, was widely praised. The *New York Tribune* singled out his closing line. Indeed, it had provoked "vehement appropriation and satisfaction," expressed through "great applause." An account in the Baltimore *Gazette*, while suggesting that the lecture as a whole "gained nothing from any charm of delivery," observed that when Huxley looked away from his notes and "dropped into a colloquial style, his voice became variable, his face expressive; he even ventured on a little gesture."[69] On this account at least, Huxley's efforts in his closing remarks to tailor his talk and collapse the emotional gap between speaker and audience appear to have been a success.

If Huxley's address, and its delivery, attracted some attention, the meaning and implications of the entire occasion generated a veritable storm of critical press commentary. Particularly shrill reactions appeared in the *Catholic Review* and the *Catholic Standard*. An account of the lecture in the *Review* spoke of the "evils" of Huxley's "system" and presented him as an "oracle . . . darkened to the light from heaven." The assessment printed by the *Standard* was yet more condemnatory and castigated the trustees of Johns Hopkins for inviting a "notorious atheist" to enunciate a paganism worse than the lowest forms found among the Greeks and Romans. This, it continued, was clear evidence that Johns Hopkins University was to be conducted along irreligious lines.[70] A string of other reports decried the lack of the usual religious observances that normally accompanied such important ceremonial occasions, not least an opening prayer. The complaint printed in the *Boston Evening Journal* was typical: "Our National Congress, our State Legislatures, our political conventions, our commonest secular and festal gatherings are accustomed to acknowledge the claims of the Deity and to invoke his blessings on their proceedings, but this great and richly endowed University . . . was formally opened and not a word of prayer and recognition of God was made. . . . Ill is the omen."[71] Gilman, who recalled this "storm" of protest from religious quarters, put it down to prejudices against a university outside of denominational control.[72] Others felt that the critics labored under a false apprehension of the nature of the event. Huxley's address had been intended, they argued, as the first of several public lectures hosted by the new university. As such, there ought to have been no expectation of ceremonial pieties and protocol.[73] The critics could not agree and argued in turn that the event had been publicly announced as the formal opening of the university. Not including a public prayer rightly offended "Christian people anywhere."[74] In the face of these strong reactions, one Presbyterian minister wondered whether "it is possible yet to redeem the University from the stain of such a beginning."[75]

What all these reactions suggest is that much of the content of Huxley's address was quickly passed over and forgotten. His substantive proposals about organizing a university degree and curriculum, about streamlining medical education and avoiding ostentatious architecture were largely ignored. What was noticed was the underlying significance of the event, captured as it was in Huxley's opening and, more especially, closing remarks.

Huxley's announcement of a new, quasi-religious vision of a scientific future that would ensure the success not only of Johns Hopkins University but also of a republican polity that would be an inspiration to the entire world rippled through the pages of American newspapers and focused the attention of friends and foes. The surrounding fuss and furor about ceremonial improprieties raised the stakes, generated both a sense of occasion and controversy and cemented Huxley's reputation as a provocateur and polemicist. All of this helped set the atmosphere for his final lectures in New York.

APOSTLE FOR EVOLUTION

If Huxley was underprepared in Nashville and, in a certain sense, overprepared in Baltimore, in New York he aimed to strike a better balance. The lectures had been long in the planning and Huxley had his argument well-rehearsed. When he arrived in New York on September 15, his three lectures on the evidence for evolution had been advertised for over a month. Organized by his publisher, the discourses were billed as the only opportunity to hear Huxley before he left for Liverpool on September 23. They were to be the climax of his visit. Huxley's friends at the *New York Tribune* took the risk of pronouncing him "the best scientific orator in England" and (with less risk) "one of the most eminent living exponents of the doctrine [of evolution]," while an editorial in the *Herald* anticipated a scientific speaker "even more eloquent and attractive [than] Prof. Tyndall."[76] Huxley's publishers were also anxious to control his image. Countering Huxley's reputation as an "infidel," a blurb attached to an advertisement of a back catalogue of Huxley's books, with blatant hyperbole, asserted that "no man has greater reverence for the Bible than Huxley, no one more acquainted with the text of Scripture."[77]

Chickering Hall, the auditorium that was to host Huxley's three lectures, was a new addition to the city's suite of public theaters. Part of the entertainment district centered on Union Square, it had been opened at the cost of $175,000 ten months before Huxley's appearance. Designed as a concert hall, it contained a large pipe organ and could hold an audience of thirteen hundred. The hall followed a standard lay out and was divided into a parquet, parquet circle and gallery. The ruby-colored leather seats, white walls and forty-eight-and-a-half-foot proscenium arch all worked together to help create a "rich ornament" on New York's Fifth Avenue.[78]

The hall was part of a larger complex that included a music store and warehouse used by the proprietors, the piano manufacturers Chickering and Sons. Fittingly, the inaugural event was a concert by the celebrated German pianist Hans von Bülow. Along with musical performances, over the next months the hall hosted lectures on history, art, literature and women's education and provided a platform for other forms of dramatic entertainment. The American Geographical Society also held its meetings in the hall. This mix of music, theater and lectures confirmed Chickering Hall as a valued space of entertainment and instruction. It also helped to reinforce a culture of critical attention that cut across the different genres of platform performances. The sounds of voices and instruments were subject to scrutiny and the embodied performances of musicians and orators were carefully examined. Audiences gathered in Chickering Hall with a diverse set of discriminating expectations about what constituted a successful appearance on stage.

Within this celebrated context, Huxley's lectures were among the most eagerly anticipated events. Priced at $5, tickets for the course of three lectures were relatively expensive.[79] As a comparison, course tickets for six lectures organized by the American Literary Bureau and delivered in Chickering Hall a few weeks after Huxley's cost $2.50.[80] On the other hand, Huxley's lectures were marginally less expensive than attending a concert by von Bülow. Reserve tickets to secure a seat to hear the pianist were priced at $2.[81]

Huxley's first lecture, delivered at 8 p.m. on Monday, September 18, tackled three hypotheses purporting to explain the history of nature. The first held that the universe had existed eternally in something like its current condition, the second that nature in its present state appeared through the operation of "extra-natural" forces and the third that nature as it now appears is a product of natural processes acting through an indefinite, but not limitless, period of time. His aim was to persuade his audience that the only tenable hypothesis was the third. Before turning directly to that task, Huxley underlined the cogency of the idea of nature's orderliness and argued that the only reasonable way to make progress in natural knowledge was to assume that the "chain of natural causation is never broken." Only the most compelling evidence could tell against this logical inference. Huxley was reinforcing his reputation as a defender of what he would later term scientific naturalism.

Huxley was also anxious, before proceeding, to defend a version of a thesis he aired in Nashville—the superiority of "circumstantial" evidence (it was "archaeological" in Nashville) over human testimony. When it came to assessing his three hypotheses, the significance of circumstantial evidence was, he urged, overwhelming. The evidence from stratigraphy left no doubt that the earth had changed dramatically through time. His first hypothesis could thus be quickly dismissed. The second required more careful handling, but not because it had better grounds. Huxley dubbed it the "Miltonic hypothesis," ostensibly to avoid claiming that the author of the account of creation found in the first chapter of Genesis necessarily subscribed to John Milton's portrayal of the creation of plants and animals. Milton's *Paradise Lost* graphically supplemented the more laconic account in Genesis by depicting the sudden appearance of "living creatures . . . Limbed and full grown," dramatically underlining their supernatural creation. Huxley was at pains, so he claimed, to show deference to those "eminent scholars" who argued that the text of Genesis chapter 1 was "consistent with" an evolving creation. In contrast, Milton's rendering of the biblical account was the one that been "instilled into every one of us in our childhood." Huxley then further undermined the relevance of "scholarly" reading of the words of Genesis (along with the text itself) by observing that he could "only stand by and admire the marvelous flexibility of a language that admits of such diverse interpretations."[82] As a report in the *New York Times* observed, in this crucial section on the supernatural creation of living forms, Huxley had employed Ciceronian paralipsis to powerful effect by, in fact, saying a great deal about the "biblical doctrine of creation" and leaving his audience in little doubt that, whatever scholars of the Hebrew text may say, the scriptural account lent its full support to the second, entirely untenable, "Miltonic" hypothesis.[83]

This rhetorical ploy might indeed be regarded, as William Peirce Randel suggested in his lively account of this episode, a tactical error on Huxley's part, playing into the hands of pressmen anxious to find an angle that confirmed the reputation of the lecturer as the scourge of religious dogmatists.[84] Huxley's intentions in using Milton, however, were not as clear or as ill-judged as Randel supposed. In Randel's reading, Huxley had merely employed his Miltonic hypothesis for the purposes of introductory illustration and had not intended to discuss, let alone disparage, the book of Genesis. It is more plausible to suppose that Huxley was deliberately drawing attention to his disdain for attempts to retain respect for biblical cosmogonies in an enlightened scientific age. Moreover, Huxley's rhetorical move not only cleared the way for presenting evidence for evolution and for a closed system of natural causes—it also allowed him to strongly hint at a shift of epistemic authority from theologians and clergy to a scientific laity. Disputes about what the Bible taught about creation were, Huxley declared, best left to "the scholars and the clergy." It is, he continued, "well for us the laity, who stand outside, to avoid entangling ourselves in such a vexed question." Huxley was redescribing his audience—which, as many reports noted contained plenty of "scholars and clergy"—as a scientific laity that he would now equip with the means to determine the truth about the history of nature. As we shall see, whatever Huxley's intentions, his lecture, as reported through the medium of copious, contentious and widely read press reports, spawned a wide range of readings and responses, in which his alleged attack on the biblical creation account loomed large.

In his second lecture, Huxley turned to "indifferent" and "favourable" evidence for evolution. Dealing first with the persistence of particular types through different geological eras, Huxley squarely faced a standard objection to evolution. This evident fact, he declared, was quite compatible with Darwin's evolutionary theory. It was only when variation offered some advantage, perhaps in a changed environment, over a parent form that evolution occurred. It was thus to be expected that certain types would persist, even across long periods of geological time. The imperfections of the fossil record were also quickly redescribed as "indifferent" evidence. The record's incomplete nature was simply a consequence of how rare it was for conditions to occur that allowed for the preservation of fossil remains. More promising in terms of evidential value were the "intercalary" forms that pointed to the existence of intermediate types between different classes of animals. Huxley concentrated on fossils that contained features shared with birds and reptiles. While not themselves examples of the "linear" forms that led from reptiles to birds, they were exactly what the theory of common descent predicted. Their ancestors may well have been in the direct evolutionary line to birds. The language of this second lecture was more technical and the argument more involved than that of the first. A comparison between several newspaper reports and the final version published in *American Addresses* (1877) suggests that the technical terminology was toned down for the live audience. Even so, Huxley's forensic analysis of fossil forms made demands on his listeners.

In his final lecture, Huxley turned to his recently discovered "demonstrative evidence" of evolution. Here again he acknowledged the help he had received from Marsh. The occasion provided Huxley with the opportunity to declare in full what he had announced with feeling in Buffalo. Thanks to Marsh's unique collection of equine fossils, it was now possible to directly trace the evolutionary history of the horse back to much earlier forms. Huxley's investigations of this history had already persuaded him that demonstrative evidence in the shape of a succession of intermediate forms had been secured. But Marsh's collection filled in gaps and allowed the story to be traced much further back in time. This was a less sensational declaration than the one he made in Buffalo. Nevertheless, after describing Marsh's equine evidence, he could declare that the "doctrine of evolution . . . rests upon exactly as secure a foundation as the Copernican theory." If the evidence for the evolution of the horse was not proof, "there are no inductive conclusions which can be said to be scientific." To arrive at that conclusion, Huxley, with due apologies for troubling the listener, gave detailed descriptions of anatomy, using a blackboard to illustrate crucial components of the "machinery" of the horse that had changed through evolutionary time. Before closing, he dismissed as irrelevant to his argument the question of the duration of geological time.

The immediate reception of Huxley's three lectures was comprehensively registered in the pages of New York's newspapers. Detailed commentary

FIGURE 2.2. A diminutive "Huxley Eikonoklastes" attacks a statue of Moses
with the bust of Milton. *Daily Graphic*, September 27, 1876, 1, detail. By
permission of the New York Historical Society.

was offered on the substantive arguments as well as on Huxley's appearance and style of delivery. Setting the scene of the first lecture, a report
in the *New York Tribune* attempted to capture the audience's first glimpse
of Huxley in the flesh: "[He] came forward upon the platform and was of
course greeted with abundant applause. He laid a copy of Milton's *Paradise
Lost* on upon the reading desk; nothing else, neither manuscript nor notes.
He leaned forward slightly over the desk and began speaking in measured
words and with a low tone of voice. Except sometimes to grasp the desk
with both hands and lean over it more intently, he did not vary his position."[85] This was a caricature redrawn: Huxley as immovable, motionless,
measured. It was a projection of Huxley's posture for the purposes of generating appeal for readers rather than listeners. Some other reports echoed

this positive image of Huxley as a resolute advocate of science. The *New York Times* columnist had Huxley speaking "under the aspect of calm judicial solemnity." Others, however, were more critical. A writer in the *Herald* found Huxley's appearance "a little disappointing." Instead of a "sturdy scientific Boanerges," the audience met a "man whose mind has overtaxed his body." He did not have the strong, clear voice of Tyndall and had a "general air and address of a not very well fed evangelical clergyman."[86] Unlike the *Tribune*'s reporter, the writer hinted at Huxley's overreliance on notes and his inept use of a geological chart, half obscured by an unused blackboard. The whole performance, it was concluded, lacked the art of generating full sympathy with an audience. The commentator in the New York *Sun* followed the *Times* in describing Huxley's style as judicial but only in the sense of "a lawyer [talking] to a bench of judges on an abstruse point of law." Huxley may be "clear and intelligible, but he does not fascinate."[87]

What did apparently fascinate was Huxley's use of Milton to attack Moses. Whatever Huxley's intentions had been in adopting the "Miltonic hypothesis," press commentary suggested that most heard his lecture as an attack on the biblical account of creation. An editorial in the *Sun* had Huxley attacking the Mosaic account of creation with a "very subtle sarcasm."[88] The headline in the *Cincinnati Daily Gazette* baldly stated, "he attacks Moses in the person of Milton."[89] A commentator in the *New York Herald* deemed Huxley underhanded. His attack on Genesis was a "suppressed sneer" and it did not become a man like Huxley "with so many titles to command attention" to stoop to such "small arts" of amusement.[90] In the New York-based *Daily Graphic*, a striking caricature of Huxley, armed with the bust of Milton and vainly trying to knock an imposing statute of Moses off its pedestal, made a similar point.

Five days after Huxley's Miltonic assault on Moses, a letter in the *New York Tribune* from a "philosophic observer" based at Princeton College—bearing the hallmarks of the president, James McCosh—wondered why someone who had already dismissed the "Hebrew Scriptures as mythical and semi-barbarous" should turn his fire on Milton. All the same, the correspondent had listened to Huxley with satisfaction. The lecturer had done good service to religion in removing the "superimposed errors" foisted on Genesis by fallible poets, scientists and divines.[91] Here was one of Huxley's "scholars" offering his reply to the lecturer's none too subtle criticisms of a reading of the Genesis creation account that made it compatible with the latest evolutionary science.

A more fully positive appraisal of Huxley's first New York lecture appeared on the other side of the United States in the *Sacramento Daily Record*. In its pages, readers encountered a strong statement of support for Huxley's rhetorical strategy. He had "avoided the trap into which his colleague Tyndall fell at Belfast" by saying nothing about the "First Great Cause." While Huxley had obliterated ancient beliefs and "left nothing worth saving of the Mosaic cosmogony," he did so without waving the "red rag" of controversy. Huxley had inaugurated a new day in which minds could be freed from the fetters of traditional theology, united in the pursuit of demonstrable truth and persuaded of the fact that there could now be "no true Science which was not also true Religion."[92]

Reviews of Huxley's second lecture, while less numerous or voluminous, were, in places, sharply critical. The report in the *Sun* described the lecture as "tedious, rudimentary and almost inaudible," with "more than one half of the audience unable to hear." The abrupt close of the lecture was, the report claimed, "greeted with slight applause." The report in the *New York Herald*, though less direct, seemed to confirm a general lack of enthusiasm. By the end of the lecture, it observed, "a very large proportion of the audience had had enough of the subject for one night."[93] Elaborating further, the report argued that the "deterrent dryness" of the material, however well delivered by the "earnest" and "eminent" lecturer, suggested that the danger to religion from Huxley's supposed infidelity was small. After all, "mankind in general does not need to be warned of the functional disorders which would follow the swallowing of a bushel of brick dust."[94]

The final lecture elicited further jibes at Huxley's reportedly dry subject matter. According to a piece published by the *New York Herald*, the "intellectual menu" served up by Huxley "was literally nothing but horse." The same article pictured a visitor dropping in on the lecture and encountering a "demonstrator in anatomy in a college of veterinary surgeons." It was incongruous then to find among the "phalanx of thinkers" listening to Huxley's discourse on horse anatomy a large number of ladies. Their presence was explained not by a fascination with the subject matter but rather by the embarrassment of having to "reply negatively at the breakfast or dinner table to the query as to whether or not they had heard Huxley."[95] Despite these jests, Huxley's own apologies for his close anatomical descriptions and his thanks to his audience for bearing with him were fully recorded. The typically more supportive commentary in the *New York Tribune* suggested that the lecture had elicited more applause than the two previous ones, and that his final sentences, "delivered with much feeling," were sympathetically received by an attentive audience.[96] The brief report in the *Sun* avoided the vituperative language that had appeared after the second lecture but ran with the headline, "A Slight Argument in the Way of Demonstrating the Law of Evolution," and ended with a quote from a clergyman on the "solitary fragment, the dubious item" that Huxley the "great expounder" of evolution had used to demonstrate its veracity.[97]

What all reports of the final lecture confirmed was Huxley's general avoidance of oratorical techniques or rhetorical embellishments. In his closing remarks, Huxley himself drew direct attention to this aspect of his craft, declaring that "I shall rejoice—I shall consider that I have done you the greatest service which it was in my power in any way to do—if I have thus convinced you that this great question that we are discussing is not one to be discussed, or dealt with by rhetorical flourishes or by loose and superficial talk, but that it requires the keenest attention of the trained intellect and the patience of the most accurate observer."[98] Even in making this declaration, of course, Huxley was deploying his own characteristic oratorical technique. He was courting his audience again by praising their patient attention. They, like himself, had embodied the very virtues he was trying to expound and inculcate. He had treated them as those anxious to know the truth and they had responded to that summons. Ultimately, he had offered and enacted a mode of speech, and a mode of listening, that would make possible a more hopeful scientific future.

With the completion of Huxley's short course of lectures in New York and his departure for Liverpool the following day, several papers took stock and offered commentary on the series and his visit as a whole. Again, comparisons were drawn between Huxley and clergymen, with one critic suggesting that he "looked like a studious, dyspeptic clergyman" and another more sympathetic observer admitting that he might "pass very well as a dissenting clergyman."[99] Several commentators expressed an overall sense of disappointment, reflecting on a scientific orator who had failed to convince or satisfy his audiences. One report several days after Huxley's departure judged, as "of more general interest" than Huxley's discourses, a rather conventional lecture on physiology delivered at the University of the City of New York.[100] Even supporters agreed that Huxley had failed to impress as a speaker.[101] Under the headline "Professor Huxley a Disappointment," D. G. Crowley of the *New York World*, despite being an "ardent evolutionist," felt Huxley's lectures lacked vigor and boldness.[102]

Huxley's New York lectures were as much a product of press attention as they were a series of arguments communicated and consumed at Chickering Hall. This was perhaps most clearly brought out in a satire published by the *New York Times* on the day of Huxley's final lecture. Playing on Huxley's jibe about the "marvelous flexibility" of the biblical creation narratives, the satirist expressed admiration for the wildly varying interpretations of a set of lectures marked by highly versatile language. Perhaps, indeed, the lectures were simply the work of a "prose-poet who writes the epic of everyday life." On the other hand, maybe it was Huxley who was playing the role of the "dogmatic and imaginative" reporter re-creating a world that produced scientists from protoplasm.[103] The crude satire contained a truth that Huxley himself acknowledged. He had worried earlier in his trip that an "ingenious critic of the future" investigating his address at Johns Hopkins University and discovering discrepancies in different reports of his lecture, ostensibly "taken down from his lips," would declare that the "pretended original was never delivered."[104] After his New York lectures, he wrote that, while in America, he had "*not* given utterance to a good deal that I am reported to have said there."[105] There was a sense in which the press reports had overwhelmed the platform performance. Others, such as Crowley, judged that the "publicist" had exaggerated the lecturer's abilities, spoiling the actual performances and creating expectations Huxley simply could not meet.[106]

Huxley's tour of America has long been known to his biographers and within the lively scholarship dedicated to better understanding his ideas, life and context. It was undoubtedly a significant episode in Huxley's own career as surely the best-known scientific publicist of his day. This was captured well by an article published in the *Proceedings of the American Philosophical Society* in 1970. Written by the American historian William Peirce Randel, the article evocatively retells Huxley's tour through creative use of extensive primary source materials. What is immediately apparent is the extent to which Huxley's American tour was tied to a growing appetite for his books, constantly stimulated by his dedicated and shrewd American publishers. The investment in his visit by Appleton and by Edward Youmans can be seen in the extravagant banquets and press receptions they helped organize. Randel also makes clear the extent to which leading representatives of America's scientific, civic and business communities courted Huxley's praise and intellectual patronage. It is also impossible to avoid the looming presence, physical and metaphorical, of America's fourth estate. Randel makes much of Huxley's reported admiration of those "centers of excellence," the newly erected skyscrapers housing New York's most successful newspaper offices, first encountered from the deck of the *Germanic* on his arrival in the city. Randel also records Huxley's close personal connections to newspaper editors in Nashville and New York. More than other accounts since, something of the backlash against Huxley is registered—against his message and sometimes his manner—even if this is not pursued much beyond choice quotations from more critical newspaper accounts. At one telling point, Randel notes that the reports of Huxley's supposed "evasions" in his first New York lecture may well have, for many, "formed the most durable impression of Huxley's visit."[107] Randel, however, supposes this was a mishearing and misreporting of Huxley's intent, as was purportedly obvious to anyone who had listened to the live lecture. In tone and overall import, Randel's narrative leans in the direction of the adulation that Huxley received from his American supporters, slipping from time to time into language treating Huxley as that "great man" whose literary and scientific abilities could never really have been in doubt.[108]

Randel's ear for telling detail and his ability to sketch out the wider historical context that formed the backdrop to the tour is taken to dramatic new levels in the chapter dedicated to Huxley's visit to America in Adrian Desmond's justly celebrated biography. Desmond's rich contextual approach, best known to readers of his biography of Darwin coauthored with James Moore, is on full display. Desmond also reports in more detail Huxley's careful study of Marsh's fossil horses and shows how Huxley put his newfound knowledge to good effect in his lectures. The horse, as Desmond reminds his readers, was, to the bulk of Huxley's lecture audiences, a symbol of American progress and pioneering spirit. Huxley, nodding to this equestrian iconography, also made the horse a symbol of sound scientific demonstration and of the irrefutable truth of evolution. In Desmond's hands, the reader is left with the impression of a tour that secured Huxley's American reputation as "the true voice of Nature," despite any challenges he encountered in harnessing the power of the spoken word to convert America to evolution.[109] Implicitly recognizing Huxley's secular appropriation of evangelical oratory, Desmond comments that in New York "the evangelism of science was beginning to produce its own Great Awakening."[110]

Randel's and Desmond's constructions of Huxley's short but significant lecture tour have not been equalled either in terms of their skillful narration or the level of detail provided. What neither scholar has done is examine the tour through the lens of American lecture culture. If their accounts confirm the central role of the press—as Desmond puts it, "all was newspaper talk"—they only incidentally notice the culture of speech that provided some of the most important measures by which Huxley's public lectures were judged. When this is taken seriously, Huxley's tour loses at least some of its sheen and takes a more chastened place within the sometimes cutthroat and always competitive world of public address. Whatever Huxley's supporters heard Huxley say—and their reports of how he said it—represents only a fraction of both the responses and reactions and the cultural consequences and contests associated with Huxley's speech performances.

As well as facing the challenges of overwhelming and conflicted press attention, Huxley had spoken in America in ways that clashed with other forms of speech that embodied and enacted a dramatically different account of social and moral improvement.[111] Three weeks after Huxley's departure, a packed Chickering Hall heard the celebrated temperance lecturer John B. Gough deliver his famous address on eloquence and orators. Performed hundreds of times to audiences on both sides of the Atlantic, Gough's lecture was closer to burlesque than anything else. Much of it was taken up imitating notable preachers and lecturers, demonstrating the elusiveness of true eloquence. Whether or not Gough masqueraded as Huxley, through mimicry he punctured the authority of many a celebrity orator. His lecture attracted little comment from the press—it was hardly newsworthy. But it represented a proximate challenge to Huxley's scientific oratory that placed truth on the lips of the trained scientist and made cultural progress rest on the patient pursuit of demonstrable facts. Huxley had adopted and adapted a standard evangelical style of extempore address to persuade audiences not only of the truth of evolution but also of the importance of key scientific virtues and emotions. It was precisely that style of address that was one of Gough's favorite targets. He rendered such formal, studied address farcical and created what Martin Hewitt has described as an "almost cinematic spectacle of action" that privileged dramatic gesture over carefully chosen words.[112] His manner and message were the antitypes of Huxley's. Gough's ethic was delivered through drama (there was "nothing of the actor" about Huxley) and his performances enacted a form of self and social improvement that pushed expert knowledge to the periphery of public concern.[113] It was populist in a way that rendered Huxley's meritocratic stance elitist. Gough's audiences, rather than sitting in silence and absorbed by his words, were raucous and fully reactive. It was not that Huxley's audiences were purely passive. They became part of the performance by imitating Huxley's patient attention to facts—a participation in a certain kind of emotion as much as a sustained effort in comprehension. But Gough's audiences performed a very different role and enacted very different kinds of emotion. In doing so, they anticipated an alternative kind of future American society to the one envisaged by Huxley.

It might be argued, then, that Huxley's clarion call for scientific righteousness—or what Huxley elsewhere called justification by verification—fell as much on stony as fertile ground. That it was a call, a "prophetic" plea, is an appropriate description. Huxley preached a new religion in America, and the performative aspects as much as the content of his lectures embodied and evoked it. In the end, as Darwin might have predicted, Huxley's oratory in America was at least as much about courting public sympathy and generating an emotional response as about instructing audiences in scientific facts. Whether or not he succeeded was contested then and remains in question now.

CHAPTER 3

RICHARD PROCTOR AND THE TEMPO OF SCIENCE

Richard Proctor was a man in a hurry. One evening, at the end of a talk on astronomy delivered at Boston's Horticultural Hall, Proctor ran down two flights of stairs and along the passage under the auditorium. He was in a friend's cab before any of his audience had left the building and had reached a nearby theater just in time for the second act of the comic opera *Dinorah*.[1] Catching his breath, he congratulated himself on his rapid transit across downtown Boston. It was typical Proctor. His career was marked by the speedy production of new material on advances in astronomy. He was among the most prolific authors in the expanding world of science for the people. He took pride in counting the number of lectures he was able to deliver within a fixed period of time. Reporting on his first visit to the United States, the *New York Tribune* pictured Proctor rushing "from town to town and city to city" on American railroads, announcing the latest findings of astronomical science These discoveries were appearing at a breathless pace, causing some to announce that a new astronomy was emerging, and Proctor was well equipped to help audiences keep in step.[2] As the *Popular Science Monthly* put it in 1874, "astronomy has now surpassed all the other sciences in the rapidity of its advancement," demanding "new men who can deal with the subject in its more novel and extended aspects."[3] Proctor fit the bill, with the added ability of making those novelties appeal to audiences as likely to attend a comic opera as a lecture on the night sky. To begin, then, it is worth asking how Proctor developed the capacity to pursue such a high velocity vocation and shuttle rapidly between digesting new knowledge and packaging it for mass consumption.

PROCTOR ON THE PLATFORM

Before Proctor began speaking to audiences across the Anglophone world, he had developed a reputation as an astronomer and successful author.[4]

After studying for a BA in theology at King's College London, Proctor read mathematics and theology at St. John's College, Cambridge. By the time he graduated, he had married and had abandoned his original aim of becoming an Anglican minister. Although he entered Lincoln's Inn to study law, the turning point in his career came when he decided to teach his oldest son astronomy. After the boy's tragic death in October 1863, Proctor settled on a quest to pursue astronomical research. His first publication, a short article on "colours of the double stars," appeared in *Cornhill Magazine* in December 1863.[5] Two years later, he published a much longer work entitled *Saturn and Its System.* The book was well received in Britain's astronomical community and was followed by his election as a fellow of the Royal Astronomical Society (RAS) in June 1866.

Just a month before his election, Proctor and his family were hit by a disastrous financial crisis after the collapse of the New Zealand Banking Corporation. Proctor had been the second largest shareholder, with a liability of £13,000. Reduced to poverty, he was forced to find ways to make money. Now deeply committed to astronomy, Proctor was reluctant to give up his avocation or, indeed, to compromise it in any way. According to one autobiographical note, he resisted applying for a position at a state-funded institution, arguing that "neither observatory work nor college teaching tend effectively to widen one's views of the great problems of astronomy; while too often it is found that official position, whether among the rank and file or in command, tends to warp both mind and character, and to diminish mental elasticity, versatility, and originality." Proctor's solution was to pour his energy into writing articles and books for pay and later into delivering profitable lectures. In Proctor's terms, this enabled him to continue "independent research" free from the baneful effects of "professional" or "official astronomy."[6]

Other changes of a personal kind were also shaping Proctor's public reputation. In 1867, not long after the start of his career as independent astronomer and author, Proctor converted from Anglicanism to Roman Catholicism, the faith of his Irish wife. This was not something Proctor broadcast widely. But it was confessed to close acquaintances, such as the anthropologist and banker Edward Clodd.[7] The word, nevertheless, got out and Proctor's adherence to Catholicism was occasionally mentioned during his first American tour in 1873–1874.[8] At the time, Proctor neither denied nor confirmed the truth of these rumors. But when he returned to the United States in October 1875, he made a concerted effort to distance himself from the Catholic Church, moving to a more agnostic position, but one which he denied led to atheism. Both before and after 1875, Proctor repeatedly asserted that God's purposes, which he did not deny existed, were entirely inscrutable. It was a form of negative theology summed up, as

Proctor frequently noted, in words found in the book of Job: "touching the Almighty, we cannot find him out."[9] This enduring religious sentiment was often accompanied with at times savage critiques of the superstitions and errors found in the Bible.

Proctor's shifting religious allegiances, and sometimes contradictory public representations of them, occurred alongside his rapidly developing reputation as a science lecturer. According to his own account, Proctor started lecturing sometime in 1869.[10] By the following year, his lectures were being widely advertised and delivered in high profile venues. His debut lecture at the Royal Institution in May 1870 is a case in point. Entitled "Star Grouping, Star Drift and Star Mist," the lecture examined the character and associations of the visible stars and argued that instead of each being a single species like the sun, they disclosed an infinite variety of forms. Using ten detailed maps of the night sky, Proctor left his audience with an impression of the universe that was vaster and more variable than anyone had previously supposed. According to an editorial in *Scientific Opinion*, his lecture was the best of the season, a fact put down to its novelty, "forcible delivery" and the use of striking photographic maps of the heavens. He was, the editorialist declared, "not only an original observer of persevering enterprise and surprising acuteness, but [also] one of the best popular teachers of astronomy which the country has ever produced."[11] Proctor had become not just a popular writer but also a widely admired speaker. In his 1871 census entry, Proctor listed his occupation as "Lecturer and Author."[12]

Proctor, as Joshua Nall and others have noted, developed the conviction that knowledge of astronomical science and its associated practices could lead to moral uplift. Lectures were particularly useful for spreading that message and for making astronomical knowledge a vehicle for self-improvement.[13] Certainly, Proctor quickly affirmed the wider moral and quasi-religious function of science lectures, partly through participation in more unconventional lecturing initiatives. From 1870 onward, Proctor featured in the program of the Sunday Lecture Society, an initiative spearheaded by William Carpenter and Thomas Henry Huxley, delivering two lectures on "The Telescope and Its Discoverers" early in December.[14] This was the beginning of a long association with Sunday lecturing societies in Britain, North America and Australia. It underlined his commitment to science lectures as a form of moral and mental culture that would help renew society across all classes, a commitment that began while he still identified (at least in private) as a Catholic and remained after he publicly distanced himself from confessional Christianity. Indeed, Proctor emerged early on as a champion of lecturing on Sundays and as a vocal critic of Sabbatarianism. This role as an advocate of Sunday lectures manifested itself not just in active involvement in lecturing societies but also through intellectual argument and political action.

The most dramatic instance of Proctor's efforts to undermine strict Sabbath observance occurred later in his lecturing career while he was on tour in Australia. As part of his Australian lecture itinerary, Proctor was due to give a Sunday talk in Sydney's Theatre Royal. At short notice Henry Parkes, the colonial secretary, blocked the lecture on the grounds that it involved ticket sales on a Sunday. The public outcry was immediate. Despite this, Proctor agreed to the lecture being cancelled because of the disastrous consequences that would have followed for the lessee of the theater. He conveyed this message from his hotel balcony to thousands of disgruntled ticket holders, noting at the same time that his Sunday lecture would not have faced any legal challenge in England.[15] The whole event made Proctor into a local cause célèbre and defender of free speech. According to Proctor, this was the reason why his Australian tour proved to be so successful.[16]

Proctor's anti-Sabbatarianism was just one expression of a wider strategy to elevate the lecture as a powerful medium for manufacturing moral culture. In his view, to make lecturing an effective lever for promoting social progress it was vital to market himself as someone with a deep commitment for disseminating new knowledge in a fresh way. Proctor, like many others, made full of print and platform culture to secure this reputation. Throughout his career Proctor generally wrote on a subject first and then presented it in the lecture hall. In his case, however, it was imperative that the lecture presentation differed from what had been written. In a brief note that appeared in *Knowledge* in 1885, Proctor claimed that his lectures had "never been printed or written." Instead, they "varied from week to week, as knowledge grows."[17] Elsewhere it was noted that one of his lectures, "The Birth and Death of Worlds," had been delivered four hundred times by 1883, "yet never twice in the same exact form, having never yet been committed to paper."[18] Proctor's appeal stemmed in large part from the novelty of his lectures and from his commitment to bringing an audience up to speed on the latest results of scientific discovery, in part to allow them to participate in that progress. According to Proctor, this lay participation in the making of science would have a positive moral as well as intellectual impact.

This commitment to delivering a fresh message was one reason why Proctor never read from a script but delivered his lectures extemporaneously. As already demonstrated, following a dominant strand in thinking about effective public speech in both Britain and America, an unscripted performance gave the impression of novelty and innovation. Certainly, Proctor was fully convinced that extempore address was essential for lecturing success. He was persuaded, for example, that the widely perceived failure of Matthew Arnold's lecture tour of America in 1883–1884 was due in large measure to the fact that Arnold read his lectures. In a word of advice to others wanting success on the lecture circuit, Proctor added the following observation: "I should suppose no one who really wished to influence his audience actually reads."[19] These sentiments strongly reflected those found in many guides to effective preaching or political speech making.

If anything, though, Proctor's commitment to extemporaneous address was more strenuously held and imbued with meaning than it was for others who simply followed the conventional and widespread advice of speech manuals to avoid reading a lecture. Proctor's use of extempore address was so ingrained that he found it impossible to perform a lecture without a full audience in front of him. The fuss and occasion of delivering a lecture became an essential component part of his ability to communicate effectively. Proctor later pointed this out by recounting an unsuccessful attempt to lecture in front of four stenographers sent by the *New York Tribune* to record a talk later to be delivered to a full audience. As Proctor described it, "the case of the aspirant to oratorical fame and the cabbages was reversed: he practised before a set of cabbages, and could only tell a real audience that 'he—he—he perceived they were not cabbages.' I had never lectured save before an audience and . . . I had to tell the four reporters I perceived they . . . were not an audience."[20]

Speaking without a script was important in another way. The invention that took place during live delivery was, for Proctor, one mode of actually doing science. In lecturing, he was not simply relaying scientific facts and theories established beforehand. Instead, fresh ideas suggested themselves during delivery and fed back into his ongoing scientific investigations. Proctor found that "new ideas outside the theories I had already formed on specific subjects have been suggested during the actual progress of a lecture, for digestion and assimilation at later leisure."[21] This reflected and confirmed Proctor's understanding of science as a nonelite, democratic and participative activity best pursued beyond the narrow constraints of professional organizations or government-funded and controlled institutions.[22] This understanding resonated with a strong sense cultivated by, or at least idealized by, lecture culture in late nineteenth-century America that speakers and audiences enacted a dialogue between equals.[23]

A final reason why an unscripted lecture was important to Proctor was that it made recording and reporting in the pages of daily newspapers more likely. It made, in other words, Proctor's astronomical discoveries newsworthy. Proctor, as Joshua Nall has demonstrated, was particularly well versed in what made a good news story, having developed a long-standing friendship with a recognized leader of the new journalism, William Stead.[24] He embraced the emphasis on the timely and the sensational that marked this new form of newspaper reporting and exploited it in full while on lecture tours. He knew more than most that lectures performed in a way that attracted press interest would have wider reach and be more financially rewarding. Proctor could also regard the newspaper as a key ally in fulfilling his vision of using astronomical science to ground and guide the development of moral culture.

At the same time, Proctor could be as wary as any of the challenges and limitations of aligning lecturing with the needs and wants of newspaper reporters. While many of his lectures were printed in full, these were not revised and issued by Proctor but came from the hands of stenographers. In America, as we shall see, the *New York Tribune* issued six of Proctor's lectures delivered in New York during his first tour. These appeared in *Tribune* extras, pamphlets of lectures sold separately from the daily newspaper.[25] The first of these, as discussed earlier, printed John Tyndall's lectures on light. Proctor's, like Tyndall's, sold well, running to the tens of thousands.[26] During Proctor's second tour in 1875–1876, the New York-based Truth Seeker Company, a society committed to the propagation of free thought, published lectures Proctor delivered at the city's Steinway Hall.[27] Proctor no doubt benefitted from the publicity, but the profits went to the publishers.

Proctor was certainly sensitive to the financial complexities of lecturing with press reporting in mind. But by never issuing an authoritative version of his own lectures, he did not have to face the problem of competing with rival accounts produced by others more quickly and in larger volumes. And by taking care to significantly rework his lectures so that they tackled subjects he had dealt with in print but in contrasting and novel ways, he was able to increase the sales of his books while benefitting from the fees paid for appearing on the platform. Unlike Tyndall, he was happy to profit directly from lecturing and argued that, by receiving substantial payments from lectures delivered in America, he was simply recouping money lost through pirated editions of his books. In one comment on this, he presented the following analogy: "A traveller who should meet in a social sort of way a gentlemanly pirate who held a quantity of what had been and should still be that traveller's property, would hardly think he was taking away other folks' money if he received in part payment a sum collected by the comrades of the pleasant buccaneer."[28] It was clear that the monetary reward of lecturing was a key motivation for Proctor. It was a way to fund astronomical science and avoid reliance on the kind of government support that Proctor felt stifled independent inquiry. This conviction, for example, led him to single out the American astronomer Ormsby MacKnight Mitchel (1810–1862), who had lectured profitably in the 1840s in order to fund his own astronomical research and to ensure the completion of the Cincinnati Observatory.[29]

If Proctor had reasons to at least tolerate the appearance of his lectures in print, he also had grounds to complain about the incomplete and distortive character of newspaper reports. Reflecting in the pages of *Knowledge* on a flawed account of a lecture by the neurologist Brown-Séquard in 1884, Proctor mused that "the enterprise of the *New York Tribune*, in taking full reports of lectures considered noteworthy, is a well-known and most admirable feature of American journalism. But it is a mistake to suppose that reports, even if actually verbatim, can exactly represent a lecturer's meaning. A speaker, by varieties of inflection, emphasis, and so on, to say nothing of expression, action, and illustration, can indicate his exact meaning whilst using language which, written in the ordinary manner, may appear indistinct and confused. Thus a most exact and carefully-prepared lecture may appear loose and slipshod in the report."

This problem was particularly acute for lecturers who spoke quickly. In that case, "a reporter is compelled to omit words and sentences occasionally, and such omissions are absolutely fatal to the effect of a lecture, regarded either as a demonstration or as a work of art." The resulting transcript becomes a "symbol" of what was said rather than anything like the substance.[30] Here, Proctor expressed a widely held view that the warmth and moral influence of the spoken word could not be replicated in print, and demonstrated his own sensitivity to the importance of speaking as a matter of performance as much as the dissemination of "dead" information.

Proctor, however, was perhaps less attentive than some orators to making his facts, as Emerson had put it, "warm and coloured and alive" through the cultivation of impassioned speaking.[31] This is arguably confirmed by the fact that, of all the prosodic features of Proctor's manner of speaking, it was the tempo of his delivery that was most often commented upon. His diction was said to be clear, but he spoke rapidly and without much variation in tone. To keep track of the deluge of information, an audience had to stay alert. America's best stenographers, skilled in the art of shorthand, struggled with the volubility. As one American journalist noted, "Proctor was the most rapid and fluent speaker that has for many years taxed the swift pencils of our reporters."[32] The brisk rate at which he spoke influenced Proctor's dominant platform posture. It was difficult for him to accompany his utterances with appropriate body language. Hands, arms, movements of any kind could not have kept pace with his words. So he stood still, arms rested on the lectern, issuing forth torrents of particulars. This did not matter much to his auditors. The bulk of a Proctor lecture was delivered in darkness, attention being directed to quick-fire images of the cosmos projected onto a large screen.

The rapid tempo of Proctor's lectures aligned with his own convictions about their function. He aimed to be understood and to instruct more than just a small portion of his audience. The information and arguments he uttered were designed to spur his listeners to scientific action and to equip them to contribute to the advance of astronomy. He was proclaiming from the platform a vision of a republic of science that, as Bernard Lightman has demonstrated, cut across a nascent professional and state-sponsored scientific culture.[33] Citizens of such a republic had much to learn, and time was of the essence. Proctor's platform performances might be described, then, as a particular embodiment of a form of scientific modernity. New information about the starry depths was arriving more and more quickly and in greater and greater volume. As an editorialist for the *New York Tribune* put it, Proctor gave the impression that there was "not time enough to tell their mighty tale."[34] Keeping up with new astronomical discoveries required lecturing at telegraphic speed.

For all that, Proctor, like Huxley and Tyndall, was skeptical about how much lectures could achieve in terms of promoting scientific understanding. Writing in 1875, Proctor argued that it was "the most difficult thing in the world" to know whether a given scientific subject had been clearly presented in a lecture. Even hearers of "such almost perfect lectures as Tyndall's or Huxley's . . . go away with the conviction that they understand the subject when they do not even know what it is the lecturer wanted them to understand."[35] Proctor's well-worn lecture on the transit of Venus had, he estimated, been heard by upwards of twenty thousand people and had been repeatedly commended for its clarity of explanation. Proctor remained unsure how much of the science had really been digested.

Whatever Proctor's own views of the purpose and power of the spoken word, the speed with which he related new astronomical facts and theories was arguably not primarily about facilitating bare-bone instruction. A case can certainly be made that the emotional effects were of more consequence. If influential contemporary accounts of prosody are taken seriously, Proctor's fast pace invested his platform performances with certain kinds of affective meaning. Generally referred to as a speaker's "movement," the tempo of delivery could be used to signal particular passions. While these were not invariant, popular elocution manuals—widely read and used in the United States as well as in Britain—associated a fast tempo with a relatively narrow range of emotions. James Rush's influential *The Philosophy of the Human Voice*, first published in 1827, included "gaiety," "anger" and "eager argument" as among the effects of a quick tempo. A slower movement was associated with sentiments such as "sorrow, respect, veneration, dignity."[36] A manual based on Rush's work noted that a significant fault found in some speakers was a "habitual rapidity" that prevented "all deep and impressive effect."[37] These associations persisted later into the century, even with changes in oratorical culture and the science of elocution. Joshua McIlvaine, professor of belles lettres at Princeton College, noted in 1871 that a fast delivery excited an audience's "impassioned sentiments."[38] In contrast, a slow pace evoked solemn and grave emotions. These assessments were independent of Proctor's own understanding of the emotional impact of his lecture performances but are suggestive of the feelings that his lectures may have provoked. There was, perhaps, a hint that the emotions and associated moral effects associated with Proctor's lectures were indeed somewhat superficial and fleeting.

Proctor had his own reasons for avoiding at least some of the paralinguistic aspects of speaking that might have countered the overall impression left by his style of speaking. In remarks published in *Knowledge* after his American lecture tours, Proctor argued that "declamation" or the "tricks of fine speaking" were entirely inappropriate for a science lecture.[39] He insisted that studied gestures drew too much attention to the "petty personality" of the science lecturer and that science was sufficiently sublime in itself not to require the tricks of oratorical appeal to emotion. This may account for the common observation that Proctor's delivery was relatively monotonous. It also suggests efforts to avoid stirring the emotions beyond those perhaps accidentally excited by fast delivery. Instead, the audience should see through the translucent lecturer to "the grand teaching of nature." As Proctor put it, "applause at the end . . . comes at a time when the lecturer recognizes to the full his utter nothingness in relation to what he has been saying. He feels that one might as well applaud the wind for the music of the Aeolian harp, as the lecturer for the grandeur of the truths of science."[40] This argument, of course, may have suited someone untrained in the exact arts of gesture or uneasy about the self-aggrandizement that was often associated with celebrity and success on the lecture circuit.

Proctor was certainly convinced that the "declamatory arts" should be avoided in science lectures. He had, he claimed, "earnestly striven to rid my delivery of every trace of declamation" and urged that any self-conscious prosody or movement in the platform performance of the science lecturer must be "unlearned."[41] Proctor, as we have seen, did have *some* awareness that his own performance was an embodied act of communication. Gesture, for example, had become for him a key mnemonic device, allowing him to include lines of poetry in his lectures. Not employing the "customary gestures" that he habitually used to accompany the recital of poetic verse led on occasion to embarrassing failure of memory. This marked out for Proctor the "wide difference between knowing a passage by heart for recital to yourself and knowing it for recital before a crowded audience."[42] It also played to the expectations of audiences, which frequently scrutinized the embodied performance of speakers on any subject and expected at least a degree of movement.

Proctor's use of humorous anecdotes and analogies also worked well with audiences that might otherwise have headed to the theater or opera house. During his tour of the United States in 1879–1880, Proctor's lectures were described by one newspaper as "one of the best features in New York entertainments." They provided a contrast to the "weakness of our best comedy," and, "in the midst of so much trash," it was little wonder that people were turning in droves to see and hear Proctor lecture. He delivered "solid information"—or facts with a moral purpose—in an attractive way, elevating as well as amusing New York's theater-going public.[43]

In the end, though, despite Proctor's episodic attention to the embodied nature of lecturing, in practice his countenance, posture and actions probably mattered less to him than the copious illustrations that he routinely used during his lectures. Certainly, there is little doubt that the oxyhydrogen stereopticon was a critical component of the total effect of a Proctor lecture. Proctor spent hundreds of hours preparing the images, a practice he was well qualified for. His star atlases offered a ready supply of striking depictions of the night sky, but many of the other images were prepared from scratch. It was also vital to Proctor that the gas lighting in the lecture hall was carefully adjusted to maximize the visual effects of his projections. The auditorium's sonic properties did matter—Proctor noted, for example, that good acoustics could make speaking and thinking much easier than a venue that placed severe strain on the voice.[44] The lighting and lines of sight, however, were of at least equal significance. It was not unknown for Proctor to use hundreds of images in a single lecture, making the task of the lanternist, and the controller of the gas lighting, a daunting one. Of all of the lecturers investigated in these pages, the importance of projected visuals was most evident in Proctor's case.

Proctor's general and growing reputation as a science lecturer no doubt rested on a combination of his early training, his personal investment in lecturing as a central part of his vocation, and his unique style of delivery. The rapid-fire utterances, extempore delivery, occasional but carefully controlled use of gesture, the humor and the exploitation of the latest display technologies contributed to their appeal to audiences keen to be enthralled as well as instructed. But to understand his tours in America, and the huge interest in his lectures, it is also important to explore how he added to his general lecturing abilities and philosophy by developing a reputation for being able to "roar" in what one American commentator described as a "trans-Atlantic accent."[45]

PREPARATIONS FOR AMERICA

In the lead up to his American debut, Proctor's standing in British astronomy and British science was, in many respects, starkly different from Tyndall's or Huxley's. There were superficial similarities. He was well known to a large public. His books and his numerous contributions to periodicals such as *English Mechanic* meant he had a large readership. His involvement in the RAS had reached new levels with his appointment as secretary in February 1872.[46] But he stood outside the elite scientific circles that had long embraced Huxley and Tyndall. His attempts at election to the Royal Society in 1873 failed, a fact not surprising given his much-publicized antipathy to scientific officialdom.

Proctor's reputation as a science lecturer placed him somewhat closer to Tyndall than to Huxley. His appearance at the Royal Institution and his involvement in the Sunday Lecturing Society had certainly put him among the best-known scientific expositors in Britain. But his popularity in Britain was not quite what it became in America. He was later to complain rather bitterly about the lack of attention paid by the daily newspapers to science lectures in London.[47] His sense of the London scene was that it had not embraced the science lecture in the way he was to observe and experience in the United States or, indeed, elsewhere in Britain and Ireland.

Another reason for his rising public profile, if not his standing among the guardians of elite British science, arose from the controversies he provoked within Britain's astronomical community. His dispute with George Airy, Britain's astronomer royal and director of the Greenwich Observatory, over the best method for observing the 1874 transit of Venus was particularly important.[48] Beginning in 1869 Proctor published prolifically on the upcoming transit of Venus, both in the *Monthly Notices of the Royal Astronomical Society* and in other periodicals aimed at a more general readership. He included criticisms of the arrangements being made for the government-sponsored expeditions to observe the transit, led by Airy. By 1872 his disagreement with Airy shifted from being largely technical to more generally political and personal. The debate played out within the RAS and in the pages of national newspapers and turned in part on what the best conditions were for supporting a truly scientific positional astronomy. Airy naturally defended his plans, with an eye to maintaining state support for astronomical research. Proctor pushed hard against such "official astronomy" and called for greater independence from the shadow of government regulation and its negative effects. When Proctor used his position as temporary editor of the *Monthly Notices* to publish a supplement on the transit of Venus again criticizing Airy's views, he faced a backlash from members of the RAS Council. As a consequence, he resigned his position as secretary a few weeks after sailing to America to begin his first lecture tour. Proctor later described this episode as an irritating interruption to his ongoing astronomical research. The RAS, he noted, was at that time "a body vexed with many turmoils, from whose perturbing influences I had trouble in withdrawing myself."[49] Nevertheless, he also realized that his widely advertised dispute with Airy helped explain why his first lecture tour proved such a success. As he put it, "my criticisms on Sir George Airy's inexact first papers on the transit of Venus . . . turned out very much to my advantage in the end." Proctor owed, indirectly at least, his "pleasant experience in America" to his criticisms of an astronomical expedition sponsored by the British state, which rivaled America's own efforts to observe the transit.[50]

Part of the pleasure was due, in his own estimation, to the "personal satisfaction" he experienced on discovering, early in 1873, that American plans for observing the 1874 transit of Venus included locations that Proctor had promoted in Britain only to be dismissed by Airy and others.[51] But, on Proctor's telling, more than mere agreement about methods endeared him to the American public.[52] The fact that he had stood up to official government science had guaranteed a warm welcome in a country that encouraged free expression of critical opinion. Other agreements between Proctor and leading American astronomers may have further enhanced the friendly welcome he would receive. Several months before Proctor arrived in the United States, the American astronomer Charles Augustus Young criticized Airy's insistence that Delisle's method of observing the transit was the most suitable. There were, Young urged in a lecture delivered at the American Institute in New York, "great difficulties" with that method, not least the problem of observing with the necessary precision a black disc against the bright body of the sun.[53] It was no coincidence that in his inaugural lecture in Boston on October 21, 1873, Proctor described Young as "one of the greatest students of the sun in this or any age."[54]

The success of Proctor's books in America also crucially contributed to securing him a warm reception on arrival. When Proctor set off to the United Sates, his books were already widely available. Some had been pirated or published without any financial benefit accruing to Proctor. If the distribution of his books before his visit had been done without proper copyright arrangements in place, Proctor nevertheless acknowledged that the very fact they were being read in America was an essential reason for the success of his first lecture tour. He noted some years later that D. Appleton had produced an edition of *Other Worlds Than Ours* in 1871. Proctor had "received no direct profit. But I gained largely—though I can scarcely thank them for that—by becoming better known than I had been in America. I cannot but attribute to this a large part of the success I obtained in America in 1873–4." Proctor also acknowledged the support offered to him by Appleton while on tour, singling out in particular Edward Youmans, "who has done more than any living man to improve the relations between literary men and publishers in the two countries."[55]

Proctor was also sensitive to the fact that the books he wrote for a wider and nonspecialist readership could attract criticism from scientific quarters and potentially damage his reputation as an authority on astronomy. Writing (and lecturing) for profit also risked reputational damage. Proctor, however, successfully parried this kind of criticism, using it to his advantage. This was particularly in evidence in-between his first and second tours, when he responded to a vicious anonymous attack on his literary output, the author of which was later identified by Proctor as the American astronomer Edward Singleton Holden, then professor of mathematics at the US Naval Observatory.[56] Writing in the *Atlantic Monthly*, Holden (if Holden it was) had charged Proctor with a form of literary theft, namely of hawking multiple books to an unsuspecting public that regurgitated the same "trash" in a new form.[57] Proctor was further charged with using religion to increase the profit margins of books, playing on the public's desire for a mix of science and theological pieties. Proctor made two responses, one in the *Atlantic Monthly* and another in a letter to the *New York Times*.[58] Among other things, his riposte denied there was anything wrong with making money from the sale of "popular works" (a "novel doctrine") and hinted that the reviewer was a competitor in the same market. He did not apologize for reworking old material for new readers, which he patently and confessedly did. But he was careful to garnish this recycling with the latest astronomical facts and theories so that the rapid production of books was seen to be driven not by the desire for pecuniary profit but by a concern with keeping pace with the accelerating advance of science.

In addition to his books and journal publications and his praise of American science, the ground for his tour was also prepared through laudatory character sketches in several popular journals. For example, Henry Ward Beecher's periodical, the *Christian Union*, noted that the "lecture business . . . depends upon lions, particular those that can roar with a trans-Atlantic accent" and pronounced Richard Proctor a "worthy specimen" and "one of the great order of brilliant scientists who can both investigate and talk."[59] Proctor, as a later previsit report in the same journal put it, was one of those "agents of international commerce in thought and emotion" whose eloquence "is as the music of the spheres."[60] The *Popular Science Monthly* added to this praise, declaring Proctor "a first class man" and a "clear, rapid and forcible speaker." His lectures, the piece continued, promised to be "the leading scientific entertainment of the season."[61] A columnist in *Scribner's Monthly* joined the chorus, noting that Proctor was among "the youngest, and in some respects, the most brilliant, of living astronomers." He had "a distinct and honorable place among the select and illustrious few who have widened the boundaries of exact knowledge and devoted great intellectual power to the elucidation of some of the grandest themes in the arcana of the sciences."[62] The same sketch praised Proctor's prescience in predicting the best method and best places for observing the transit of Venus in 1874 and noted his zealous and courageous opposition to the astronomer royal's alternative plans.

These pretour commendations were vital, as Proctor jostled with a number of other celebrity British authors and speakers about to embark on visits to America. The author Wilkie Collins and "the greatest living elocutionist," J. C. M. Bellew, were both about to speak in America for the first time.[63] Proctor also faced comparison and competition with the agitator and atheist Charles Bradlaugh, the transatlantic temperance lecturer J. B. Gough and the sensationalist explorer and gorilla hunter Paul Du Chaillu. Proctor, however, was well placed to outperform these celebrated speakers. He was not alone in thinking that he succeeded in America in the season that "Bellew and Wilkie Collins both failed there."[64]

Before he crossed the Atlantic in October 1873, Proctor seemed ready made for America. In lectures and writings he had publicized his love of America and American science and had articulated a vision of scientific culture that arguably resonated more in the United States than it did in the United Kingdom. He was ready, when the invitations came, to "rush from town to town" making, some claimed for the first time, the rapid advances of science "generally appreciated on the [American] rostrum."[65] Unlike Tyndall or Huxley, he was willing to lecture to lyceum audiences and signed up with the American Literary Bureau in order to maximize lecturing opportunities.[66] The scene was set fair for his American debut.

PROCTOR'S AMERICAN DEBUT

As with other visiting lecturers, one important impetus came in the form of an invitation to give the prestigious Lowell lectures in Boston. Proctor committed himself to delivering twelve lectures at the old Marlboro Chapel, a public hall used by the Lowell Institute, through October and November 1873. Announced in Boston's newspapers, Proctor's series proved to be an immense success. The tickets were all gone within thirty minutes of being released.[67] At his inaugural lecture, given on October 21, Proctor laid out the overriding objective of the whole series. His aim was to "furnish information to comparatively well-educated people as to the details of astronomical science, and to obtain recruits in amateur astronomy." To do this he had to demolish the "prevalent idea that the science can only be pursued to advantage in government observatories." To his Boston audience, he insisted that "important researches" [sic] frequently relied on "private observation."[68] Toward the close of the series he again reminded his audience that it was quite possible for amateur observers to comprehend and contribute to the growth of astronomical knowledge. Opening his lecture on "the means by which the size and distance of the heavenly bodies are determined," Proctor drew attention to George Airy's claim that "it was not possible to describe the means employed in the transit of Venus so that they could be understood by a popular audience."[69] Proctor sought, once again, to prove Airy wrong.

When Proctor first took to the stage at the Lowell Institute, reports—as was customary—offered a sketch of his appearance and lecturing style. Comparisons were made with Tyndall, with one report noting that Proctor "harmonizes more with our idea of the average Englishman. He is a short, thick-set man, with a full face, and parts his hair in the middle." His style, the report continued, was "clear and lucid" and his manner "very pleasant."[70] His inaugural lecture took the sun as the subject and introduced the audience to the "tremendous operations going on in this wonderful orb." Proctor discussed the sun's distance from the earth, size, mass, generation of light and heat, using familiar analogies and with reference to the latest discoveries. For instance, he demonstrated the vast distance between the earth and the sun by picturing a "baby with an arm long enough to reach the sun." He would, Proctor calculated, "have to live to an old age and die before he would know that his finger was burned there."[71] Provoking laughter, this farcical image further endeared Proctor to his audience.

Another way in which Proctor commended himself to his Boston audience involved frequent favorable mentions of American astronomy and arguments that American scientists had anticipated European discoveries or had fallen on the right side of various astronomical debates. In his second Lowell lecture on the sun, Proctor pointed out that American astronomers had correctly argued that the corona was a product of solar activity and not, as some European astronomers had supposed, partly a consequence of the earth's atmospheric processes. Singling out William A. Norton and Charles A. Young, Proctor declared them "entirely in the right."[72] He also championed the astronomical investigations of Maria Mitchell, mentioning her discovery of a comet in 1847 in one lecture and arguing in another that she deserved the honor of having a crater of the moon be named after her.[73]

The remaining ten lectures covered a great sweep of astronomical subjects, ranging from the planets of the solar system, their evolution, comets, stars and nebula and ending at the grandest scale of all, the universe itself. Proctor constantly returned to the possibility of life on other planets or elsewhere in the universe. In his third lecture, for example, he looked at the case for life on the sun (citing but not approving John Herschel's theory of solar life) and on the "inferior planets."[74] Although he pointed out that life might exist on the solar system's inner planets, he made it clear that conditions on each of them made for extremely harsh environments. The subject, popular with the public given the long history of controversy and cultural and religious questions associated with it, was absorbing and allowed Proctor to compete with John B. Gough, who delivered a lecture the very same evening on London's street life just a few doors away from the Marlboro Chapel at Boston's Music Hall.[75] While Gough had spoken many times on

London street life, this lecture on "night scenes" was new.[76] The Music Hall, which could hold up to three thousand people, filled to capacity to hear it.[77] Proctor's lectures, unlike Gough's, were free. But his full capacity audience of about two thousand was also attracted by exciting new material that had its own intrinsic interest.

Proctor used a range of visual aid techniques to make his lectures appealing and to maintain interest as his series continued, including diagrams and drawings. In lecture six, for example, he displayed artistic depictions of the night sky as it would appear on the moon. He also utilized the power of photography and lanternslides, although he reserved the use of this media to just one lecture. He made sure, however, to announce this to his audience in advance, noting in his eighth lecture that images by the American lunar photographer Lewis Rutherfurd, "the best ever taken," would be projected using a magic lantern.[78] The photographs were shown in the eleventh lecture, and Proctor underlined again that they were the "finest in existence."[79]

As well as stimulating his audience's interest through promises and projections of high-quality images, Proctor kept them engaged by addressing questions raised by his hearers. In his fifth lecture, for example, he responded directly to a query raised by one member of his audience who had written to the *Boston Daily Advertiser* to state an objection to Proctor's allusion to the sun's gravitational pull in his first lecture. Proctor took time to correct the "error into which the writer had fallen," perhaps caused by a "slip of the memory."[80] In responding, Proctor not only secured his credibility as a masterful expositor of astrophysics but also reacted to his audience in a way that made them participants in the production of his own discourse. Once again, this positioned him as a champion of amateur scientific inquiry and set him apart from lecturers who largely discouraged direct audience participation of this kind.

On completion, Proctor's first Lowell lectures were proclaimed a success by the press and, despite the occasional challenge to some of his claims, reinforced his reputation as a superlative lecturer on science. As the *New York Tribune* put it shortly after the series, Proctor had come "at the zenith of his fame" and his success brought assurance of his ability "to pure and elevated thought combined with novel and entertaining instruction."[81] The lectures, along with a condensed parallel series delivered at the Essex Institute in Salem, Massachusetts (an abridged version of his Lowell talks), provided a powerful platform from which to launch a wider tour that took in Albany, New York, Brooklyn, Cincinnati, Indianapolis, Chicago, Philadelphia and Washington. They also introduced his first American audiences to some of his characteristic techniques: humor and charm, arresting images, a receptivity to dialogue and a projected confidence that his listeners could keep up with astronomical discoveries and even make significant ones of their own.

Proctor's first stop after Boston was Albany, where he delivered a single lecture on "The Wonder of the Star Depths" on December 4.[82] A reception at the Lotus Club in New York then marked Proctor's final public appearance before the New Year. The reception provided him with another opportunity to praise American science and astronomy. The club, founded in 1870 by a group of New York journalists, critics and businessmen to encourage the arts and literature, had established a tradition of organizing lavish banquets for visiting celebrities from Europe. Its committee "watched the wharves" for any distinguished visitors arriving in New York to extend an invitation.[83] Proctor was the first scientist to be honored with a Lotus Club banquet and, as he revealed in his speech, it was a prominent club member, its secretary John Elderkin, who had been instrumental in making possible his first American lecture tour. Proctor, aware of the club's tradition of lionizing—and criticizing—British literary figures (Wilkie Collins had dined with its membership three months earlier), flattered his American hosts with a defense of their own astronomers in the face of British conservatism. Proctor declared himself more American than British in his attitudes toward scientific authority and strongly criticized British astronomers such as Sir George Airy and Norman Lockyer for failing to recognize the superior results produced by astronomical science in America (once again, he mentioned Maria Mitchell).[84] In part this was a continuation of feuds that had led to Proctor's resignation as secretary of the RAS. But it strengthened his pro-American credentials and underlined his judgment about the failings of English official astronomy.

The Lotus Club public banquet continued to stimulate newspaper commentary and correspondence over the following weeks. Maria Mitchell, taking some of the wind from Proctor's sails, wrote to the *New York Tribune* to exonerate Airy from Proctor's accusations that the astronomer royal had, out of sheer superciliousness, tried to prevent the award of a gold medal for her comet discovery.[85] Proctor's reply, while diplomatically praising Mitchell's "graceful tribute" to Airy, repeated his original accusation, noting that Airy's objection had been overruled by a committee majority. He used the opportunity to suggest that English astronomers, and scientists more generally, had been far too slow to recognize American achievements.[86] Equally, American scientists had been too retiring in publishing the results of their discoveries in a timely fashion. They had not kept pace with the necessary rush to establish priority that marked science internationally. Speed was of the essence.

The close engagement with American scientists continued as Proctor geared up in early January to deliver two parallel lecture series in New

York and Brooklyn. The New York series was run under the auspices of the Young Men's Christian Association and held at Association Hall. It was inaugurated with a reception to which leading representatives of American science were invited, along with nearly two thousand other guests.[87] The reception committee included Henry Jackson Morton (the chemist, celebrated lecturer and president of the Stevens Institute), along with Edward L. Youmans, F. A. P. Barnard (then president of Columbia College) and other American scientific luminaries and philanthropists. Proctor delivered the first lecture, on the sun, to a capacity crowd (hundreds did not find a seat) and was introduced by Morton, who had provided expert help preparing the visual illustrations projected in rapid succession in a darkened hall.[88] Again, observers singled out Proctor's ability to clearly and quickly present novelty. One editorial published the following day, noted Proctor's fluency, "freshness of style," and popular presentation of the "latest phases of discovery" as guarantees of the success of the entire course.[89] The *New York Tribune* promptly produced an extra that printed the full texts of all six lectures, alongside a lecture series recently delivered by the celebrated Swiss American scientist, Louis Agassiz. Like tickets for Proctor's lectures, the extra, produced on an eight-cylinder press, sold quickly. Several weeks later a reprint was issued after unanticipated demand (but the experiment of using the new printing method was abandoned for the sake of a "more attractive" edition).[90] It is interesting here to underscore the fact that Proctor was, in person or in print, standing alongside two of the most celebrated American science lecturers of the day. It is curious that newspaper reports did not rush to draw direct comparisons between Proctor and these American science lecturers despite the clear juxtaposition. His competition, as with other visiting British science lecturers, was more often considered to be the other foreign lecturers on the circuit that particular season. But Proctor's proximity to Morton and Agassiz will have done his own reputation no harm. Proctor, of course, could take nothing for granted as his tour quickly moved on.

The Brooklyn lectures began the next day, on January 9, organized by the Mercantile Library Association. The *Brooklyn Eagle* declared the series superior to the New York one on account of the better hall (the Academy of Music) and because most of the lectures would be delivered before Proctor went back across the East River to finish his course at Association Hall.[91] While reports were fulsome in their praise of the Brooklyn lectures, one member of the audience at least thought Proctor in too much of a hurry. Writing to the *Eagle*, they asked that Proctor might in subsequent lectures "deliver with less rapidity, especially when facing his illustrations." It was their view that "very many" of his hearers had missed portions of the lecture due to the sheer speed of Proctor's utterances.[92] The tempo of Proctor's lectures had now been noted and would fast become *the* characteristic mark of his style of delivery.

When Proctor gave the same six lectures in Cincinnati, the complaint reemerged, with even more force. His closing lecture, delivered on March 4, was, according to a report in the *Cincinnati Commercial*, spoken "with even more volubility than on former occasions, and so inundated his audience with facts, statements, theories and deductions about the moon past, present and future that it was exceedingly difficult for the ear and understanding to keep pace with him." According to the same report, the lecture was also the "least interesting" Proctor had delivered and, worse yet, he had neglected to include mention of the work of the Cincinnati-based astronomer and mathematician Daniel Vaughan.[93] These complaints contrasted with the more celebratory accounts that appeared in the rival *Cincinnati Daily Gazette*. Even so, they risked Proctor's reputation as a champion of American astronomy and a skilled and fascinating lecturer.

The quickening pace of Proctor's lectures was mirrored by the rate at which he gave them. During his first two months in the United States, he gave one lecture every three days. Between early January and April 8 (when he departed for England) he lectured almost on a daily basis. He traveled as far west as St. Louis, relying on trains to transport him on time for his next performance. When he delivered his final address at Association Hall in New York on April 7, he had lectured 102 times in total.[94] This took its toll both on Proctor's health and his reputation.[95] His lectures in St. Louis, Indianapolis and Columbus and in other more western cities were repeats of ones given in Boston and New York. While minor deviations in content were noted by some, it was possible for newspapers—as in Cincinnati—to forgo the employment of a stenographer and simply reprint the text that had appeared in the *New York Tribune* extra, risking Proctor's standing as a lecturer committed to novelty and freshness of delivery.[96] In Columbus, professor of mathematics and civil engineering at the State Agricultural College Robert White McFarland publicly declared that Proctor's lectures had been "mostly composed of old material, bad in style and delivery and incorrect and erroneous in those parts not old."[97] It was perhaps in light of these sorts of judgments that Proctor agreed, before he departed, to give six additional lectures in New York, some of which included new material not found in his original Lowell lectures and all of which had not been heard in New York. This costly strategy appears to have worked, and his final lectures drew large and enthusiastic audiences and much praise from the press.

Whether or not by design, the subject matter of Proctor's final lectures led him to questions of a more metaphysical and religious kind. This was most dramatically the case in his farewell lecture at Association Hall on "the infinities around us." Proctor admitted that his subject matter took him beyond astronomy strictly speaking but proposed that astronomy more than other sciences suggested the topic. For much of the lecture, unusually delivered without any visuals, Proctor explored the concepts of infinite space, matter and time. He argued that, though inconceivable according to the human senses, the realities of these infinities were the sound deductions of science and philosophy. That they were inconceivable provided an argument against the "materialist" who denied the existence of a God of infinite wisdom, power and beneficence. Mere inconceivability provided no grounds to deny certain realities. But the overall tone of Proctor's lecture suggested not apologetics for religious conceptions but rather an argument for what he maintained was a scientific and sound view of an infinite universe. As on other occasions, he generally avoided direct commentary on specific religious concerns and, with a handful of exceptions, his tour had invited little theological controversy. What was clear, however, was that his main task had been to rapidly record for his hearers the marvels of new scientific discoveries. This was a task that, for Proctor at least, could no longer be effectively done by avoiding the contrasting tempos of religious change. When he returned to the United States eighteen months later for a second tour, his own changing religious views came under much closer scrutiny, particularly as he directly attacked the braking effect of conservative religion on accelerating scientific progress.

FRAMING PROCTOR'S CONVERSION

Proctor arrived in New York on October 13, 1875, for his second lecture tour, again managed by the American Literary Bureau. The tour began on October 18 with a lecture in Philadelphia (on the sun) before continuing with another set of twelve Lowell lectures. For the most part, little in these lectures hinted at any controversy over changing religious views. On the contrary, they were, as Proctor himself put it, "the driest on my list."[98] He was therefore amazed at the interest in the lectures, reporting that long queues had formed to secure tickets for the Lowell course. Others may have been surprised as well. The *New York Tribune* simply noted that Proctor "began his series . . . with his old lecture on the transit of Venus."[99] Although addressing familiar subject matter—comets, the planets, the transit of Venus—Proctor was still able to draw large audiences.

Press interest in Proctor's opening lectures, however, was not especially high. While Boston newspapers carried relatively brief abstracts of some of his Lowell lectures, there was little additional reporting elsewhere. This, however, changed dramatically with the announcement that Proctor would speak on four successive Sunday evenings alongside his final few Lowell lectures. The chosen topics were racier, designed to catch the attention particularly of readers who would already have encountered his views on the announced titles: "Astrology and Superstition," "Birth and Death of the World," "Other Worlds and Other Suns" and "Religion and Astronomy."[100] In a note inserted in newspapers before the lectures began, Proctor explained that the lectures would "indicate the relation of modern astronomy to the great questions at present agitating the scientific and religious worlds." Perhaps playing on expectations already raised by previous reports of his recent change of mind, Proctor explained that "the views I now entertain on such subjects as the plurality of worlds, cosmic evolution, the supervision and control of the universe are altogether unlike those which I indicated in my . . . earlier works." He viewed these matters "differently now than . . . a year or two ago."[101] This was surely designed to pique the interest of the public, given the popularity and intrigue surrounding questions of life on other planets and the fascination of those who had followed Proctor's influential arguments on that subject.

His first lecture at Horticultural Hall on Sunday, November 7, did not disappoint. Proctor declared himself on the side of John Tyndall, fully embracing the views expressed in the physicist's infamous Belfast address delivered the previous year. This did not, Proctor insisted, amount to atheism or to an attack on religious sentiment. But it did mean adherence to truth and a view of the universe incompatible with an interfering deity. The lecture attracted considerably more attention than his prestigious but dry Lowell lectures. The *Albany Evening Journal* declared that Proctor, who before occupied a "middle ground" in science and religion, was now "in full sympathy with the most advanced evolutionists."[102] The *New York Tribune*, previously uninspired by Proctor's "old" lectures being delivered at the Lowell Institute, published an extended editorial that announced him a "convert of science" and printed his lecture in full on the front page.[103] The editorial declared that he now accepted not just the evolution of living forms but also the generation of life from inorganic matter. Evolutionists would "welcome him with open arms."[104] Two days later, the *Cincinnati Daily Gazette*, a paper that had followed Proctor's lectures in that city closely during his first visit, announced that Proctor, "previously a Roman Catholic . . . has radically changed his views." This, as shall become evident, was an exaggeration, but was not without foundation.

It was, however, a declaration that sat uneasily with Proctor. In his opening lecture at Horticultural Hall he insisted that he had held the same views about the origins of all forms of life in matter as those presented by him to a New York audience in April 1873. He could not understand why Tyndall had to endure a vicious assault on his supposed atheism when Proctor's claims had raised no objections whatsoever. Proctor could only suppose that his religious opinions were so widely taken to be the same as those held by Louis Agassiz and Michael Faraday that he simply had not been properly heard. It was not his "duty" to disclose his religious views, which he confessed now aligned with Tyndall's. Religion operated in the domain of emotion, science in that of knowledge. Religious sentiment was "a phase of the energy of nature" and one completely separate from that of science. Despite these statements, the editorial in the *Cincinnati Gazette* intoned that "his course will . . . be greatly deplored by his Catholic friends, as he was one of the most prominent of the scientific representatives of their faith." Not only this, but the Gazette also proclaimed that Proctor had changed his views "in all branches of natural science," and these apparently sudden developments could not be assessed without the famous lecturer "describing more minutely the various steps he has taken."[105] Proctor, it seems, was struggling to keep control of the narrative about how precisely his views on religious as well as astronomical matters had changed and how the two may or may not be connected. But his hearers, and those avidly following the progress of his second tour in America, were being fed a picture of Proctor as mercurial, a portrait surely enhanced by his rapid-fire approach to delivering scientific and metaphysical material in the lecture hall and by his commitment to keeping his audiences abreast of developments in astronomical science.

As news of Proctor's apparently altered views circulated widely, the reports became more dramatic. The *Portland Daily Press*, for example had Proctor moving from holding the same opinions as the antievolutionist geologist John William Dawson to becoming an outspoken proponent of "the formation of life from inorganic matter."[106] The *Christian Union* put the change in similarly stark terms: "Two years ago he was a Catholic in religion and an opponent of Darwinism. Now he is no longer a Catholic, and avows himself a champion of the development hypothesis."[107] Though Henry Ward Beecher, as the journal's editor, warned his readers not to rush to the conclusion Tyndall and Proctor were opposed to Christianity properly understood, the impression given was of someone who had experienced a sudden deconversion. Proctor was quick to act in an attempt to suppress the now widespread impression that his views, at least on science, had undergone "sudden change." The real change, Proctor insisted, was not to his science but rather to his rejection of a supposed compatibility between the dogmatic claims of religion and the positive findings of science. Two years ago he was "satisfied" that religious dogma and science could be reconciled. After the insistence of theologians that Tyndall's science was utterly inconsistent with the teachings of the Catholic Church, Proctor gave them up in order to adhere to the truth disclosed by science.[108]

In commenting on his change of mind, a further editorial in the *New York Tribune* suggested that the "acrimony" of the debates about science and religion in Britain in the wake of Tyndall's address had forced Proctor's hand. Things in the United States were different: "In this country there are numbers of thoughtful men who clearly perceive that a belief in the highest truth of science is not of necessity in conflict with the profoundest religious faith."[109]

An article published in the *Catholic Telegraph* made the same point but related it to the compatibility of Roman Catholicism and science in particular. The anonymous author was at pains to point out that evolution was compatible with the Catholic faith, as long as God's sustaining and cooperative agency in creation was accounted for and properly understood. Citing Augustine, the article urged that it was quite possible to hold that God had created "all things in the elementary germs of life" that were spontaneously formed when he first declared "let there be light."[110] These "germs" then developed through time into myriad living organisms as described by evolution. This view, already widely circulated through the medium of Catholic periodicals and in the work of the Catholic zoologist St. George Mivart, ought to have allowed Proctor to remain true to the faith.[111] The only exception was the direct formation of the body (and soul) of Adam, and even in that case the church had never formally condemned the theory positing the indirect creation of the first man.[112]

No record of a response from Proctor has been found, but it was clear from his previous commentary that he had indeed come to reject the case for compatibility between the findings of science and Roman Catholicism, as defined by Tyndall's Catholic critics. But he also denied that he had come to America on his first visit as an advocate for Roman Catholicism. Responding to a comment in the *Daily Graphic* that he had indeed done so and was now a proponent of atheistic science, Proctor categorically denied both. He had "never made or sanctioned any public statements about my religious opinions." Any previous rumors that he had were due to his name being added to a list of "senators" of a new Catholic University in London without his knowledge. This did not mean that on his first visit he had already cut ties with his Roman Catholic faith. But what he now publicly insisted on was that "the Roman Catholic priesthood considers my views—even the views indicated in the earliest of my books—about evolution inconsistent with Catholicity."[113] Proctor had come to agree with them and had rejected the Catholic side of the equation as a consequence. At the same time, however, Proctor was careful to deny that he was now appearing as an atheist or antireligious propagandist. On the contrary, he regarded the fact that "the human race demanded a religion" as "full of promise for the future."[114]

Proctor did not recognize any construal of accord between science and confessional religion. Writing several months after his second tour, he noted "the influence . . . of religious scruples upon scientific progress and research in America. Here I must admit that I was somewhat disappointed. I expected to find America a long way in advance of England. But with some noteworthy exceptions, especially in the West, America seems to me to be behind England in this respect."[115] If American newspapermen were accusing Proctor of rapidly changing his views and siding with those who attacked cherished religious beliefs, Proctor hit back and accused Americans of allowing theological "scruples" to slow the advance of science. Here he echoed views held by a number of prominent commentators on science and religion on both sides of the Atlantic, namely that the conflict lay between theological dogma and science rather than with religion understood as an ineradicable, deeply felt but ineffable aspect of human consciousness.[116]

In spite of these attempts at presenting himself as consistent as to science, opposed to (Catholic) dogma but accepting of religion, Proctor did not hold back from launching attacks on cherished religious beliefs. Proctor, of course, well knew that stirring up controversy about religious concerns was a tried and tested method of generating press attention and drawing crowds into lecture theaters. The three remaining Sunday evening lectures at Boston's Horticultural Hall all tackled subjects that had explicit connections with theological concerns. The first two, on "Other Worlds and Suns" and "Astrology and Superstition," did not tackle in any sustained way the most controverted questions. Nevertheless, Proctor took the opportunity to decry the distortive effect of religious or superstitious views that were too firmly held. The former had held back an unprejudiced look at the plurality of worlds. Even the Scottish physicist David Brewster, an enthusiastic proponent of a solar system and universe teeming with life, had been guided too much by a desire to show that the light of the sun had not been "wasted" on barren planets.[117] There was, Proctor declared, no way to discern particular divine purposes in the universe. Whether such purposes existed lay beyond the scope of science and human knowledge in general. The moral from the lecture on astrology and superstition, on the other hand, was that men of science could now "undeceive" people and reveal the falsity of many common beliefs. Not that the task was an easy one. Proctor finished the lecture by bemoaning the fact that "ninety-nine out of a hundred" prefer to remain deceived.[118]

It was the fourth and final Sunday lecture, on November 28, that caused the greatest reaction. To a crowded hall, Proctor made the case that the science of astronomy, despite having been obstructed by religious forces and superstitious beliefs, was now fully established. His account of the history of astronomy included the "torture" of Galileo and the burning at the stake of Giordano Bruno for his belief in the plurality of worlds. Those scientific martyrs had helped pave the way for a science of the heavens that finally dispelled influential misinterpretation of scriptural texts. Whatever aspersions this narrative cast on the formal structures and actions of the Christian churches, Proctor remained mainly conciliatory for the first half of the lecture. The second half, however, was more obviously polemical. Proctor pointed to the "discrepancies" between the two creation accounts, expressed horror at punishments sanctioned by Moses, traced the Jewish Sabbath to Egyptian "superstitions" and dismissed as incredible the astronomical miracles recorded in the Old Testament. Proctor launched, in other words, a hailstorm of objections to scriptural narratives and biblical beliefs that would have been fully at home in a freethinkers' Hall of Science. At the lecture's close, he emphasized the importance of the uniformity of law and, following Tyndall, denied the physical efficacy of prayer. All of this, as was his style, was delivered at a breathless pace and may well have left listeners startled by this outburst of religious skepticism.

In a report of the lecture, the "occasional correspondent" for the *New York Tribune* noted that the audience had gathered in expectation of a "more extended review of the conflict between science and religion." Proctor, at least in the latter half of his lecture, did not disappoint. According to an accompanying editorial in the same newspaper, much of what he said "grated upon the ears of many hearers."[119] It was, the editorial continued, less Proctor's "separate expressions" that indicated his "antipathy against . . . those who draw their weapons from the Biblical narrative" than his overall "tone." It was the manner in which he had commented on Moses and miracles that had demonstrated to his hearers "how wide is the gap between his present views and those held generally by men of the religious faith to which he formerly adhered." When the Truth Seeker Company—a New York publishing firm managed by the freethinker DeRobigne Mortimer Bennett—published an extended abstract of the lecture, its antireligious tenor and associations were further reinforced.[120]

The Sunday evening lectures were undoubtedly the most controversial and widely reported of Proctor's second tour. Proctor did not repeat them as a set and did not repeat his lecture on religion and astronomy anywhere else. When he gave five lectures at Steinway Hall, New York, starting on December 13, his tone was somewhat more conciliatory. In his opening lecture on the "growth of worlds," Proctor impressed on his audience that the "infinitude" of the universe was compatible with the existence of an "infinity of purpose of personal will."[121] Against what some reports of his second tour suggested, he categorically denied that the universe was without purpose. He only meant to claim that knowledge of that purpose was not available to the human mind.

His second lecture evinced a religious sentiment, if not any warmth toward traditional Christian accounts of creation. The death of one planet, in the cosmic saga Proctor presented, was but a sign of new life on another,

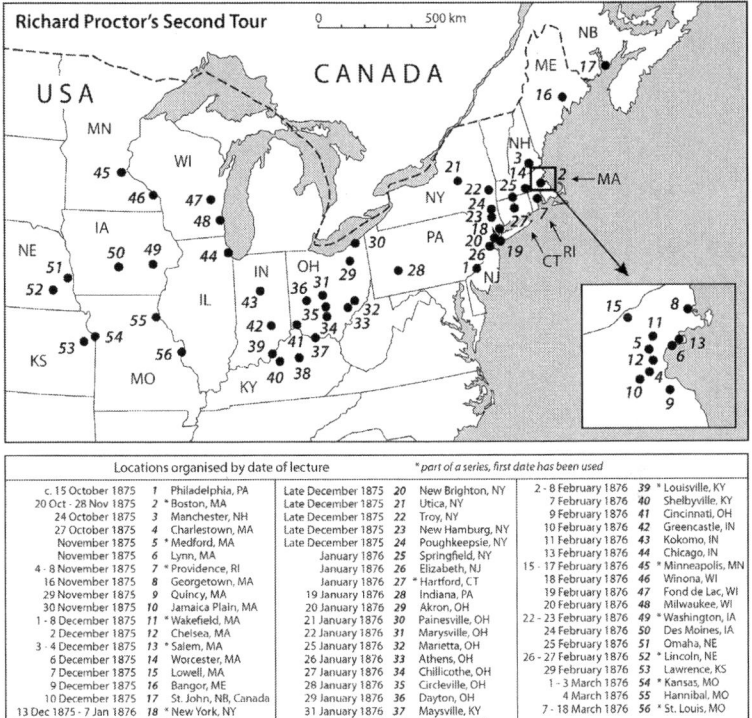

c. 15 October 1875	*1*	Philadelphia, PA	Late December 1875	*20*	New Brighton, NY	2 - 8 February 1876	*39*	* Louisville, KY
20 Oct - 28 Nov 1875	*2*	* Boston, MA	Late December 1875	*21*	Utica, NY	7 February 1876	*40*	Shelbyville, KY
24 October 1875	*3*	Manchester, NH	Late December 1875	*22*	Troy, NY	9 February 1876	*41*	Cincinnati, OH
27 October 1875	*4*	Charlestown, MA	Late December 1875	*23*	New Hamburg, NY	10 February 1876	*42*	Greencastle, IN
November 1875	*5*	* Medford, MA	Late December 1875	*24*	Poughkeepsie, NY	11 February 1876	*43*	Kokomo, IN
November 1875	*6*	Lynn, MA	January 1876	*25*	Springfield, NY	13 February 1876	*44*	Chicago, IL
4 - 8 November 1875	*7*	* Providence, RI	January 1876	*26*	Elizabeth, NJ	15 - 17 February 1876	*45*	* Minneapolis, MN
16 November 1875	*8*	Georgetown, MA	January 1876	*27*	* Hartford, CT	18 February 1876	*46*	Winona, WI
29 November 1875	*9*	Quincy, MA	19 January 1876	*28*	Indiana, PA	19 February 1876	*47*	Fond de Lac, WI
30 November 1875	*10*	Jamaica Plain, MA	20 January 1876	*29*	Akron, OH	20 February 1876	*48*	Milwaukee, WI
1 - 8 December 1875	*11*	* Wakefield, MA	21 January 1876	*30*	Painesville, OH	22 - 23 February 1876	*49*	* Washington, IA
2 December 1875	*12*	Chelsea, MA	22 January 1876	*31*	Marysville, OH	24 February 1876	*50*	Des Moines, IA
3 - 4 December 1875	*13*	* Salem, MA	25 January 1876	*32*	Marietta, OH	25 February 1876	*51*	Omaha, NE
6 December 1875	*14*	Worcester, MA	26 January 1876	*33*	Athens, OH	26 - 27 February 1876	*52*	* Lincoln, NE
7 December 1875	*15*	Lowell, MA	27 January 1876	*34*	Chillicothe, OH	29 February 1876	*53*	Lawrence, KS
9 December 1875	*16*	Bangor, ME	28 January 1876	*35*	Circleville, OH	1 - 3 March 1876	*54*	* Kansas, MO
10 December 1875	*17*	St. John, NB, Canada	29 January 1876	*36*	Dayton, OH	4 March 1876	*55*	Hannibal, MO
13 Dec 1875 - 7 Jan 1876	*18*	* New York, NY	31 January 1876	*37*	Maysville, KY	7 - 18 March 1876	*56*	* St. Louis, MO
14 Dec 1875 - 10 Jan 1876	*19*	* Brooklyn, NY	1 February 1876	*38*	Lexington, KY			

FIGURE 3.1. Map showing locations of Proctor's lectures during his second tour.

so that "age after age the worlds of eternity will live in glorious beauty." Death, in the end, became in Proctor's telling, "a higher sign of life."[122] His third and fourth lectures contained similar appeals. In his encomium to the fourth, for example, he assured his audience that "if we knew the reality, instead of seeing a small part of the Universe, we should find a meaning that would agree with our ideas of an almighty power."[123] That power, of course, remained entirely inscrutable.

In the fifth and final lecture, he took his audience back to the very origins of matter, an area of inquiry where science and religion came into close association. In the visible universe, the clear and "absolute end" was the gradual equalization of temperature, which spelled an inevitable and final death. But that visible realm was plausibly one in which the sun and planets were mere atoms of a larger "invisible universe," where a "future life" was possible and where "thought and recollection still remain in existence."[124]

The lectures were repeated in Brooklyn in a parallel series (an approach adopted during his last visit) and then taken around the country as far west as Lincoln, Nebraska, and as far north as Minneapolis (where his audience numbered seventeen-hundred—a fact that startled those who had not kept up with the astonishing growth of a city known only twenty years previously as an "Indian reservation").[125] In the main, Proctor gave one-off lectures in smaller centers but spoke three or four times in places such as Chicago, Louisville and St. Louis. In total he delivered 142 lectures, 40 more than during his previous tour, and while no final figures are available, no doubt made a significant amount of money (likely running into thousands of dollars).[126] Reports of his lectures after he left New York in early January 1876 became less frequent and rarely filled more than a paragraph or two of a local newspaper. Whether by design or not, Proctor avoided controversy for the rest of his tour, relying instead on hundreds of new images projected onto canvas to entice and startle his audiences.[127]

The publicity engendered by Proctor's high-profile lectures on subjects that intersected with controversial religious concerns came at the same time that his book *Our Place Among the Infinities* was released in America by D. Appleton. Published on December 22, it was based in part on talks he had previously given in America, but with additional material added before Proctor signed off the proofs in September.[128] His *Science Byways* also appeared around the same time and was widely reviewed by American newspapers and periodicals. All in all, it was a successful tour, and one extended longer than first planned despite his wife giving birth to twins near its beginning (reportedly making Proctor the father of eleven children).[129] Proctor, however, was still chasing the reduction of debt (with so many offspring no wonder) and struggling to overcome the financial catastrophe that had struck ten years earlier. There was every reason to return a third time to the United States for another yet more extensive tour. He had successfully played the religious card, at least in terms of the publicity it had generated. He had kept pace with some of the most recent controversies in science and religion and deftly and swiftly presented to audiences hungry to learn what one of the most celebrated science lecturers of their age made of them. They had not been disappointed, at least in terms of the interest Proctor's shift toward scientific agnosticism had created. If Proctor had to deny any sudden change of views and emphasize the consistency of his thought through time, he nevertheless benefitted from reports and rumors of dramatic changes in his religious position. But to create a similar level of interest, on his return for a third tour, Proctor had to find another angle to pull in large audiences to hear the latest findings of astronomical science.

PYRAMIDS AND THE END OF THE WORLD

In early July 1879 the American Literary Bureau received a letter from Proctor suggesting a third tour, starting that autumn. Proctor noted that he had not intended to return until 1882, when the transit of Venus would be clearly seen in the United States. But "recent domestic sorrow" (the tragic death of his wife) meant he was anxious for a change of scene.[130] This time his tour would take him into Canada and then right across the United States, finishing in San Francisco. He would then sail to Australia and New Zealand to tour there. His lectures would all be new, accompanied with fresh and specially prepared paintings and illustrations. In addition to recounting recent progress in astronomy, there was one topic that Proctor wanted to bring to his audiences. "I refer," he wrote "to the great pyramid. I do not promise to confirm [Charles Piazzi] Smyth's theory that in that building the end of the world in 1881 is unmistakably indicated."[131] Proctor was now a master of creating suspense and intrigue. Audiences knew they could expect a torrent of the latest thoughts on controversial astronomical subjects delivered at high velocity and with spectacular accompanying images. Tapping into pyramid mania was a master marketing ploy. If Proctor did not "promise" to confirm Smyth's apocalyptic theory, neither did he (initially) deny its basis in fact. That, though, was soon to change.

The anticipation that Proctor would discourse on the pyramids and the prophecy of the world's end was reinforced, perhaps deliberately, in the more immediate lead up to his third tour. An editorial in the *New York Herald*, for example, insinuated that Proctor may indeed believe that "the great pyramid . . . is a silent prophecy that the world will terminate in 1881."[132] On his arrival, the same paper gave a more sober account of the topics Proctor would cover, printing an interview with him at the Westminster Hotel, New York, where he was staying. The *Herald* discussed Proctor's program of talks, along with some of the new discoveries, mostly by American astronomers, that he would present in his lectures.[133] On that occasion, the pyramids were not mentioned, being held in reserve for later in the tour. This kind of publicity likely benefitted sales of Proctor's latest book, *Rough Ways Made Smooth*, published on December 20, 1879, and favorably reviewed, at some length, in the *New York Herald*.[134]

The series of four lectures that he offered in American cities through the autumn of that year and on into the spring examined in turn "The Poetry of Astronomy," "The Immensity of Space," "The Vastness of Time" and finally "Other Worlds and Other Suns." The topics were familiar, but with new images, fresh facts and a lingering curiosity about Proctor's opinions on controversial subjects, the lectures proved enormously popular.

The first lecture in New York, on November 10, was delivered to an overflowing Chickering Hall, with many more disappointed not to gain entry. According to at least one report, the lecture was almost pious in tone, showing how, to quote the account in the *New York Tribune*, "the exact teachings of science . . . are well calculated to impress the mind with the greatness of the Creator."[135] Scriptural quotations were used, not, on this occasion, to query biblical beliefs but to underscore the religious resonance of the latest astronomical research. A brief mention of the pyramids was made but only to argue that they had been designed to support astronomical observations. It might be reasonably supposed that Proctor was teasing his audience.

The following three lectures, in contrast, eschewed conventional pieties. The lecture on the vastness of time, for example, presented a vision of a future earth barren of life, a world that had died from natural causes and the passage of time. If this predicted death was millions of years in the distant future, in the context of the universe it was but a moment away. In a satirical editorial response to the lecture that appeared in the *New York Herald*, the lasting impression from the lecture was of an almost unimaginably old and vast universe that made humans, whether descended from Adam or from protoplasm, "infinitesimally small."[136]

Proctor's fourth and final lecture on other worlds and suns returned to one of his signature topics, the plurality of worlds. Walking in his "brisk fashion to the front of the platform," he first apologized for not offering a more technical account of physical astronomy, noting that his aim was to avoid the "*technique* incident to the classroom" and to offer something more attractive for the "cultured ladies and gentlemen" that attended his lectures.[137] That did not mean he had abandoned his commitment to casting his audience as earnest pupils of astronomical science. Proctor did this by beginning his discourse by addressing questions posed to him by letter from members of his audience. Having done this, he once again painted a vision of a universe in which vast tracks of space were devoid of life but which also contain millions of worlds with the right conditions for life to appear and thrive. Proctor then swiftly moved through several topics, including eclipses on Saturn, craters on the moon and spectroscopic analysis of the sun. The final denouement was a passage on the "immense numbers of suns," closing with the dream vision composed by Thomas De Quincey from verse by the Romantic German poet and writer Jean Paul Richter, perhaps accompanied by some modest gestures from the more usually restrained lecturer. De Quincey's evocation of the astronomical sublime and "the persecutions of the infinite" provided a finale for Proctor's four lectures.[138]

As was his custom, Proctor spoke during his first four discourses in a mostly darkened hall, illumined only by the images he had projected on the screen behind him. His disembodied voice rapidly transported the auditors across the vast reaches of space, impressing them with a sense of infinitude. It was a technique he used in the parallel series run in Brooklyn. But when he stood up on the Sunday evening following the end of his course to give a "sacred lecture" on the "religion of astronomy," he departed from this technique and used a pointer to draw his audience's attention to the diagrams and pictures arranged behind him.[139] On this occasion his voice was, according to one report, "even more monotonous than usual." This was off-set by the variety of his thoughts, which "poured forth with almost pause-less rapidity." Lecturing in the light invited comment, too, on his posture, with the *New York Tribune* report noting that, for the lecture's duration, he "stood with both arms and a part of the weight of his body resting on the reading desk."[140] This was not because he used notes (he had none), but simply reflected the rather static stance he adopted, matching his colorless vocal performance. This did not mean his lecture was judged a failure. At least one listener suggested Proctor's voice was "beautiful and sounded like the far off [*sic*] tones of the music of the spheres, and yet it seemed as if the man did not belong to his voice."[141] This ethereal quality was judged to be in perfect harmony with the celestial subject matter.

The emotive force of the lecture no doubt added to its cultural signifi-cance. But there were also those who saw it as an important occasion be-cause it was an opportunity to learn more about Proctor's much-discussed religious views. A comment in the *New York Times* published on the day of Proctor's Sunday lecture presented it as an occasion to test whether or not Proctor was "orthodox," a reason, perhaps, for why the lecture promised to be "unquestionably the most interesting of this brilliant course."[142] Proc-tor's secular sermon certainly proved to be immensely popular, attracting "one of the largest and most critical audiences ever assembled in Chickering Hall."[143] It also generated copious comment from reporters and correspon-dents in New York's newspapers.

The dominant note was one of disappointment. Proctor talked very lit-tle about the burning religious questions that at least some auditors hoped he would address. He studiously avoided speculations about the divinely ordained reasons behind various astronomical phenomena. He eschewed direct inferences about purposeful design like those suggested by those "primitive popularizers of astronomy" Thomas Dick and David Brewster. Evoking the "splendors of an infinity of space filled with the glories of un-numbered suns," Proctor ruled out "direct teleology," including the idea that humans were the goal of the created cosmos. While admitting the possibility a "creative purpose" behind an infinite universe, he "paused in conclusion" to state that "science could no longer light the way." Quoting once again his favorite verse from the book of Job, "touching the Almighty we cannot find Him out," Proctor provided his usual biblical warrant for agnosticism.[144]

If Proctor appealed to the book of Job to deny the possibility of knowl-edge of divine purpose, he also undermined the Bible's veracity by accusing Moses of covering up the astronomical origins of various ritual and litur-gical practices set out in the Pentateuch. Proctor once again argued that the Jewish Sabbath and sacrificial system were both of Egyptian origin and had arisen within a context in which astronomy was itself a religion. The reason Moses gave for Sabbath observance—that God rested on the seventh day—was, Proctor declared, patently absurd. The idea of an infinite being requiring rest defied understanding. It made far more sense to trace the idea of the seventh day back to the lunar calendar of Egypt and Assyria. In passing, Proctor also called attention to contractions in biblical accounts of the ritual requirements demanded of the Israelites during the Exodus. In the end, Proctor left the impression that the "superstitions" of the Egyptians, their beliefs in the influences of the sun and planets and their worship of them, were more reasonable than the claims made in the Pentateuch.

All of this effectively repeated what Proctor had said four years before. Some, however, detected, or feigned to detect, a softening of tone. Following the lecture, an editorial in the *Northern Christian Advocate* declared that "the surrender of Professor Proctor to the materialists three or four years ago was not a complete surrender." Even if he still opposed definite dogmas, "he reserved the right of belief in those grand basal truths which to a sensitive mind are the clearest intimations of Nature." [145] As far as the *Advocate* was concerned, Proctor had confirmed that final causes could reasonably be assumed and that it was quite legitimate to affirm that creation had a purpose. Other religious commentators were less sanguine. Two letters published in the *New York Herald* took Proctor to task and one demanded a public debate.[146] Several days later, in Brooklyn, the Reverend Thomas Mitchell delivered a free lecture attacking "the fallacies of Professor Proctor."[147] In this way, Proctor's lectures on topics that ventured on religious ground directly intersected with pulpit as much as lecture culture.

On the day following his lecture on religion and astronomy, Proctor gave a matinee lecture on the great pyramids that allowed him to continue to critique popular religious conceptions associated with astronomical topics. This was among the most reported of his talks during his tour of 1879–1880. Smyth's account of the pyramids and structures as made possible by direct supernatural revelations relating to metrology and astronomy had been keenly discussed, digested and derided since his book on the subject appeared in 1861. Smyth himself was in earnest and his status as an astronomical scientist of repute meant that his millenarian predictions, based on his detailed investigations of the Great Pyramid of Giza, were widely broadcast in print and from pulpits.[148] Proctor was, of course, fully aware of Smyth's work and considered it delusionary. But tackling pyramids was an ideal way to attract an audience and drive home a message about the religious implications of astronomical science, ancient and modern.

Proctor, in the end, avoided mention of Smyth, concentrating instead on his theory that the pyramids were designed to facilitate astronomical observations. Proctor's avoidance of Smyth's evangelical pyramidology did not prevent his lecture from drawing the ire of the Reverend Joseph Wild, a Brooklyn Congregational minister well known for his support of British Israelism, the thesis that the ten lost tribes of Israel were the ancestors of the people of Britain.[149] In Wild's evening lecture to the "Lost Israel Identification Society" at his church, he argued that Proctor had changed his mind about the great pyramid since he last spoke on the subject and now acknowledged that it had been designed with certain scientific ends in mind. Wild disputed this theory, arguing that the pyramid's structure and content pointed not to its use as an observatory but as a repository of divinely inspired knowledge. The pyramid, Wild contended, held "figures and dates linked to [Biblical] prophecy" and was justly called, by the Prophet Isaiah, "the pillar and witness of God."[150] Such dramatic claims contrasted with Proctor's cautious and cool assessment but were warmly welcomed by Wild's many supporters. There were limits to the "warmth" and interest Proctor could inject into his more religiously skeptical lectures and Wild was a seasoned platform performer who worked to benefit from that fact. This was competition of a different kind for Proctor, but one that he could take advantage of, if only to rebut, just as he did with Smyth's sensational claims.

Pyramids, however, faded for a time from view as Proctor continued his third tour. After a brief lecture visit to Montreal, Proctor traveled back to New York a few days before Christmas. There, he gave an extended interview to a *New York Herald* reporter on the recent inventions of Thomas Edison. Proctor did not disappoint his readers with his fulsome praise of the invention of the electric light bulb by the "wizard of the wire." Speaking as a recognized "scientific authority," Proctor praised Edison for inventing something that would doubtless transform domestic and city lighting and act as a spur to more general scientific progress.[151] He dismissed the possible dangers and fears around the electric light (that it would, for example, be damaging to the eye) and looked forward to Edison's promised demonstration on New Year's Eve and to an electric future free from the hazards and limitations of gaslight. Already in print as a champion of America's celebrated but also, in some circles, much abused, inventor, Proctor again stood behind American applied science and simultaneously confirmed himself as an advocate of a technological modernity.[152] Proctor here used the journalistic interview to good effect, reinforcing his reputation as someone with a deep kinship with American values and a keen ear for fashionable scientific causes.

Having completed his stays in New York, Proctor continued west, going first to Boston, and then on to Indianapolis, Cincinnati and various smaller centers before arriving in San Francisco early in April. As he traveled, he delivered some or all of the same lectures that he had given in New York and other Eastern Seaboard cities. The reported reactions to these lectures were almost uniformly positive. One reasons for this was a sense of civic pride. Large and enthusiastic audiences in New York set a precedent and citizens of other cities were encouraged to demonstrate an equally lively local intellectual culture. Reports in Cincinnati newspapers repeatedly cited success in New York as a reason to attend upcoming lectures in their own city.[153] A newspaper in Indianapolis lamented the fact that the municipality could only manage two lectures from the great astronomer.[154] Such civic rivalry and comparison helped propel Proctor west, the fees from his lectures more than covering the cost of train tickets, accommodation and other expenses.

The heightened interest also aided his growing celebrity, and newspapers clamored to know Proctor's opinions not just about the latest astronomy but on any number of subjects of current interest. In an interview for the *Cincinnati Gazette*, as well as being questioned about the latest astronomical discoveries, Proctor was asked for his views on women's education (he was in favor but hinted at certain inherent limits to female participation in science). Proctor's own letters, written as he traveled across America and most often sent to the *New York Herald*, added to this trend. Proctor's views on the latest opera performance, disputes about literary matters, the shortcomings of New York cab drivers, the biblical star from the east and much else were given space in the *Herald* and often reprinted in shorter form by other papers. If other visiting scientists were more reserved and at times reluctant to allow interviewers into their places of residence as they toured, Proctor played the celebrity to the full and helped to make possible a productive conjunction between a reputation for scientific expertise and popular personal appeal.

There did continue to be challenges, of course. Proctor's pontifications on the latest cultural fad or debate stood in some contrast with the content of his lectures, which, according to some reports and despite Proctor's ambitions, contained little that was new. If that was not held against them (it was often Proctor's style and approach that were held to make the lectures worth hearing), it did render them old news.[155] One result was that his visits to places he had never been to before took on an increased importance, generating new audiences and new controversies—a strategy which, to borrow a concept from the geographer David Harvey, might be described as a "spatial fix."[156] His arrival in San Francisco in particular generated a degree of civic fanfare and public interest and his lectures there attracted significant attention from local newspapers. Billed as his last in America (perhaps, if Proctor was to be believed, his last ever), they generated considerable excitement. The lectures he gave were not, in fact, repeats of what he had presented in eastern cities. Instead, newspapers announced a course of three lectures on the sun, the planets and the stars. Held at the hall of the Mercantile Library Association, the lectures were advertised for the evenings of the fifth, seventh and ninth of April.[157] The large crowds that gathered in the hall to hear Proctor were treated to an arresting sequence of images projected using limelight on a large screen that "occupied the entire width and height of the platform."[158] Bigger was generally better when it came to such visuals. For much of the time, the hall was darkened, with Proctor giving a running commentary on the images displayed. Dazzled by diagrams and star charts, audiences caught the merest glimpse of "probably the greatest astronomer of his time."[159] What they heard at length was the monotone delivery of astronomical facts issued at high speed from the lips of a seasoned celebrity speaker.

It was a delay in the departure of the steamship *Australia* that allowed Proctor time to hastily organize extra lectures, one in Oakland on Saturday April 10 and the other in San Francisco the following day. For his Oakland lecture, Proctor chose the pyramids as his subject. The reporter who described it in the *Daily Alta California* suggested that Proctor's tone made it clear he considered Piazzi Smyth's pyramidology ridiculous. The "lecture as a whole was bitter and sarcastic" and the view of the pyramids adopted by Proctor "eminently material." The disquiet among the audience that Smyth had been treated to "ridicule and paltry jest" suggested some offense had been given and taken.[160]

No such negative reporting resulted from his final and farewell Sunday evening lecture in San Francisco on "The Birth and Death of Worlds." Before beginning his discourse to a packed Grand Opera House, Proctor expressed gratitude for the "unprecedented encouragement" he had received in California to lecture, marking his time there as the high point of his longest lecture tour yet. Finishing his well-worn and fast-paced narrative of cosmic evolution, tracing all that makes us human back to the "giant mass of gaseous vapour" that birthed the earth, and pointing forward to the inevitable death of our own planet, Proctor took his leave and, the next morning, made his way to the port to sail for Australia.[161]

Two or three hours before Proctor departed on the delayed steamship, he wrote to the *New York Herald* to sum up his third tour.[162] As usual, he was keen to record the basic statistics of his three tours, enumerating the number of lectures delivered in each. In his most recent tour, he had given 136 lectures, 6 fewer than four years before. In total, he had lectured on 380 separate occasions in the United States. It is possible to extrapolate from these figures and estimate that Proctor performed in front of upward of one hundred thousand people on each of his tours, allowing for those places where he lectured several times. On his final tour, his gross income, according to an article in the *New York Times*, was $50,000, with an average income per lecture of $367.[163] After expenses, Proctor's profit amounted to $15,000, a considerable sum and one likely matched by his previous tours. What is also striking is that the average income per lecture was more than what was generously offered to Lowell lecturers (of the order of $200 to $250 per lecture). Proctor's tour had also boosted sales of his books and had encouraged his publishers in both England and America to release new editions of his *Myths and Marvels of Astronomy*. As one report put it, his presence in America had created a "Proctor boom," such that his writings "were never more widely read than they are today."[164] Yet it is worth noting that it is likely more people heard Proctor's lectures than read his books. And the lectures almost certainly earned him more money, more quickly, than book sales.

Of course, the impact of the lectures was about much more than benefits (no doubt accompanied by certain burdens) accrued by Proctor. Hall proprietors and staff, various organizations, Proctor's agent—for his third tour, Benjamin Webb Williams of the Williams Lecture Bureau—and American science in general all profited.[165] Newspaper owners, editors, reporters and stenographers also took advantage of Proctor's popularity. Stories about the astronomer's (purported) change of mind about the relations of religion and science or his views of aspects of American life made good copy. Full reports of his lectures sold newspapers, especially in places where thousands attended his lectures and hundreds more could not obtain tickets. Indeed, his tours were arguably as much a product of intense newspaper attention as anything else. And it was not his books that created that level of public interest but his lectures. There was something about the nature of these live events that created the conditions for this degree of visibility and intensity of discussion.

Proctor's lectures, through their remediation into newspaper columns, became news. Their fast-paced delivery resonated with the frenetic tempo that marked the generation of journalistic print. The text of the lectures, whether printed in full or in abridged form, no doubt departed from what was actually said. This was certainly Proctor's view and he consistently decried the "slipshod" reporting of his lectures.[166] Short of quickly producing an authorized version, there was little Proctor could do about this. It was a problem made more difficult by his rapid delivery. For all that, however, the intense interest of journalists aided the success of his tour.

The speed at which Proctor published books and delivered his lectures was material evidence of his ability to keep abreast of rapidly changing scientific developments. Religious dogma, and the recalcitrance of its defenders, could not but appear as a barrier to such high-velocity scientific progress. His pace of delivery embodied the progress of science and, by contrast, the friction created by dogmatic religion. At the same time, Proctor was anxious to impress on his audiences that his views on cosmic and biological evolution had developed organically and progressively. His commentary in print and from the platform about science and religion remained remarkably consistent. There was no ultimate conflict between science and religion; there was no reason to deny ultimate purpose even if scientific knowledge of this was impossible.

Toward the end of his third and (as he then believed) final tour of the United States, Proctor addressed this subject in full, writing a long letter to the *New York Herald* outlining again the long-term consistency of his views about astronomy, evolution and religion. He recalled how, during his second tour, false reports of a dramatic conversion to evolutionism had created a firestorm of controversy and misunderstanding. Proctor's own attempts to clarify his position seemed only to create further confusion, even leading to reports in London newspapers that "caused pain to dear ones at home who could not conceive what dreadful doctrines I had begun to promulgate." Once again, then, he tried to clarify his views and show they had remained substantially unchanged across all three of his visits to the United States. How he did so was telling. He began with mention of a number of his earlier books before turning to newspaper reports of a lecture delivered in January 1874 to show that what he said then about evolution and about science and religion was effectively the same as what he had maintained before and after, in print and from the platform. The most important piece of evidence, however, was from the text of a lecture delivered on April 1, 1874. Unfortunately, the newspaper report was marked, in Proctor's estimation, by a "certain incorrectness and wildness of statement." He had "spoken at that time too fast to be reported verbatim, skillful though the reporters were."[167] To overcome this difficulty, his own version, published a few months later in the *Contemporary Review*, was to be used instead.[168] The sense was roughly the same as in the newspaper report, but the words were distinctly different. Whether or not Proctor's version or the report was a more accurate account of what Proctor had uttered at Association Hall in New York cannot now be recovered. What is clear is that the meaning of his live utterances on controversial issues could never be controlled by print and ran ahead of even Proctor's swift pace.

CHAPTER 4

ALFRED RUSSEL WALLACE, ANTICELEBRITY

On February 16, 1886, Alfred Russel Wallace wrote to the English essayist and poet Edmund Gosse to ask for advice about lecturing in America. A little over a year earlier, Gosse had visited the United States to deliver a series of Lowell Institute lectures and meet with some of the country's best-known writers. According to Gosse's own account, his tour was a success, cementing his reputation in America as an essayist and litterateur of international repute. Oliver Wendell Holmes Sr., who attended Gosse's Lowell lectures, had told Gosse that "he had never known a stranger to make such a conquest" and that "we are all a little in love with you."[1] Gosse, it seems was the right person to ask about the best recipe for a successful tour. Wallace had several questions and concerns to raise. Should he "employ an agent to manage the tour" and give "only one or two lectures in each town" to guarantee the "largest return"? This was especially important to Wallace, who had bills to pay and would only "risk the fatigue of a tour" if it was going to be lucrative. Wallace also worried about topics—he thought perhaps something on natural history and "social questions" might work. Then there was his voice. His throat, he told Gosse, "was delicate . . . and very sensitive to chills." What, he wondered, was the "Eastern" winter like? Was Boston in November "a tolerably good month"?[2]

Wallace was right to inquire about agents, timings, money, topics and health. But he missed some of the crucial ingredients behind what W. D. Howell called the "Gosse boom" in America.[3] Gosse had combined an intense season of "lecturing almost every night" with constant "parties and receptions." He met "over 600 people" and enjoyed invitations to the White House and private conversation with Walt Whitman and Oliver Wendell Holmes Sr.[4] His wife, who accompanied him, attracted the eye and commentary of reporters nearly as much as Gosse himself. He had charmed his audiences with humor, warmth and a "clear, musical voice."[5] He made sure to lecture with aplomb in New York, despite having to squeeze his visit there into a hectic East Coast lecture schedule. Though a shadow later fell over the publication of his Lowell lectures (famously mauled by the critic John Churton Collins), Gosse's reputation in the United States proved resilient and his short visit benefitted both his celebrity status and his bank account. Wallace, a rather reluctant lecturer and socially shy, could not hope to replicate Gosse's approach. Even so, Wallace was not ignorant of, or uninterested in, the art of public speaking. Like Tyndall and Huxley, throughout his life he paid significant attention to preachers, politicians and other professional orators and believed in a general need to use the lecture platform to shape public opinion, not least on matters scientific and political. His observations, opinions and experiences of lecturing provide an important backdrop to his more immediate efforts to prepare to speak well and succeed on the lecture circuit in America.

WALLACE'S LESSONS IN SPEAKING

Wallace's earliest experiences of lectures came during his work as a surveyor in Wales. His growing interest in science and his involvement in the Kingston and then Neath mechanics' institutes meant lecture attendance became part of a wider diet of intellectual pursuits. He later recalled developing clear ideas about what constituted an effective science lecture and tried his own hand at creating one. In his octogenarian recollections of his time in Neath, Wallace recalls that a talk on Linnean nomenclature that he had listened to as a nineteen- or twenty-year-old starting to develop an interest in botany prompted an early attempt to write out a lecture. Not only did the speaker, a "local botanist of some repute," champion the Linnaean system over and against a natural system of classification (something Wallace thought mistaken), the lecture was "so meagre, so uninteresting, and so utterly unlike what such a lecture ought to be, that I wanted to try if I could not do something better." As Wallace sat and wrote his autobiography, having in front of him the script of that early attempt refreshed his memory and prompted the thought that "although crudely written and containing some errors, I still think it would serve as a useful lecture to an audience generally ignorant of the whole subject, such as the young mechanics of a manufacturing town."[6]

It is not clear if Wallace ever delivered this script in front of an audience. His reminiscences suggest that his first lectures were given to audiences of working men at Neath Mechanics' Institute after he had returned to the town following the death of his brother in 1845 (dated by Wallace as February 1846 but since revised). In Leicester, he had taught pupils at the Collegiate School in his position as a junior assistant. Over the next two seasons, he lectured at Neath Town Hall on the invitation of William Jevons, owner of a local iron works and cofounder of the mechanics' institute. His memory of those first lectures was of a nervous young man finding his voice and developing the skills needed for clear scientific exposition.

According to Wallace, the lecturing he did throughout his life was largely motivated by financial necessity. His mature thoughts on the matter suggest that he regarded his writings as a more effective tool for inculcating scientific habits of mind and communicating scientific content and concepts. This was not because of any blanket privileging of pen over platform. Rather, he felt that his own "deficiencies" as a thinker and speaker prevented him from offering effective lectures. In a typically self-critical reflection, he confessed to poor "verbal memory" and a struggle "to reproduce vocal sounds." These, Wallace lamented, "put me at a great disadvantage as a public speaker. I can rarely find the right word or expression to enforce or illustrate my argument." As far as writing was concerned, however, this apparent defect was judged advantageous because "the absence of the flow of words which so many writers possess has caused me to avoid that extreme diffuseness and verbosity which is so great a fault in many scientific and philosophical works."[7] More generally, as someone of weak physical constitution and a lack of "assertiveness," Wallace preferred the private space of writing to the public platform. Wallace did qualify this assessment of the worth of his own lecturing, noting that, even if his lectures were motivated by a search for financial security, they had provided the basis for several of his more important books, which otherwise would never have appeared. He also mentioned his speaking tour in America as a positive by-product of the search for financial security. For all that, from 1890, Wallace determined to refuse all invitations to lecture, believing, as he had long felt, that "I could do more good with my pen than my voice."[8]

When Wallace looked beyond himself, however, his support for lectures as an important mode of communicating science and a scientific disposition is clear. His advocacy on behalf of mechanics' institutes, a movement that made the lecture a central instrument in science education and the virtues thought to come along with it, demonstrated Wallace's commitment to the lecturing cause. The building he designed for Neath Mechanics' Institute centrally featured a lecture hall. Wallace's first substantial publication, which appeared in 1845 but was likely written several years earlier, was a five-page apologia for the institute as a place to educate the people in science. Among other things, Wallace argued that "the materials for diffusing information being thus prepared by the institution, it becomes desirable to create a taste for its acquirement, and to promote an interest in scientific and literary pursuits. . . . lectures are of great avail; and whenever persons can be found willing to undertake the delivery, they should be immediately requested to do so."[9] Although Wallace used fairly conventional language to describe the "diffusion of knowledge" that mechanics' institutes encouraged, his short essay emphasized the full participation of members in the production and propagation of science. If anything, despite an opening disclaimer about downplaying the "higher branches of science," Wallace pushed for scientific instruction and scientific activities that were less about equipping an artisanal class with practical knowledge than they were about enrolling them in scientific enterprise for its own sake. While his examples of those who had risen from obscurity to scientific stardom included George Stephenson and James Watt, he also pointed to William Herschel, who had added "immensely to our knowledge of the structure of the mighty universe." Wallace adverts an openness, then, to members of mechanics' institutes making botany, astronomy or geology their special study, and offering lectures on those less applied and often "abstruse" subjects.[10]

Wallace's advocacy of science lectures was expressed in other ways. A year after his essay appeared, while resident again in Neath, he signed a petition along with two others appealing to parliament to repeal parts of a statute that imposed heavy fines on anyone connected with unlicensed rooms used for lectures, debates, libraries or reading news. The petition deemed the difficulties of securing a license "a hinderance, vexatious and uncalled for, to societies established for the advancement of knowledge and the progressive improvement of mankind."[11] Based on the neutral ideal that regulated the activities sponsored by mechanics' institutes, the appeal defined lecturing in strict opposition to the kind of seditious speech that was the target of the statute under review.

Wallace's convictions about the importance of lectures as a vital mode useful for the dissemination of science and a scientific disposition were reinforced through his own experience of listening to celebrated speakers. In his autobiography, Wallace mentions two lecturers who had stayed in his memory because of their power and appeal as speakers and their influence on his early intellectual development. The first was Spencer Hall, an advocate of mesmerism whom Wallace had encountered while in Leicester. As Wallace remembered it, the lectures not only persuaded him of the truth and utility of mesmerism but impressed him because Hall, in his manner, "was not at all that of the showman or the conjurer."[12] The sobriety of the lecturer was an important part of his rhetorical power, making a potentially sensational and suspect subject matter sound and look scientific and temperate. Hall, as well, communicated an account of scientific instruction and scientific pursuits that resonated with Wallace's mature views of the matter. As Alison Winter has shown, Hall's lectures on mesmerism supported a democratic vision of science that was formed in opposition to a culture of elite expertise and that emphasized the capacity of urban-based artisans to become key scientific actors.[13]

The other lecturer whom Wallace recalled being impressed by in this decade of his life was James Silk Buckingham. According to Wallace, "the impression made upon me then was, and still is, that he was the best lecturer I ever heard, the most fluent and interesting speaker." When Wallace heard Buckingham in London, likely early in 1844, he had a long-established reputation as a celebrated orator.[14] The exotic subject matter—part of his "oriental travels"—was coupled with what one contemporary called Buckingham's "conversational oratory," a style that was at once accessible and "visionary."[15] His fluency, natural and chaste gesture and variety of tone and modulation were all also celebrated. It was the beau ideal of the public speaker and one that Wallace took to heart even if he felt he could never himself rise to those heights. But Buckingham, as well as being an enviable communicator, was also an advocate of causes dear to the young Wallace. As a sitting MP, he had backed civic museums and libraries among many other reformist causes and championed a political philosophy that shared features in common with the social vision of Robert Owen.

Lectures such as those given by Hall or Buckingham were not the only kinds of public speech that Wallace relished as a young man. The sermons he heard while attending church and chapel services in Welsh villages also left a distinctive and lasting impression. In his recollections, he comments that he "was greatly struck with the grand sound of the [Welsh] language and the eloquence and earnestness of the preachers." Their "energetic" performances in the pulpit did "full justice to their rich and expressive language, and even without being able to follow their meaning it is a pleasure to listen to them."[16] That Wallace could not follow the content but could appreciate the atmosphere and energy of the sermons shows a clear appreciation of the affective power and dimensions of spoken address.

It was not only Welsh preachers that Wallace remembered. While he was in Leicester, Wallace listened "almost every Sunday" to the sermons of John Brown, "one of the most impressive and eloquent preachers I have ever met with." What struck Wallace most about Brown was that, unlike many Church of England clerics at that time, he "preached extempore, and he did it so admirably that it was a pleasure to listen to him."[17] Although there was on these occasions not the same language barrier, it seemed this pleasure was entirely independent of the content of the sermons. Much of what Brown declared Wallace found simply incredible, being by then a convinced sceptic about conventional religion. Just as he did not have a musical ear but could be profoundly moved by "grand, pathetic or religious music," so Wallace could delight in the pleasures and power of eloquent speech without believing himself capable of replicating it.[18]

Wallace's enthusiasm for all of these preachers and lecturers demonstrates the sorts of virtues he felt were important for the successful science lecturer. It is not surprising then, to find him observing that when he heard Thomas Henry Huxley lecture for the first time, "[Huxley] did not read the paper, but, with the help of diagrams and sketches on the blackboard" made the technical subject matter "perfectly intelligible and extremely interesting."[19] If Wallace felt he could not equal Huxley in terms of fluency and interest, he had, at least, a model to aspire to when he found himself compelled by financial insecurity to instruct large crowds on science or on political economy.

Wallace's participation in lecture culture, whether as a listener or as a speaker, was interrupted by his long sojourns in the Amazon and Malay Archipelago. Apart from a fifteen-month interlude between travels, it was not until his return to England in April 1862 that lectures again became a feature of his intellectual life. In the face of his struggle to find permanent employment, delivering lectures became one means among many to earn money. With a few exceptions, it was not until 1867 that Wallace began to seek opportunities to lecture for a fee. Up until then, speaking to an audience often involved the reading of a paper before a learned society. In the first three years after his return, he gave several papers to the London Zoological Society, addressed the Royal Geographical Society, and the Anthropological Society of London. The sectional meetings of the British Association also offered opportunities for Wallace to deliver papers that, while always accessible because of Wallace's commitment to clarity, were largely aimed at scientific specialists. A good example is Wallace's lecture on "the varieties of man in the Malay Archipelago." First given at the Newcastle meeting of the British Association on September 1, 1863, it was repeated on December 1 for the Leeds Philosophical and Literary Society and again in January 1864 at a meeting of the Ethnological Society of London. The rendition in Leeds was unusual but demonstrates the relative ease with which Wallace could address different audiences, not least because of the nontechnical nature of his scientific discourse.

During this period, it was his prolific pen far more than his voice that was the chosen instrument for communicating his views of tropical nature and human society. This was a pattern that continued throughout his life. In 1867 Wallace did seek and receive invitations to lecture to provincial literary and philosophical societies. His series of three lectures on tropical nature were delivered over the space of a week in November to the Newcastle Literary and Philosophical Society. With a full hall and a warm local welcome, these lectures were a success.[20] Wallace's chaste and lucid style was on full display, supplemented by a number of diagrams and preserved specimens of tropical plants and animals. They were, as was typical for Wallace, fully scripted affairs.[21] Wallace wrote out his lectures in full, down to the opening "ladies and gentlemen," and delivered them verbatim, apart from the occasional extemporaneous explanation of a diagram or chart. With these lectures written out, it was not difficult for Wallace to pick up the same scripts and deliver them again the following year to other societies such as the Birmingham and Midland Institute, the Leeds Philosophical and Literary Society and Bedford Mechanics' Institute. The lectures were then incorporated into Wallace's book, *Tropical Nature and Other Essays*, published in 1878.

These early efforts to offer accessible lectures for a fee did not, despite their apparent early success, translate into Wallace becoming a common fixture on Britain's lecture circuit. In terms of paid lectures, there was something of a hiatus through the 1870s. It was not until the later years of that decade and into the 1880s that Wallace took again to lecturing in various urban centres around Britain and Ireland. In 1878, for example, he traveled to Bristol, Halifax, Cardiff and Bradford to talk on the distribution of animals, a paper originally delivered to the Royal Geographical Society in June 1877. He later contributed to science lectures for the people in Manchester (1879) and Glasgow (1882). In 1880 he gave a series of lectures at Rugby School and crossed the Irish Sea in 1882 to deliver lectures for the Belfast Natural History and Philosophical Society and the Royal Dublin Society. Lectures at the Birmingham and Midland Institute and at the Edinburgh Philosophical Institute followed in 1883 and 1885 respectively. These public lectures were celebrated as a boon for the society that hosted them and brought Wallace the writer-naturalist onto the public stage. They also earned income. Between 1878 and 1885 Wallace received about £300 in fees, generally charging between £15 and £20 per lecture.[22]

The early 1880s were also a period in which Wallace began to offer public lectures on political subjects. As president of the Land Nationalisation Society from 1883, he was obliged to offer an annual address, but he also spoke on the subject outside of the society's base in London. Wallace also found himself addressing various institutions on other economic and political topics. In June 1886, for example, he lectured on "the depression of trade: its causes and remedies" in Glasgow and Dundee.[23] If in this lecture he circled back to land nationalization as perhaps the most important remedy for widespread poverty, he nevertheless covered a range of causes and solutions to economic downturn and inequality. By this time, he had an eye on America and spent a portion of his lecture accounting for the deep depression that had taken hold across the Atlantic, which he understood to stem from railway manias and the corruption of municipal politics.[24]

Although Wallace's repertoire of lectures expanded over time, there was one subject strongly associated with Wallace that he rarely spoke on. Wallace's interest in spiritualism, which began in 1865 or earlier, became a passion and mainstay of his intellectual life.[25] In 1870 he did read, on at least two occasions, a long paper critiquing objections to miracles, which was, in part, a defense of the truth of spiritualism. At least one of those readings, at a soirée held in the Beethoven Rooms on London's Cavendish Square, was to a sympathetic, semiprivate audience. In August 1876 Wallace also spoke in defense of the possibility that spiritualist claims were true, after a paper on the subject by the physicist William Barret was presented to the anthropology section of the British Association. But in the main, Wallace did not lecture on the subject, despite the growing public interest in it. Spiritualism was not, of course, a topic likely to be welcomed at literary and philosophical societies. There were places where a lecture on spiritualism would have been welcomed (even if it may not have paid), but Wallace for the most part did not avail himself of those opportunities.

Whatever Wallace's reasons for not lecturing on spiritualism in public halls, in October 1876 he did very publicly address the subject in London's Bow Street Police Court, during the trial of the American spiritualist Henry Slade. The sensational trial was brought to court after the young zoologist Ray Lankester and the medical student Horatio Donkin attempted to expose Slade during a séance.[26] Slade was well known for performing automatic writing on a piece of slate held under a table fold. At one séance, Lankester snatched the slate from Slade, discovering writing already on it before the automatic message was predicted to appear. The trial, which ran on for weeks, ended with Slade being convicted of conspiracy and fraud under the Vagrancy Act. For the first weeks of the trial Wallace kept his distance but eventually took the witness stand on October 28, offering a straightforward defense of the spiritual cause behind the writing that appeared on slates during Slade's séances.[27] His own sittings with Slade were presented in forensic detail, as Wallace attempted to demonstrate that the writing on the slates could not have been produced by anything other than a spiritual force. It is likely that Wallace's words on this occasion were more carefully scrutinized and more closely reported than any of his lectures had been. The *Times* reported them verbatim, and several other newspapers provided detailed summaries. The pressure of speaking in the courtroom was a foretaste of the kind of close public attention and detailed newspaper coverage that he would experience in America. It also dramatically increased Wallace's fame and notoriety as spiritualism's leading defender, a reputation that would eventually play in his favor toward the end of his American tour.

READYING FOR AMERICA

Even before Wallace began contemplating a North American lecture tour, he was making his views widely known there. Eighteen months before his visit, Wallace wrote an article for Boston's *Sunday Herald* on the harmony between spiritualism and science. His writings on spiritualism had appeared in America before, but this article brought his scientific spiritualism to a mass audience. In it he outlined what were now well-settled convictions. Spiritualism stood in the place of novel scientific claims in other periods, facing skepticism and hostility from conservative forces. The criticisms leveled at it were too often made by those who had not investigated it thoroughly. When the evidence was examined, even the strongest "prepossessions" against it would be overcome. The existence of immaterial minds provided grounds for subscribing to personal immortality in a way that was credible in a scientific age. Spiritualism opened up an "unseen universe" to science and offered a basis for the renewal (and transformation) of more conventional religious beliefs. It removed the burden of skepticism from the historians of religion, allowing them to accept the reality of supernatural events in the past without affirming the conventional interpretations placed upon them (witchcraft, for example, need not be dismissed as diabolical). Above all, spiritualism gave suffering moral meaning by making it a necessary condition for a more noble and elevated future existence. This spiritual evolution, fueled by a "constant struggle" with material imperfection and the play of natural forces, was essential for the development of the highest virtues, not least "those highest moral qualities summarized as 'love' by St. Paul, and 'altruism' by our modern teachers." In the face of the moral nihilism of "modern philosophy" and against the blindly accepted dogmas of established religions, spiritualism provided the only rational basis for ethical living.[28]

Writing articles for newspapers was one way to make his name known in America. It also seemed to suit Wallace to write rather than speak about spiritualism. But lecturing in the United States had the potential to generate a good deal more attention and, if done well, a healthy income. Wallace, however, was reluctant to embark on a speaking tour. When an invitation to lecture in Sydney, Australia, arrived just before Christmas in 1885, the prospect of lecturing across the United States before moving on to New Zealand and Australia presented itself. Wallace, then sixty-two, only contemplated it because of the large sum he still owed on his house in Godalming, some forty miles southwest of London, and because he wanted to "leave something for his family." If such an arduous tour meant "clearing 1,000 [pounds]" Wallace would endure it.[29]

In the New Year, Wallace wrote to the American palaeontologist O. C. Marsh and to C. W. Ernst, editor of the *Beacon*, a Boston newspaper sympathetic to socialism. Both assured him they would seek out lecture opportunities and both wrote to Augustus Lowell of the Lowell Institute and to Daniel Coit Gilman of Johns Hopkins University, alerting them to the possibility of Wallace's visit. Their efforts paid off, and Wallace received an invitation from Lowell in early February to give eight lectures at the institute in the autumn. He quickly accepted, but with some trepidation. To a friend, the chemist Raphael Meldola, he remarked, "Now I am in for it!"[30]

The invitation from Gilman, however, was much slower to arrive. Both Ernst and Marsh reported to Gilman that they had made efforts to secure an invitation for Wallace to deliver lectures at the Lowell Institute. Marsh also encouraged Gilman to approach the Peabody Trust to suggest Wallace as a lecturer at the trust's institute. Gilman's delay suggests doubts about the matter, the seeds of which may have been planted, perhaps unintentionally, by Ernst. In his letter, Ernst had noted that while he "had the greatest admiration [for] Wallace's scientific attainments and style," he could give no assurances about his "social graces." To investigate the latter, he had written "to a cousin of mine in London about Mr. Wallace's personal tastes."[31] This caution, along with Ernst's anxiety to avoid the impression he was "making suggestions" to President Gilman, significantly qualified his otherwise enthusiastic endorsement. Marsh's letter of recommendation was terse, noting only that he had already suggested a course of Lowell lectures and that "anything [Wallace] says on natural science will be good."[32] This ran against Wallace's own expectations. He was keen to lecture on political as well as scientific topics, not least because the former were now a central part of his repertoire.[33]

Two or three weeks later, perhaps prompted by Gilman, Henry Newell Martin, professor of physiology at Johns Hopkins, set out to discover whether Wallace would be a success as a lecturer. One surviving reply was sent by Martin's former teacher, Thomas Henry Huxley, who offered his own opinion of Wallace's lecturing abilities. Qualifying his view by first stating that he had never heard Wallace "address a large assembly" and that he "hated listening to lectures," his letter nevertheless damned Wallace with faint praise. While the "substance" of Wallace's lectures, "even if it should be about spirit rapping and writing, would no doubt be of interest," Huxley had "grave doubts whether his style of speaking is such as to lay hold of a large general audience."[34] Given the esteem in which Huxley as lecturer was held among elite scientists in America, this rather pointed remark profoundly threatened Wallace's chances of success. In the end, Huxley's insinuations did not prevent Gilman from extending an invitation to Wallace to lecture at the Peabody Institute. But two cajoling letters from Wallace's agent, one sent in March and another in May, were necessary to secure it.

Other parties interested in Wallace's lecturing abilities were also seeking information about his reputation as a public speaker. A note appearing in the *New York Herald* in April, for example, communicated the contents of a "private letter" from the Edinburgh Philosophical Institute that highlighted "the clearness" of Wallace's style, "the distinctiveness of his utterance" and, contra Huxley, his "ability to hold his audience."[35] Though positive, this understated praise lacked the hyperbolic style that usually characterized announcements of a much-anticipated visit of a British orator. It certainly was not enough to generate the level of fascination and curiosity associated with other speaking tours.

Faced with this challenge, Wallace took the advice of British lecturers with experience of American platform culture and hired a lecture agent. His first thought was the Boston-based Redpath Lyceum Bureau (suggested to him by the American journalist and political economist Henry George) but, on the advice of the British naturalist and science popularizer Reverend J. G. Wood, he secured the services of another Boston agency, the Williams Lecture Bureau.[36] The Redpath Bureau, founded in 1868, was the bigger and better-known of the two.[37] Nevertheless, Williams had experience with British scientists and had been the agent for Richard Proctor as well as Wood.[38] Proctor, of course, had been a success. Wood's speaking tours in America, on the other hand, had been a failure. Whether or not Wallace was aware of Wood's lack of success when he decided on Williams, that failure would return to haunt him.

As well as settling on an agent, Wallace spent the months leading up to his tour composing five lectures for the Lowell Institute (he had already delivered the other three in various venues during the preceding years) and preparing four additional lectures on political economy. He was determined to ensure his Lowell lectures were vividly and accurately illustrated. To help accomplish this, Wallace wrote to several leading scientific authorities requesting images or, in the case of Francis Galton, help with the construction of diagrams to graphically depict the law of deviation applied to species variation.[39] Illustrators and photographers were also approached, and the large collection of resulting slides and diagrams—packaged in a six-foot-long waterproof canvas case—accompanied Wallace to America.[40] In advance of the tour, Wallace tested out his presentations on audiences nearer home, displaying the diagrams in a lecture given at the Essex Field Club in Loughton (of which Wallace was an esteemed honorary member) and exhibiting his slides to the masters and boys at Charterhouse School (just a short walk from his home).[41] With lectures written, illustrations prepared and the first part of his itinerary in place, Wallace set sail from London on October 9, 1886.

PLATFORM POLITICS IN NEW YORK

When Wallace arrived in New York on October 23, he checked into the hotel recommended by Wood and, in the evening, met the veteran journalist Albert Gallatin Browne Jr. The following day Browne called again, along with the labor candidate for mayor of New York, Henry George. Wallace had previously met both men in England and they provided the contacts he needed to launch his lecture tour. Wallace, who had been deeply impressed by George's *Progress and Poverty* (1878), found the champion of labor rights and enemy of landowners and capitalists "full of work & excitement with his candidature."[42] This was a promising start even if it lacked some of the fanfare and excitement that accompanied other visiting lecturers. Meeting Browne, with his connections to the world of New York journalism was, on the surface, shrewd. But confabbing with George immediately threw Wallace into the throes of a three-horse race for mayor and aligned him with one political camp.

The politicking soon intensified. On his third day in New York, Wallace made his first public appearance, speaking at New York's Chickering Hall during a meeting of the Columbia College Alumni Henry George Campaign Club. The report of the meeting published in the *New York Tribune*, a paper loud in its support of the Republican candidate, Theodore Roosevelt, did Wallace no favors. The impression left was that Wallace was speaking not as a Henry George supporter but as a "friend" of the Democrat nominee, Abram Hewitt. Introducing Wallace as president of the English Land Nationalisation Association, the report printed only a fragment of his "written speech." Wallace, it reported, argued that "the workmen were making a false start and were falling into a prepared Republican trap." According to the same report, George, who had entered the hall just as Wallace had begun to "read" his speech, offered an extemporized reply dismissing Wallace's argument as that of the "old Democratic party." Those arguments, George declared, obscured the "real issue" that society was being ruined by a "gang of thieves" accumulating wealth at the expense of laborers.[43] Wallace appeared not only as a political washout (and political foe to Roosevelt) but also as someone with little public presence or personal charisma. It was to be expected, perhaps, that the *New York Times*, another champion of Roosevelt, did not even mention Wallace's contribution to the Chickering Hall event.[44]

A strikingly different (and presumably more accurate) account appeared in the pages of the New York *Sun*. In that report it was not Wallace who had opposed George but one Charles Miller, an advocate for Abram Hewitt (not incidentally, the candidate being cheered on by the *Sun*). Wallace, a "tall, thin man with a long snowy beard" simply made "a few remarks on the land question in England." Miller, on the other hand, launched an attack on George's political philosophy. Anything Miller had to say was, not surprisingly given the composition of the audience, "swamped in a roar of hoots, and groans and hisses." And when George stepped on the platform, "the crowd simply went mad, tossing their hats in the air and shouting themselves hoarse."[45] It was to Miller that George replied. Wallace's voice was almost entirely lost amid the political brouhaha.

Wallace's appearance at Chickering Hall contrasted sharply with those of Huxley and Proctor before him. Wallace had spoken briefly in the hall as a support act to the main attraction, and even that rather ineffectually by his own and others' reckoning. To be misidentified by one of New York's major newspapers as an unpopular political figure was even more of a disaster. It was perhaps a good thing, then, that, with his first Lowell lecture set for November 1, Wallace did not remain long in New York. In the greater scheme of things, however, this might reasonably be considered among Wallace's gravest (if inadvertent) mistakes. To not even appear before a New York audience on the subject Wallace was best known for—evolution—was a serious strategic blunder. The very mixed and contradictory reporting of his brief intervention in a political meeting told against him. If those reports starkly reflected the political bearings of the newspapers in which they appeared, more positive ones did little to create a public appetite for hearing Wallace in the city that arguably set the tone for many a lecture tour. The more specific fact that the *New York Tribune*, deliberately or not, blatantly misrepresented the political bearing of Wallace's remarks and appearance stands in dramatic contrast to the fawning and fulsome reports of Huxley's lectures at Chickering Hall over a decade earlier in the same paper. What Huxley had done in private in his letter to Henry Martin, the *New York Tribune* did in public—presented an entirely average and underwhelming public speaker with little ability to command a large assembly.

Wallace does not appear to have appreciated the importance of New York as either the springboard or final landing place of a successful tour. There are, however, hints that he did try to garner press support and interest in New York through time spent with Albert Browne during his sojourn in the city and his later encounter with the Washington correspondent of the *New York Herald*, Charles Nordhoff. Browne, a lawyer and journalist famous for his reports during the Civil War, was at the time on the staff of the *Herald*. Yet there was from that newspaper nothing like the coverage of Wallace's visit that marked the tours of other British celebrity scientists. And despite Nordhoff's effusive praise for Wallace when they met three months later in Washington—Wallace wrote at the time that Nordhoff had told him that "he had been hoping all his life to see me—& that he considered it a great honour to have me at his table"—this too does not seem to have translated into extended public praise in the pages of the *Herald*.[46]

A "GREAT ENGLISH NATURALIST" IN BOSTON

After five days spent in New York and exploring its environs, Wallace traveled to Boston to prepare for his Lowell lectures. There were eight Lowell

TABLE 4.1 ADVERTISED DATES AND TITLES OF WALLACE'S LOWELL LECTURES

Date	Title
Monday, November 1	*The Darwinian Theory*
Thursday, November 4	*The Permanence of Oceans and the Relations of Islands and Continents*
Monday, November 8	*Oceanic Islands*
Thursday, November 11	*Continental Islands Present and Ancient*
Monday, November 15	*Relations of New Zealand and Australia*
Thursday, November 18	*The Origins and Uses of the Colors of Animals*
Monday, November 22	*Mimicry and Other Exceptional Developments of Color*
Thursday, November 25	*The Origin and Uses of Colors of Flowers and Fruits*

lecturers for the 1886–1887 season, one of whom overlapped substantially with Wallace.[47] On Tuesday, November 9, Rodolfo Lanciani (1845–1929) began a course of twelve lectures on recent archaeological discoveries in Rome.[48] Wallace, however, was at the top of the roster and first among equals on the Lowell platform.

Wallace's arrival was noted in several newspapers, with a list of his eight Lowell lectures appearing in their advertising columns.[49] In a short notice, the *Boston Journal* presented Wallace as "the most distinguished [naturalist] now living in the world." A much fuller report and interview with Wallace appeared in the *Sunday Herald*, which announced him as "the English Naturalist" and "England's greatest living naturalist." Wallace's countenance was described in detail. He was "not at all a typical Englishman," with a "worn rather than heavy [white] beard." His "bearing" was that of an "ordinary citizen rather than that of a scientist." That indistinctiveness was compensated by a "strong individuality." His audience could anticipate a lecture from someone whose "presence is so good and whose enunciation is so clear for an Englishman that he ought to be easily heard." The question was whether that description would be enough to excite demand for tickets and for hearing and then reading Wallace.

While much of the interview concerned Wallace's views on land nationalization, universal suffrage and Home Rule in Ireland, the first part allowed him to introduce his forthcoming lectures. As he put it, they would consider two topics—island life and the coloration of plants and animals—in order to demonstrate the wide application of Darwinism. His lectures, then, would be a defense of Darwinism as a biological theory against its various detractors. When Wallace was asked about religious opposition to Darwinism, he suggested that there was "not much, except to the part regarding man." This "antagonism," he surmised, was caused by the "extreme agnostic party," members of which included Haeckel, Huxley and Tyndall, who aggressively argued on "general grounds" that humans did not possess a spiritual nature that survived death.[50] It remained to be seen whether Wallace would introduce his own controverted position on that subject into any of his eight lectures. His argument, by then well-known and much discussed, was that natural selection, or any evolutionary mechanism whatever, could not provide a full explanation for the cognitive and moral capacities of humans and that appeal must be made instead to a "Higher" or "Overruling Intelligence."[51] For Darwin, and those who followed his more naturalistic account of human evolution, Wallace's view was unwarranted and untenable. Others, anxious to reconcile an evolutionary account of life with religious convictions about the special status of humans, warmly welcomed it.

What was given only the barest of mentions in Boston's leading dailies was Wallace's strong convictions about spiritualism. This fact was pointed out in a lengthy piece on Wallace published in the *Banner of Light*, Boston's weekly spiritualist paper. Where one "parasite newspaper" belonging to the city's "creed parties" had mentioned Wallace's writings on spiritualism, it summarily dismissed them as being convincing only to members of the spiritualist "sect." To rectify the situation, the article in the *Banner of Light* referred to Wallace's defense of spiritualism that had appeared in the *Sunday Herald* the previous spring. It took pains to stress that Wallace believed that evolution was "the great fundamental law of the universe of the mind as well as of matter," and that "the whole *raison d'être* of the material universe . . . is to serve the grand purpose of developing human spirits in human bodies." This positioned Wallace as an advocate of the "New Dispensation" of modern spiritualism and gave readers something to listen for in his upcoming lectures.[52]

Wallace entered Huntington Hall at 7:45 p.m. sharp on Monday, November 1, and began his lecture without preface or introduction. The *Boston Advertiser* reported that despite his reliance on notes, the whole talk resembled an "informal conversation" delivered in a "very simple" style. The crowded hall was composed of an audience that "in intelligence and in interest . . . would have given pleasure to any speaker." The brief summary of his lecture told readers very little of its content beyond the fact that it laid out the basic principles of "the Darwinian theory," rebutted "the many theories" opposed to it and argued that the mental powers of humans could "never have been developed from lower animals."[53] This final claim was not widely commented on despite being one of Wallace's more controversial theories. Wallace had not managed to deliver it in a way that energized public interest. It was only the *Banner of Light* that made something of it, claiming that in making such a statement Wallace had "showed himself a Spiritualist."[54]

Wallace's second Lowell talk received the most coverage of the three public lectures delivered in Boston on the evening of Thursday, November 4. Even so, reports were brief, the longest being in the *Boston Herald*, a précis republished verbatim in the *Banner of Light*.[55] Wallace's subject—the permanence of continents and oceans—introduced his listeners to one of the fundamental controls on the history and geography of life. Little of the material was new but, with the help of geological charts, he offered a characteristically clear account of the latest theories about the earth.

The continued popularity of Wallace's lectures was signaled by the fact that the audience for his third lecture on oceanic islands remained large even though President Cleveland was addressing a gathering to celebrate the two-hundred-and-fiftieth anniversary of Harvard College the same evening. As one thousand costumed undergraduates paraded by torchlight through Cambridge, Wallace expounded upon the flora and fauna of islands inaccessible to all life except that which could travel huge distances across oceans.[56] But if his audience at Huntington Hall continued to come out, the newspapers gave little space to record his explanations of island life. The *Banner of Light* rushed through the third, fourth and fifth lectures in a single column published several days after they were delivered. Other papers omitted mention of his third, committing their column inches to the presidential visitation and the torrent of words spoken by national and local notables at the Harvard anniversary celebrations. Wallace's fourth and fifth lectures only appeared in the form of short and unexciting abstracts. The *Boston Evening Transcript* inserted the briefest of notices of the fifth lecture next to a full column account of the evangelist Dwight L. Moody's revival meeting, which had taken place the same evening less than three miles away at Sanders Theater.[57] This textual juxtaposition, fairly typical in terms of newspaper accounts of lectures and sermons, is a reminder that the kind of talks Wallace was giving almost invariably were only one of a number of often very different options for those wanting to participate in Boston's lively culture of public speech. In more material geographical terms, Wallace also found himself again at a distance from the center of lecture culture, this time on an urban scale. The Lowell lectures certainly had intellectual prestige, but their move to Huntington Hall in 1879 raised concern that they would no longer be performed at the heart of Boston's entertainment district.[58]

The final three lectures marked a change in subject and in approach. Wallace shifted from island life to coloration in plants and animals as his chief illustration of the explanatory power and wide application of natural selection. To illustrate the relations between color and the "habits and economy of animals," Wallace used a number of stereopticon views.[59] As he noted to his wife Annie several days earlier, this marked the beginning of his "lantern lectures," a change that he no doubt hoped would further strengthen the tour's appeal.[60] His American addresses, he confessed to his daughter, did "want 'flummery,' but the 'slides' serve the purpose in some of them."[61] The brief reports of the final lectures that did appear were, however, perfunctory and did little to give a sense of any general excitement or widespread interest. Wallace avoided any grand statements about the cosmological implications of evolution and stayed close to his script. His account of plant fertilization in his final lecture, while putting the power of Darwinian explanation in the limelight, did not touch on the metaphysical matters that were hinted at in his opening lecture.

When all was said and done (or reported and done), Wallace's Lowell lectures were not, by most measures, a failure. While Huntington Hall had not been entirely full for every lecture, hundreds still came out every Monday and Thursday evening to hear the undoubtedly famous naturalist. One commentator even felt that they were the most successful Lowell lectures for some time. According to an editorial in the *Boston Herald*, Wallace's course had been "the most notable presented . . . for many years. . . . Nothing approaching it in importance has occurred in the history of the Lowell Institute since Tyndall came to America under its auspices." Indeed, Wallace had on one count surpassed Tyndall in pitching his lectures at the right level for the intelligent audiences that the Lowell Institute was famous for. Where Tyndall had risked making his lectures too "elementary," Wallace had assumed his audience already knew something of his subject. It was, the editorial concluded, "Wallace's modest bearing and gentle, winsome manner" that had "done much to gain him the affection, as well as respect, of his hearers."[62] It is worth emphasizing that it was this rather than other aspects of Wallace's lectures that the report finished with. "Bearing," "manner," "affection," and "respect" were not the same as a sound grasp of evolution by natural selection but were nevertheless of clear importance in assessing whether or not a lecture had been a success or failure.

In the end, though, this glowing write-up masked some of the reasons why Wallace, whatever his own virtues or the virtues of his discourses, had failed to pack Huntington Hall or capture the attention of a wider newspaper-reading public. Although reports of at least some of his lectures had appeared in the pages of Boston's leading dailies, they supplied only the most cursory remarks about Wallace as a speaker. The *Herald*'s warm words did not match the coverage even that paper had given. And the fact that the *Banner of Light* was the only paper to provide reports of all his Lowell lectures reinforced Wallace's association with the spiritualist cause despite his general avoidance of that subject at Huntington Hall. Those reports, published as they were in a weekly, appeared long after the lectures' delivery, reducing the kind of positive feedback effect of accounts printed the morning after an evening discourse.

Wallace was aware of the underwhelming coverage of his Lowell lectures. Most immediately, he was sensitive to the fact that the Harvard celebrations and the contemporaneous congressional elections had left little room for reports of his talks. As the lack of attention continued, and as lecture invitations failed to materialize, Wallace offered reasons other than his well-known advocacy on behalf of spiritualism. These included a lack of interest in lectures on natural history (people, he supposed, preferred talks on "Art, or Travels, or Battles, or something more exciting"), the surfeit of American scientific lecturers and his association in the public's mind with J. G. Wood.[63] Writing to Annie a day before his fifth Lowell lecture, he reported that "my agent, Mr. Williams tells me, in confidence, that Mr. Wood has injured my chances very much, because he too was cued up as a 'great English Naturalist' & the people were disgusted with him, both for the little that was in his lectures & his indistinct speaking—so that they are afraid of having more lectures from 'great English Naturalists'!!!"[64] Others also noticed the meagre press coverage. Two months into his visit an editorial note in the *Nation*, a New York newspaper, asked why Wallace, "co-discoverer with Darwin of the doctrine of natural selection," was not attracting the same public interest that had been lavished on Tyndall and Huxley in the previous decade.[65] Around the same time a commentator writing for the *Minneapolis Tribune* wondered why Wallace had not been invited to Harvard's anniversary celebrations.[66] The same reporter speculated that Wallace's support of spiritualism was the chief cause. But publicity about the lack of publicity was not going to help Wallace secure the level of invitations he needed to keep his tour on track.

The story was much the same when it came to the occasional talks he gave once he had delivered his first few Lowell lectures. Apart from a short paper read during the annual meeting of the National Academy of Science, these tended to be in smaller centers, attracting only local interest. The first was an informal talk on land nationalization given to the New England Women's Club on November 15. Lectures a week or so later at Williams College (on the colors of animals) and under the auspices of the Meriden Scientific Association (on Darwinism) were well enough received. Some reports, however, suggested disappointment among at least a portion of his audiences. The *Meriden Journal*, for example, declared the lecture a "very disappointing one." The content was not sufficiently "popular" and Wallace's delivery, while fluent, was marred by the "decided elision of his g's."[67] Such indistinctness was, unfortunately for Wallace, reminiscent of the failings of the Reverend J. G. Wood.

SPIRITUAL EVOLUTION IN BALTIMORE

After Wallace completed his Lowell course on November 24, he spent time at the Peabody Museum and in the company of Othniel Charles Marsh. His journey to Baltimore, where he was due to deliver five lectures, included a stop off at Poughkeepsie, New York. There, he gave an evening lecture on oceanic islands at Vassar College, to a "good audience," one that included many of the "300 lady pupils" that Wallace had met during the day.[68] From there, he took the long train ride to Baltimore, arriving just a few hours before his inaugural lecture at the Peabody Institute.

The Peabody was one of America's leading bodies dedicated to the promotion of science and the arts. First proposed in 1857 by the American financier George Peabody, its striking white marble building was opened in 1866 and later extended to incorporate an art gallery and music academy. The $1 million donated by Peabody also funded a sumptuous library and commodious lecture hall with a capacity of about fifteen hundred. As well as hosting musical concerts, the hall was the venue for the "school of lecturing," one of the institute's central activities. At least in the original conception, the lectures were to be on "the most useful science and arts" and were to be "collegiate in manner." In addition, a series of more popular lectures on the arts, sciences and literature "delivered by the most distinguished lecturers" were held at the hall and followed something like the pattern of the Lowell Institute's courses.[69] The tickets, however, were not free but had to be purchased from the institute.

Despite the grand designs, as a space for lecturing the Peabody hall was less than ideal. As with many similar halls, ventilation and heating were a constant challenge. Acoustics, too, were a problem and it was observed early on that those with voices of normal volume struggled to make themselves heard. Loud voices, while reaching the entire audience, also tended to reverberate, and the echo effect could be off-putting for auditor and speaker alike. Efforts to improve the acoustics included the erection of curtains hung from the ceiling and a sounding board. The latter allowed a speaker to be heard and reduced reverberations but confined lecturers to standing in one place, making experimental demonstrations or a livelier delivery difficult. Lines of sight were also rather poor and thought unconducive to scientific lectures. While the floor of the hall gently sloped away from the platform, it was still difficult to see any experiments performed live. The trustees of the institute acknowledged the need for a new hall specifically designed for science lectures and built in the form of an amphitheatre.[70] This vision, however, had not been realized.

By the time Wallace came to give his lecture, at least some of these snags had been fixed. While his diagrams and slides would have been visible to all, it nevertheless took sustained vocal power to ensure a hearing throughout the hall. Like his Lowell lectures, the Peabody cycle attracted a well-educated audience schooled in the art of listening to high caliber speakers. Again, like at Lowell, his lectures were part of a larger series delivered by some of America's most distinguished academics and orators. The year Wallace was there, subjects included travels in the Near East, Gustave Doré's illustrations of Dante's *Divine Comedy*, a lecture on the Brooklyn Suspension Bridge, marriage and divorce laws, and "great British orators." None of them, Wallace's included, attracted a full hall but ticket sales (priced at 25 cents for one lecture or $1.50 for a season ticket) were nevertheless healthy.[71]

The reception of Wallace's lectures was not widely registered, at least in the region's leading newspapers. The *Baltimore Herald* gave a brief account of the opening lecture on Darwinism and offered a sketch of the speaker. Describing Wallace as "a man of short and stout build, with white hair and a beard," it also noted that his voice had "little animation." His scientific reputation and humble countenance offset this lack of vocal assertiveness and attracted a "large and attentive audience."[72] Wallace delivered a total of four lectures, derived from his Lowell course but in a condensed form. As well as covering Darwinism, he tackled the theme of color in general along with mimicry and coloration in plants. These were the lectures that included vivid images on lantern slides, which, in Wallace's estimation, added to their popular appeal. In many ways, Wallace fit the space he was lecturing in. With the adjustments to the hall's acoustic properties and the use of good-sized diagrams, Wallace's lectures, despite his rather dry delivery, would have been accessible to all who had gathered to hear and see them. Still, apart from one or two positive reviews, the lack of any extended public record of his lectures suggests that they were enjoyed but did not provoke controversy or stir up wider public discussion.

One person who did respond privately and positively to Wallace was Daniel Gilman. Whatever fears he may have harbored about the reception of Wallace's Peabody lectures seem to have been overcome by the experience of hearing the renowned naturalist present his vision of the theory of natural selection.[73] Gilman was especially moved by Wallace's final words, which placed the formation of human spirit at the very heart of nature and evolution, making meaningful what would otherwise be without significance. Anyone familiar with Wallace's writings, even as briefly expressed in his widely read article on modern spiritualism, would have recognized the message, one that transformed his lectures on Darwinism into a natural history of the "the Soul . . . the real man, of which the physical body is but the temporary manifestation." Evolution was ultimately a spiritual affair and had moved toward "the ultimate production of the noble and perfect human form." This spiritual evolution occurred "through struggle and effort, and by unceasing warfare against physical and moral evil, for a higher and more permanent existence."[74] This was Wallace's keynote, and his distinctive brand. It was the outworking and elaboration of his conviction that human intellect, consciousness and morality were evidence of a higher and more spiritual reality. In making these claims the dogged defender of natural selection against its many detractors reversed the Copernican revolution and directed the whole realm of nature, and the whole arc of evolution, toward the making and welfare of humanity. Gilman, who years earlier had privately expressed his discomfiture at the complete absence of any religious allusions in Huxley's address at Johns Hopkins, seems to have found a kindred spirit in Wallace. But his sentiments were kept private. Wallace, in public at least, had not really found his voice or captured the imagination of a public often hungry for a vision of evolution that left scope for spiritual meaning in nature and above all in human society.

THE LEAN MONTHS

His Peabody course concluded, Wallace returned directly to Boston, where his agent was based. The news was not good. To Wallace's annoyance, Williams had not secured any more lecture engagements. Consequently, Wallace gave himself three more weeks in Boston after which he would travel to Washington to see out the winter. While Williams made further efforts to elicit invitations by advertising Wallace's offer in periodicals like *Scientific America*, Wallace made himself busy by closely observing the organizing principles on display at the nearby Museum of Comparative Zoology and the Peabody Museum of American Archaeology and Ethnology. He also spent time socializing with key scientific and literary figures in Boston, including Alexander Agassiz (zoologist and son of Louis Agassiz), Asa Gray and Oliver Wendell Holmes Sr. Attending meetings of the Boston Natural History Society kept him in close contact with the city's scientific community. Two days before leaving for Washington, he caught the first of Sir John William Dawson's Lowell lectures on the geological history of plants. Dawson, then principal of McGill University in Montreal, was well known as an arch critic of Darwinism. In his lecture, however, he avoided direct mention of evolution, cataloging instead traces of the earliest forms of vegetable life.[75]

As well as cultivating scientific friendships and spending time listening to lectures, Wallace immersed himself in Boston's lively spiritualist community, visiting the offices of the *Banner of Light* and attending several séances. The most notable of these were hosted by Hannah Ross, a noted Boston medium. Wallace's travel diary, normally used to make a very brief record of his activities, included detailed accounts of the Ross meetings. By the time Wallace attended for a third time, he had been in contact with the American psychologist William James and had persuaded him to participate. Wallace was clearly comfortable with talking privately about his spiritualist convictions and with highly influential figures like James and Holmes. But he did not make any public statements on spiritualism in the United States until he came to the defense some months later of Ross, who was by then accused of serious fraud after a dramatic exposé in the *New York Times*.[76] It would not be until he reached San Francisco in May 1887 that Wallace would be persuaded to lecture on the topic.

Wallace had not long arrived in Washington from Boston when he had to travel again, this time for a brief visit to New York. The American Geographical Society had invited him to address its January meeting at Chickering Hall. Apart from his brief remarks made in the same venue in October, this was Wallace's only public appearance in one of the key lecturing centers in the United States. But besides a brief report of his address on "Ocean Islands and Their Relations" in the *New York Tribune* and a longer synopsis in the *Bulletin of the American Geographical Society*, Wallace's lecture attracted little attention.[77] Worthy in itself, the presentation to the Geographical Society did very little to spark sustained interest in his platform performances.

The lean months in Washington, at least as far as lecturing went, were only broken up by two talks on anthropological subjects given in February. The first was an apparently informal affair, held in a "private Drawing room" for the Women's Anthropological Society of America.[78] The society had been founded two years earlier and met biweekly in one of the reception rooms of Columbian University.[79] Later incorporated into the Anthropological Society of Washington, among other things it worked to promote the participation of women in a discipline that was particularly resistant to female involvement.[80] Wallace, although in general sympathy with that aim, appears to have regarded his talk as a duty more than anything else. In a letter to his daughter Violet, Wallace remarked that "he had to give [the Society] a little lecture" but complained in the same note that "all this time I have no lectures," which was to say, none that paid.[81] However, his sketch of the great anthropological problems of race and language was, for the society itself, of some importance, so much so that a later report on the society's progress published in *Science* singled it out as given "at a special meeting."[82]

Three days later, Wallace spoke again, on the invitation of the American ethnologist John Wesley Powell, to members (and guests) of the Anthropological Society of Washington. At the law lecture theater of Columbian University, Wallace addressed the topic of "social economy versus political economy." This time his performance attracted press attention, largely because of its provocative, and to one Washington reporter, suspect argument. It also drew a large and interested assembly. A report in the *National Republican* described the audience as "classical, scientific and learned," with "many ladies in attendance." Wallace's manner as much as his message attracted comment. The speaker "read from a manuscript for over an hour. There was a clearness about his voice which, though not strong, was easily

heard. He made no effort at oratory or gesticulation." This, however, did
not detract from the fact that "there was a bold, fearless, and at the same
time radical method of presenting his views . . . characteristic of men of his
standing."[83] From this view, the listeners, absorbed by the man speaking,
either had eyes fixed upon him or were scribbling notes, capturing Wallace's
words on paper.

Wallace's message was about social economies of (the small) scale.
He argued against the mass production of goods in large factories, called
for cottage industries, decried the inevitable inequities of individualistic
capitalism and advocated for the state ownership of "all great public im-
provements." This brought applause and a "rising vote of thanks" from his
audience.[84] But it also attracted caustic commentary. An editorial in the
Washington Critic the following day described his vision of "miniature machine shops" as "a nursery romance" and declared that "one of England's
foremost thinkers . . . was so far ahead of the average . . . as to have left reason and good common sense far in the rear."[85] Wallace's memory of the talk
was that while it was "altogether too revolutionary for many of my hearers,"
it prefigured the rise of socialist ideas in the decades that followed.[86]

Wallace did not lecture again for the remainder of his stay in Washington. He continued to shuttle between the capital's scientific and spiritualist
communities, pursuing his love of botanizing and séances. On his final
full Sunday, he attended the Metropolitan Memorial Church to hear John
Philip Newman preach on Christ and the angels. Newman was among
America's most celebrated preachers, most famous, perhaps, for his account
of Ulysses S. Grant's final hours (Grant had died two years previously).
The Metropolitan Church was American Methodism's national cathedral
and seated twelve hundred people.[87] In the sweep of his grand sermon,
Newman took in the plurality of worlds (the universe, Newman declared,
was "densely populated" with spiritual beings) and discussed the "graded
positions of the celestial hosts."[88] Weaving in some of the pressing scientific
questions of the day—to the plurality of worlds was joined the age of the
earth—Newman's sermon had all the appeal of a lecture by Richard Proctor and the spiritual energy of a more strictly religious evangelical address.
Wallace, however, was unimpressed, at least with the content. In his diary
he complained that Newman had ignored the "spirits of men and women"
and had ascribed all "spiritual influence to angels and archangels."[89] Wallace, though a connoisseur of Methodist preaching, made no comment on
the delivery and noted only that the church had been packed. It did not
seem to register with Wallace that part of Newman's draw was his ability
to stir decidedly religious emotions and speak to metaphysical issues that,
for many, were of deep personal importance (whatever their diverse views
on the subjects in question). Wallace had the convictions and the material

to do just that but had not really touched the religious nerve that so often enlivened, for better or worse, much public discourse.

Wallace left Washington the following Sunday and traveled north to Niagara and then on to Kingston, Ontario. His ten days in Canada were justified by invitations to give four lectures, one in Kingston on March 6 and three in Toronto. They contrasted to what had gone before in that, at last, some of the trappings of celebrity and the frisson associated with religious questions began to appear. All four discourses were widely reported in local newspapers, and Wallace's name even found its way into an advertisement for new "spring hats."[90] In Kingston, Wallace's promised visit had prompted a master from the town's Vocational Institute to elicit the views of local clergy on the relations between Christianity and evolution. Wallace's lecture on Darwinism received the compliment of a lengthy report in *The Daily Whig*, which give space to a detailed description of the content and of Wallace as a speaker. Noting his kindly "countenance," his clear enunciation "with not a word lost" and his spare use of gesture, the report repeated what many others had said before. Wallace, once again, resisted the kind of high-flown oratory he had heard in the Metropolitan Memorial Church in Washington and spoke plainly, clearly, about the facts of evolution.

Wallace's two lectures at Convocation Hall at Toronto's University College were organized under the auspices of the college and the Canadian Institute, the city's premier scientific society. Wallace chose to speak on Darwinism and the "origin and use of color in nature." His opening lecture was delivered on March 10 to a crowded hall, having been widely advertised in Toronto's newspapers and among the college undergraduates. The lecture was chaired by the college president, the Scottish archaeologist and university administrator Daniel Wilson. In his introductory remarks, Wilson presented Wallace as the codiscoverer of "the great doctrines of evolution and the survival of the fittest." As a report in the *Varsity*, the University of Toronto's weekly newspaper, put it, "Wallace took as his text the closing paragraph of the Introduction to Darwin's 'Origin of Species.'"[91] Defending the thesis that species were not independently created but had evolved mainly through the operation of natural selection, Wallace argued that it was as well established as ever. Moving through well-worn objections, Wallace concluded with his own argument that the "mental structure" of humans could not be explained by the otherwise powerful workings of natural selection. To his Canadian audience he repeated his now oft-stated conclusions: "mind was more fundamental than matter; soul and spirit is the real man" and that the "very reason and purpose . . . of the whole material world had been the production of the . . . human form by and through which the spirit of man might be . . . prepared . . . for a higher and more permanent existence."[92] All this was, yet again, delivered in a "clear and well-modulated voice" ("silvery" according to another report) and "without gesture of any kind."[93]

Two weeks before Wallace delivered his lectures, a brief note in the Canadian periodical *Arcturus* had noted that the codiscoverer of evolution "differed from Huxley, Haeckel and the more advanced school" in arguing against a naturalistic explanation for the human intellect. But this reassurance, made good by Wallace himself, was not sufficient to prevent a backlash against the visiting lecturer. At the time of Wallace's visit, programs of lectures at the college were closely watched by opponents of moves to absorb denominational colleges into a federation governed by the University of Toronto. The guardians of various kinds of theological orthodoxy scrutinized the lecture programs for evidence that the college, and the university of which it was a part, was a seedbed of infidelity.

Wallace's lecture on Darwinism did not escape notice. One correspondent, an "onlooker," wrote to the *Toronto Mail* to express their objection to Wallace's lecture being delivered at the Convocation Hall. Wallace was not the target, but he did not escape collateral damage. Wallace, the letter writer claimed, was "by no means a scientist of high order" and had simply served up the "cold victuals" of a theory that had already been exhaustively discussed. There was nothing new in Wallace's lecture. The real complaint, though, was that University College had hosted a lecture that could too easily be used by opponents of Christianity to undermine the faith. The "coquetting and dallying of the Provincial University with Darwinism" was all the more objectionable when that institution was on a mission to destroy the independence of denominational colleges and "concentrate all the more advanced scientific teaching . . . in the university professoriate." If now the threat was stayed under the leadership of Daniel Wilson, there was nothing to prevent "more adventuresome spirits" taking over and cultivating the "pride of intellect . . . into greater strength."[94] This kind of complaint was entirely predictable, and Wallace became a foil for a local discussion that raged on after he left.

For his part, Wilson acted to diffuse the condemnation of Wallace and the college but also privately disclosed his own doubts about the former. While acknowledging in his journal that Wallace "unquestionably takes a high rank among men of science," personal encounter led him to "class him very far below Darwin." Wallace left on Wilson an "impression of a selfish man somewhat coolly trading on his name in a money-making tour" and someone who had made "spiritualism his religion." He also found Wallace's account of the emergence of humanity's moral and intellectual nature "utterly untenable," making members of the "lower races . . . seemingly soulless."[95] Wilson, a "low church" Anglican sympathetic to evolution but skeptical of evolutionary explanations of the human mind, was concerned that Wallace's ideas threatened the doctrine of the divinely ordained unity, equality and dignity of humanity.[96] It is not clear why Wilson interpreted Wallace in this way or why he did not instead, as many with similar religious convictions did, regard Wallace as an ally in the struggle against more thoroughly naturalistic explanations of human distinctives such as mind and morality. It may, perhaps, have been Wallace's insistence that there was little evidence of mathematical, artistic and musical development among "lower savages."[97] Wilson may also have deemed Wallace's argument that the human body was but a temporary and material expression of an immaterial soul a departure from Christian orthodoxy, perhaps associating this account with a gnostic denigration of the corporeal and a denial of the hope of bodily resurrection.

As it happened, however, Wilson's key role in brokering an agreement for a regional federation of colleges meant that he kept such critical thoughts to himself. As president of University College, he wanted to demonstrate to critics of the federal scheme, both religious and secular, that the college could host someone of Wallace's stature and welcome open discussion of a profoundly significant scientific theory. At the same time, Wilson was not beyond using the kinds of religiously motivated criticisms leveled at Wallace to censor more aggressively secular critics of absorbing religious colleges into a state-funded university.

All told, despite a flurry of local fuss, Wallace's visit was largely lost sight of in the ongoing tumult surrounding university and religious education. His lecture on "the origin and use of color in nature" the following day, and his repeat of Darwinism to three hundred young men at the Ontario Veterinary College the day after were successful enough in their own way.[98] But Wilson's rather cool response to Wallace reflected a wider sense of unease in certain quarters of Toronto's college communities. If on this occasion the lectures did help fuel ongoing theological and related institutional disputes, they were never center stage and were remembered or forgotten depending on their usefulness in a local but intense culture war. Wallace, wisely perhaps, did not overstay his welcome and left Toronto on March 17, with little prospect of further lectures to fund his remaining time in America.

WALLACE COMES OUT IN SAN FRANCISCO

Wallace's lectures in Canada, which earned him a little over $200, were his last for over a month. Returning to Washington, DC, by way of the Niagara Falls, Wallace continued to enjoy the scientific and spiritualist communities in the capital. An audience with President Cleveland, then midway through his first term in office, was an indication of the circles Wallace was moving in. The company he kept included leading lawyers, journalists, scientists and philanthropists, along with well-known figures in American spiritualism. None of this translated into lecture opportunities and Wallace had all but given up on Williams delivering anything more than the odd invitation. On March 30, a few days before he departed Washington to travel west, Wallace met with Colonel William F. Morse, the American agent of the London impresario Richard D'Oyly Carte, to arrange lectures for the next season.[99] Wallace was apparently unaware that Morse, who he mistakenly thought was an agent of the Redpath Lyceum Bureau, had managed Oscar Wilde's recent American tour and had overseen tours by the journalist Archibald Forbes and the celebrity lawyer William Bannatine. Despite that fact, Wallace was nevertheless persuaded that Morse could turn around his fortunes. Even so, he decided not to pursue the possibility of a further tour organized by Wilde's canny manager.

Around the same time, Wallace's fortunes picked up a little, and he made the decision to travel west, funding the cost (which, if he made it to San Francisco, would be as much as crossing the Atlantic) through lectures in Cincinnati, Bloomington, Sioux City and three urban centers in Kansas. It was not until he reached Kansas that he was confident enough of lectures in San Francisco to risk paying the fare that would take him to Salt Lake City and then along the Central Pacific Railroad to the West Coast. Two of his Lowell lectures provided the bill of fare along the route. He delivered his lecture on Darwinism three times, first at the chapel of Indiana University on April 26, then in Sioux City at the Academy of Music on May 2 and finally, one week later, at the State Agricultural College of Kansas, Manhattan. A synthesis of his sixth and seventh Lowell lectures on the origin and uses of the colors of animals and then on mimicry in animals provided material for a lecture delivered to the Cincinnati Society of Natural History, at Belmont College (again in Cincinnati) and then repeated for the Sioux City Scientific Association, the University of Kansas in Lawrence and in the Opera House in Salina. This allowed Wallace to use his diagrams and lantern slides, which he hoped would draw larger crowds. The only other lecture Wallace gave on his journey to San Francisco was on oceanic islands, a talk that made up a series of three for the Sioux City Scientific Association.

Wallace's reception during his brief visits to cities en route to the West Coast was generally positive and his lectures earned him almost $500. In Cincinnati, his lecture to the Society of Natural History earned him an unexpected $50 despite difficulties with an inept lantern operator and a faulty lamp. One newspaper report noted his "drawling" speech (which, the reporter averred, was typical of his countrymen's efforts at oratory), something Wallace himself noted but, in a letter to his wife, put down to being unable to read his script.[100] In Bloomington, where he lectured on the invitation of the University of Indiana's president David Starr Jordan, Wallace had to hastily create new diagrams after the ones that traveled with him failed to arrive. Even when they did keep up with his winding progress west—as in the case of his Sioux City series—Wallace spent much of the afternoon prior to his lectures in setting them up.[101] The challenges only increased the further west Wallace went. In Salina, Kansas, after paying for the hall, printing and a (poor quality) magic lantern, he lectured to an audience of about one hundred and earned a rather paltry twenty-six dollars.

About a week before his lecture in Salina, Wallace took the decision to travel on to California.[102] His brother John, who lived in Stockton having emigrated to California nearly forty years before, had assured him that arranging one or two lectures there was possible. The promise of at least some income allowed Wallace to commit to the journey. Traveling via Salt Lake City, he visited the famous Tabernacle and was amazed by the acoustics that allowed six thousand people to hear "a speaker . . . [talking] in an ordinary conversational tone."[103] This, of course, was Wallace's preferred style of address and he was to use it again in San Francisco to much applause.

The attention Wallace received when he arrived on May 23 at the Baldwin Hotel in downtown San Francisco was unprecedented.[104] His brother had booked accommodation on the first floor that included a living room to receive visitors. Unlike elsewhere, Wallace was staying at an opulent hotel with the reputation and capacity to attract reporters and other visitors. After a visit to Pioneer Hall on Fourth Street where he was due to give two lectures, Wallace received a reporter from the *San Francisco Examiner*, who called on and interviewed Wallace at length. The interview "embraced a hundred topics" and the reporter left with the impression of a "brilliant and easy conversationalist" and of a "hale and hearty English gentleman with mild blue eyes and a benevolent countenance."[105] The printed version included Wallace's views on his codiscovery of natural selection, the land question and his impressions of the railroads and denuded and neglected landscape he had traveled through. Wallace thought the interview was "very well done considering the short time."[106] It also meant that his introduction to a San Francisco public was in stark contrast to the one that had played out so badly in New York. For perhaps the first time in his tour, the ingredients for success were being gathered together.

Over the next two days Wallace was busy meeting some of the city's leading citizens, including Edward Holden, president of the University of California, Joseph LeConte, professor of geology at the university and the lawyer and friend of Henry George, James Maguire. In what time he had remaining he sourced a stereopticon and explored San Francisco. His first lecture in the city was delivered on May 25 at the newly built Pioneer Hall. Erected through the patronage of the California Pioneers after the destruction by fire of a hall of the same name, the building materially embodied the spirit of the city and was a major civic landmark.[107] Able to seat one thousand, the luxuriantly decorated hall was unrivaled in the city as a venue for public lectures.

When Wallace took to the stage, he looked out on what the *San Francisco Examiner* described as "one of the most intellectual gatherings ever witnessed in San Francisco." Joseph LeConte, one of America's leading advocates of evolution, provided some brief words of introduction and heralded Wallace as the "greatest living champion" of the theory, an epithet that had none of the negative associations of being described as a great English naturalist. With "that modesty characteristic of great minds," Wallace ignored the puffery and began a two-hour long lecture that reportedly held the "absorbed attention" of his illustrious audience throughout.[108] At last, it would seem, Wallace's lecture was drawing the kind of journalistic praise so key to attracting crowds of listeners and generating wider media attention.

Reports appeared the next day in several of San Francisco's leading dailies. The headlines variously proclaimed Wallace's theme as "Man and Monkey," "Darwinism" and "Evolution."[109] The subheads picked out some of the more provocative points: "The ultimate purpose of the wonderful progression of nature"; "The body is evoluted. The mental nature, he believes, was differently acquired."[110] Wallace had rehearsed the lecture many times, and it had been critical in funding his travel to the West Coast. This time, though, the metaphysical as much as the scientific content made the headlines. Wallace once again examined the evidence for Darwin's theory of natural selection, detailing and, through "great diagrams," displaying the high frequency of variation in plants and animals.[111] Wallace also rolled out his well-worn argument about the difference between successive changes produced by domestic breeding and breeding in the wild. From there he turned to the struggle for existence, and, according to one journalist, "all present assumed even a more attentive attitude than previously as if afraid to lose one syllable of what was said." Yet for all its apparent interest, it was "too long for insertion."[112] Instead, like other reports, Wallace's closing passage on the evolution of the body and the unexplained emergence of mind received full treatment. The reports differed somewhat and at least one mangled part of his argument. But the main lineaments were there in all. There was now no doubt that the human body was a product of natural selection. But the appearance of the human mind cannot be explained by the theory of natural selection. It is dormant in uncultivated "savages" and of no survival use. It is, however, more fundamental than matter and will survive the death of the body. The human mind supplied the "very reason and purpose" of nature and of evolution, which had provided the "form by . . . which the spirt of man might be prepared . . . for a higher and more perfect [or permanent] existence."[113] Wallace's final flourish was to declare

Darwinism entirely compatible with this noble philosophy.[114] Journalists took note and made this the central feature of their reports.

The second lecture, given two days later, was, comparatively speaking, underreported. Only the *San Francisco Examiner* carried a full account. The lecture's long title, "The Origin and Uses of Colors in Animals in Relation to Their Habits and Surroundings, Mimicry and Exceptional Modes of Color" echoed the length of time it took to deliver. The two-hour discourse was nevertheless imparted from the "lips of a master" in "good voice" and was "illustrated profusely with stereopticon views." The appeal of his first lecture, buoyed by fired-up newspaper accounts, helped Wallace to sustain interest and positive press attention even though the second discourse risked overtaxing his listeners.

The lecture nevertheless hit some high notes. Wallace took as his starting point the immense variety, richness and aesthetic charm of color in nature. He noted that, with the advent of Darwinism, it could no longer be assumed that coloration in plants and animals lacked purpose. His task was to explore the extent to which color in animals was "due to general law and how much to the action of natural selection under the law of utility." The latter, Wallace argued, was evident in animals whose color and patterning provided concealment, protection or recognition. The former, understood in terms of chemical composition and organic structures, operated unimpeded in animals in which bright and diverse coloring was not injurious. Along the way Wallace took issue with Darwin's theory of sexual selection and argued that striking ornamentation and color in males was due to more vigorous development of parts. This, through the "segregation of chemical or organic molecules" increased colored patterning as a natural by-product. After exploring why tropical animals were typically marked by more vibrancy, Wallace finished his account of the "marvel and mystery" of animals colors by quoting lines by the "poet-artist" John Henry Dell, "on a peacock's feather."[115] Dell's verse left Wallace's listeners with the wonder of nature's colors, "obedient to plan" and to a more profound "poem" of which the peacock's feather, for all its arresting hues, was but the "lightest word."[116]

Wallace's two lectures were given as the "great English scientist" (note not naturalist) and "co-discoverer of the theory of natural selection." Trading under these arresting descriptions, the lectures confirmed his scientific reputation and standing. But there was another element at play in San Francisco that in the end benefitted Wallace and helped generate interest in his presence in the city. More than in any of his other stop offs, Wallace appeared as a leading advocate of spiritualism. This was due to the efforts of James J. Owen, editor of the *Golden Gate*, San Francisco's weekly spiritualist newspaper, and another prominent spiritualist, Dr. Albert Morton.[117]

Wallace, of course, had not hidden his support for spiritualism throughout his tour, whether in private gatherings, involvement in séances or through his defence of Hannah Ross. Perhaps most significantly, Wallace's views had been widely circulated through a special symposium on the possibility of postmortem existence published on Easter 1887 by the *Christian Register* and later reissued in pamphlet form.[118] But although Wallace had offered the occasional talk on his views on land nationalization, it was only in San Francisco that he agreed, finally, to lecture on spiritualism. Responding to Albert Morton's appeal to speak on the subject, Wallace observed that he had *"never* lectured publicly on Spiritualism, partly because I feel I can do more good by *writing* on the subject and also from the fact that I have no power of expression or of style to do justice to it."[119] But whatever his long-standing misgivings, Wallace acquiesced to Morton's invitation.

The lecture, which Wallace agreed to deliver at the city's Metropolitan Tabernacle, a large auditorium owned by a Unitarian congregation, was widely advertised among San Francisco's spiritualist community. The *Golden Gate* called for all public meetings of spiritualists announced for Sunday evening, June 5, to be "adjourned" to encourage attendance at Wallace's talk.[120] The lecture on spiritualism also attracted greater numbers than his two science lectures and generated unprecedented levels of press attention. Most dramatically, the *San Francisco Examiner* carried the headline, "WALLACE COMES OUT."[121] This was, of course, an exaggeration. The *Golden Gate* had already provided a forensically detailed account of a séance Wallace and his brother had attended the morning before his lecture on coloration in animals.[122] But outside spiritualist circles, the lecture topic may indeed have been a startling revelation.

Wallace's Sunday lecture, which drew a crowd of about one thousand, took as its subject the question found in the book of Job, "If a man die, shall he live again?"[123] It was a favorite among spiritualists, who either answered in the affirmative or denied that men die at all. This "question of questions" provided Wallace with the opportunity to recommend spiritualism and detail the evidence that lent it support. He began, however, with a more general concern that the materialism of the age, buttressed by a scientific mindset that reduced all to molecular motion, endangered the very possibility of moral culture.

The opening section was written out in full in Wallace's own script.[124] Carefully crafted, and with a compelling cadence, it had an eloquence that later sections lacked. Wallace, preaching a spiritualist sermon, began with the words from Job, declaring that "this is the question of questions, which in all ages has troubled the souls of men." In his own age, "modern science, instead of clearing up the difficulty and giving us renewed hope . . . advances powerful arguments against the affirmative reply."[125] This, Wallace proclaimed, was a road to disaster. Utilizing the rule of three, Wallace went on to argue that if all men, all children, "come to believe that there is no life beyond this life . . . there would cease to be any adequate motive for justice, for truth, for unselfishness."[126] Even now, he insisted, selfishness was on the rise. The tempering effects of conventional religion were fading, and the possibility of an altruistic social order was, in turn, under severe threat. "Hell upon earth" beckoned.[127] To those who denied this dismal prognosis, who pointed to the "blameless and even philanthropic lives of many agnostics and materialists," Wallace had this to say:

> This is not to the point. These men are unselfish in spite of their beliefs, and are illogical for being so. For surely <u>reason</u> and <u>self-interest</u> must ultimately determine the conduct of men; and, under absolute materialism it is impossible to give an adequate <u>reason</u> for unselfishness, it is impossible to prove to men that it is <u>for their own benefit</u> to give up their own happiness—the satisfaction of their own appetites and passions—for the sake of others. Under a strictly materialistic regime altruism is illogical and absurd, and could not continue, except in rarer and rarer cases, to influence the conduct of mankind.[128]

Wallace, despite his denials, knew how to preach. His residual memories of Methodist ministers and Welsh preachers may well have echoed in his mind as he wrote these lines while staying with his brother in Stockton.[129] The message and the mode of address were not so different from the sermons he had heard as a young man. Adherence to materialism meant moral decay. His answer to the state of things did, of course, contrast dramatically with this gloomy picture. He moved from threat, moral peril and social degeneration by inserting an empathetic "But" in his manuscript, turning his audience's faces toward a bright new hope, the emergence of a spiritualist movement in spite of the gathering gloom of scientific materialism.

For whatever reason, the passage countering the argument that there are benevolent materialists did not appear in any of the reports of his lecture, including the verbatim account published in the *Golden Gate*. Wallace had marked it as page 3a in his manuscript copy and had perhaps decided not to include it once he began speaking or had not given it to journalists in advance of delivery. But they summed up well the spirit and sentiment of his opening argument. In any case, at this point his lecture shifted to a genealogy of the "revulsion of feeling" against the claims of spiritualism. He traced this antipathy to the "witchcraft mania" of the sixteenth and seventeenth centuries and to the parallel rise of modern science.[130] Because of the horrifying persecution of witches and the findings of scientific inquiry, people were uninclined by habit of mind to lend any credence to spiritualism. Designating witchcraft as mere superstition had provided a basis to condemn those who put innocent practitioners of harmless but naïve occult practices to death. And science had demonstrated that nature's workings required no explanations other than material ones. These moves, while understandable, were fundamentally mistaken. To demonstrate this, Wallace spent the reminder of his lecture arguing for the scientific credibility of twelve types of physical and mental occurrences that could only have a spiritualist explanation. Each pointed to the evident truth that mind could act independently from matter and could thus perdure without a material substrate.

He illustrated his "Twelve distinct classes of phenomena" by his own experiences of them.[131] At this point, Wallace began to improvise. His notes reduced to brief prompts pointing to his encounters with mediums and entranced individuals. The report in the *Golden Gate* recorded in full Wallace's demonstrative proofs of the truth of spiritualism. The evidence they provided allowed him to outline two attributes that marked all spiritual phenomena: their uniformity and their essentially human character. In doing so, Wallace worked to rebut claims that they were manifestations of nonhuman spiritual beings, benign or otherwise, or that they could be explained away by more material explanations. As he moved toward his conclusion, he argued that spiritualism had superseded (or fully incorporated) Christianity and could stand its ground against the skepticism of scientific naturalists. More than this, it offered sure grounds for the kind of unselfish actions and altruistic society that Wallace believed was crucial for the future happiness of humanity. It even suggested an answer to the problem of evil and suffering. The formation of moral virtue and personal character before death would, he argued, be the foundation or seedbed for a better and happier existence in the disembodied life to come. Wallace left this lofty vision in the minds of his listeners by closing with two contrasting poetic depictions of life, the first by Grant Allen and the second by a "trance-medium" channeling the spirit of Edgar Allen Poe. The former enunciated a "gospel of doubt and despair" with its embrace of the finality of death and the replacement of "living gods" with "blind laws."[132] The latter, by contrast, offered the "joyous teaching," of that "glorious morning . . . when the soul no longer fetter'd to the feeble form of clay/ to a high harmonious union soars, elate with hope, away."[133]

Wallace's hour-and-a-half-long Sunday evening lecture on spiritualism was, by many measures, the most successful of his tour. It was printed, and reprinted, in full in spiritualist newspapers (in Los Angeles, Boston, and as far away as Melbourne and London) and in pamphlet form.[134] It paid better than any other lecture he had given, earning him $146.[135] It did not, of course, entirely persuade everyone. An editorial comment inserted in the issue of the *Golden Gate* that carried the full address expressed "a little disappointment that the distinguished lecture did not take a more scientif-

FIGURE 4.1 Albumen print of Alfred Russel Wallace by the photographer
Isaiah West Taber (1887). NPG x5112, © National Portrait Gallery, London.

ic view of the subject."[136] What was meant by this was not clear, but the
observation, likely made by the prominent spiritualist who had introduced
Wallace to his assembled audience, registered a certain reserve about the
effectiveness of the lecture as a defense of spiritualism. Other papers, not
aligned to the spiritualist cause, generally passed no comment about the
plausibility of Wallace's arguments, simply reporting them in abbreviated
form.[137] Yet from an interview printed in the *Golden Gate* between one of
its representatives and an editor of another San Francisco newspaper it was
apparent that at least some of Wallace's hearers were aghast that the great
naturalist who gave "two or three lectures on scientific subjects" should
"[go] off in another about this intangible subject—even carrying us into
table rapping and the like."[138] According to this hearer at least, Wallace had

failed to render spiritualism scientifically credible. But at last Wallace was being widely talked about.

Whatever the scepticism of some, along with all its other benefits the lecture on spiritualism provided an opportunity that Wallace had not yet availed himself of while in America. On the evening of the lecture Wallace spoke to the editor of the *Mining and Scientific Press* about securing a portrait for a brief feature on Wallace published in the journal on June 18.[139] The next day Wallace visited the studios of the celebrated photographer Isaiah West Taber. Taber's photograph of Wallace was the only one taken during his tour, and, as well as being inserted in the *Mining and Scientific Press*, was used as a frontispiece to a pamphlet version of his lecture on spiritualism and as a carte de visite sent to, among others, the celebrity conservationist John Muir.

All this occurred as Wallace's tour wound to a close. After a trip to Yosemite, Wallace repeated his lecture on spiritualism in his brother's hometown of Stockton. He did not lecture again until the end of July. By then he had traveled back across the United States and was staying in East Lansing, Michigan. He spoke there, at the State Agricultural College, for the last time on Darwinism and coloration in animals. Although invitations had come in to speak in September on spiritualism from Boston, Chicago and other centers, Wallace had already decided to return home.

In many respects his time in San Francisco ended his tour in the way it could, and perhaps should, have begun. As Wallace himself noted in a letter to his wife written a few weeks after his appearance at the Metropolitan Tabernacle, spiritualism "paid better than natural history."[140] The portrait of him by Taber began to circulate widely. Invitations, which had only trickled in before, started to appear more frequently with the promise of better profits. He had, at last, found his voice, or a way of communicating that produced energy and excitement. But Wallace was finished. He wanted to get home in time to catch some welcome British summer weather and to avoid a "stormy passage" back across the Atlantic.[141] His tour, in the end, if not a failure, fell short of his own and others' expectations. After all expenses, his net profit was only £350.

The rather meagre takings from the tour were a consequence, in part, of Wallace's reluctance to play the role of a celebrity scientist. On arriving in America, he already felt he faced an uphill struggle to attract attention despite his famed "codiscovery" of the theory of natural selection. By character and by reputation he was modest and understated. His position, as Frank Turner put it, "between science and religion," opposing scientific naturalism and conventional Christian belief, meant it was hard to frame his tour using standard tropes and cultural divisions.[142] His decision not to lecture on spiritualism—a subject that certainly did generate intense public fascination and debate—meant that he largely avoided topics that made such tours newsworthy. Although throughout his tour there were hints of the regalia of celebrity, for the most part Wallace's tour might better be described as one conducted by an anticelebrity. There was little of that "demand for perpetual performance" or "constant scrutiny" faced, or enjoyed, by Huxley, Tyndall and Proctor and recently identified as key characteristics of modern celebrity culture by Robert van Krieken.[143] Wallace may have wished for more invitations to speak, but it was not until the closing chapter of his visit that he began to employ the technologies of celebrity that made a speaking tour lively and lucrative. He wanted, or rather needed, to make money largely for the sake of his family and his own situation but there was little evidence of an appetite for the level of attention that went with profit making. His style of speaking, frequently described by American newspapers as clear, if not dynamic, was symptomatic of Wallace's failure to attract sufficient public interest to make his tour more than the sum of several fairly dry lectures on evolution. He was not considered an orator, and when his manner was noticed at all, his lack of gestures, his "modest" bearing and his reliance on notes rendered his talks, considered as performances and cultural events, of limited moment. Where Proctor had compensated for his own lack of gestures and oratorical prowess through arresting imagery, rapid-fire delivery of newsworthy science and the stirring of debate about the rather drastic religious consequences of scientific advance, Wallace generally stayed on the safe side of controversy and wholly lacked the volubility and volatility of the much more excitable, and exciting, Proctor. Wallace was nondescript and, as a speaker, did not fascinate. Unlike Tyndall, who made his bodily presence as much as his scientific and metaphysical convictions felt in his lectures, Wallace struggled to attract the eyes, ears and hearts of his auditors. And compared to Huxley, whose American lectures were loved and loathed with a passion, Wallace was damned by faint praise. As will become apparent, the next and final lecturer in our pantheon, the Scottish evangelist and evolutionist Henry Drummond, proved to be entirely more adept than Wallace at exploiting the opportunities and influence afforded to celebrity speakers and exploited in full an appetite for a heady mix of evolution and religion. From the very start, Drummond offered, without very much gesture but with an almost hypnotic stage presence, a combination of religious revisionism and evolutionary speculation skillfully packaged and adapted for success on a revivified American lecture circuit.

CHAPTER 5

EVOLUTION'S EVANGELIST

The American Addresses of Henry Drummond

A few days before Christmas 1889, the publisher Hodder and Stoughton launched *The Greatest Thing in the World*, a small and delightfully produced booklet containing a sermon by the Scottish evangelist and evolutionist Henry Drummond.[1] With its "deckle edged paper . . . white covers and gilt tops," all designed by Drummond, it was the ideal Christmas gift written by an author with a household name.[2] The text for the sermon was 1 Corinthians 13 and the subject was love. Within a few months, 185,000 copies of the British edition had sold, an American edition published by the New York firm James Pott & Co. had appeared, and a German translation was underway.[3] The text itself was based on a stenographic record of a talk Drummond had given two years previously at Northfield in Massachusetts. It was a sermon Drummond had preached many times over many years. The American evangelist Dwight L. Moody had heard it while in London in 1884 and was determined to get Drummond to his summer convention at Northfield to deliver it there. But Drummond had been preaching the same message to thousands since Moody's first evangelistic crusade to Britain and Ireland in 1873–1875. When it was carefully recorded and then printed with Drummond's agreement shortly before Christmas 1889, many of those who read it could hear in their heads the voice of one of the best-known speakers of the day. The booklet might be taken as an epitome of Drummond's care with words—spoken, then printed with the same intense commitment to the aesthetics of style, form and content. And even though on the surface an entirely religious text, it was for Drummond the expression of a profound unity between natural or scientific and spiritual or theological worlds. It was no accident, either, that the formative context and original medium for the sermon was oral. This was Drummond's pattern and was critical to his success both as speaker and author. As with the other chapters in this book, Drummond's own experiences and thoughts on the spoken word provide an appropriate place to begin appreciating his commitment to effective speech in service of religiously meaningful science and science-infused religion.

SCIENTIFIC EVANGELIST IN TRAINING

Public speaking began for Drummond at the school he attended in the Scottish town of his birth. While at Stirling High School, Drummond won the reading prize on several occasions. Further opportunities to develop his speaking abilities presented themselves during his time as a student at Edinburgh University. As a member and then president of a debating society, the Philomathic, Drummond addressed fellow students on a range of topical subjects, from the education of women (he was in favor) to the Irish Church (Drummond called for its disestablishment).[4] He became a valued and popular member and was widely admired for his debating abilities.

Drummond also made the most of other, less conventional, opportunities at the university to hone his communication skills. He included, for example, animal magnetism and mesmerism among his extramural interests. By the late 1860s, "electro-biology" was moving from being a pursuit with considerable if contested intellectual influence to being considered by at least some leading scientific figures as marginal to mainstream science.[5] Drummond, however, was convinced of its truth, believing, in common with many of his contemporaries, that it pointed to a divinely ordained "life giving, communicable and curative power."[6] His own experiments in hypnotism demonstrated to him and to others that he could profoundly affect those in his company using mesmeric techniques. While he apparently later rejected mesmerism, ostensibly because of a concern that it gave him undue influence over other people's actions and decisions, some who knew of his youthful dalliance with it described his attractions as a speaker as plausibly mesmeric. His student friend, James Stalker, for example, noted that Drummond "had the most perfect, effortless command of every audience which I have ever seen in any speaker. It was like mesmerism."[7] This sentiment resonated with a commonplace judgement that the science of mesmerism provided a means to understand the mechanisms by which minds were changed about matters of profound political or metaphysical importance. That Drummond's flirtation with mesmerism was well known helps explain a recurring observation made by those who heard him speak, namely, that it was his eyes that confirmed him as a speaker of uncommon power. Only Wallace among the other scientist lecturers that this book has explored had given mesmerism any kind of attention. Drummond, though, was alone in experimenting with its use, perhaps in more subtle ways than usual, for increasing the power of the spoken word to move (or some might say manipulate) an audience.

Drummond spelled out his student ideas about effective communication in more mundane terms in his valedictory address as president of the Philomathic. Comparing the strengths of lectures, reading and conversation as ways of acquiring knowledge, Drummond argued that the lecture was the "best means." Its reputation may have suffered, but only, Drummond argued, "because there are no good lecturers." In a way that echoed Thomas Huxley's assessment of the value of lectures, Drummond suggested that the advantage lay in the "sympathy which it creates."[8] While his views on lectures did not stay static, the emphasis on "sympathy" remained central to his assessment of the power of the living voice. There was something about speech that made it, in Drummond's view, an especially potent means for altering the affections and redirecting intellectual attention and development.

In 1871 Drummond left the university without being awarded an MA. He did have some successes—notably in geology—but like many of his contemporaries he did not complete all the components necessary to secure a degree. At the age of nineteen he entered New College in Edinburgh to train as a Free Church of Scotland minister. There, Drummond further developed his interest in science and was exposed to a conservative form of biblical criticism. His ideas about communication also evolved. In an address to the Theological Society, delivered in the third year of his course in divinity, Drummond outlined a science of "spiritual diagnosis" that would aid the minister or evangelist in changing the religious and moral compass of those under their care. Although Drummond believed the dimensions and movements of the human "soul" were infinite and escaped the limits of human knowing, it remained vital to systematically study patterns of spiritual growth or degeneration. Among Drummond's key contentions was the importance of studying individuals, moving away from reliance on preaching to many through "pulpit eloquence" and standard rhetorical techniques. Learning how to draw in and hold a crowd was relatively straightforward in comparison with influencing an individual soul. What mattered most was learning how to "fascinate the unit by our glance, by our conversational oratory, by our mystery of sympathy."[9] If this found immediate application in work with individuals, it also described Drummond's approach to public speaking. He eschewed rhetorical techniques and the conventional arts of oratory for a form of delivery more akin to someone talking directly and plainly to an individual soul. This was not only a prioritizing of the individual over and against the collective but also a way to treat the collective as an individual. And while it echoed some common British views about what constituted effective (and affective) speech—spare in gesture, accessible in language and direct in style—it included other elements that, when conjoined to more standard approaches, made Drummond's platform performances virtually unique.

Drummond's thoughts on spiritual diagnosis were aired just before he found himself caught up in Dwight L. Moody's first British campaign. It was this, perhaps more than any other experience, that launched Drummond's career as a public speaker. He began by volunteering in the "inquiry rooms" set up for those convinced of their need for conversion by Moody's forceful evangelistic sermons. After being noticed by Moody, Drummond was asked to organize follow-up meetings in the towns and cities that Moody subsequently visited during his two-year crusade. Drummond soon found himself addressing large audiences that had been caught up in the religious atmosphere of Moody's stirring appeals. He learned to speak in public halls to crowds of hundreds and then thousands. But he also gleaned much from Moody himself of the art of compelling public speaking. In observing Moody closely, Drummond learned from one of the period's most successful evangelists, a masterful marketeer for evangelical religion.[10]

Participating in Moody's "campaign" was also an apprenticeship in effectively rendering speech into print, trying to retain the freshness and the atmosphere associated with live talk. One of Drummond's responsibilities was writing up and editing Moody's sermons for publication.[11] Along the way, Drummond witnessed Moody's courting of the press, his way of speaking for stenographers as much as the audiences that gathered to hear his popular sermons. Drummond thus became intimately familiar with the whole infrastructure of a celebrity speaker able to address rows of journalists in both custom-built and makeshift auditoria in Britain and America. It gave him the opportunity to form his own judgements and develop a unique style and celebrity quite different from Moody's, but in ways fundamentally influenced by the strategies the latter used to secure and exploit popular appeal.

As Drummond's career developed, the opportunities for speaking to different audiences on a range of subjects and under contrasting conditions increased. In 1877 his appointment as a lecturer in natural science at the Free Church College in Glasgow necessitated the preparation and delivery of lectures on scientific subjects for divinity students. Drummond took this dimension of his new job seriously. His college lecture notes from this period show meticulous attention to word choice, sentence structure and overall composition, and demonstrate his practice of writing out his more formal lectures fully in advance of delivery even if he later relied on an outline alone.[12] As with much else in Drummond's thinking and acting, he did not operate with the sharp distinctions between classroom lectures and popular talks in large public halls. This is not to suggest that he was insensitive to the need to adjust his speech for the occasion or audience, but it demonstrated an underlying commitment to a kind of oral communication that registered strongly with the emotional as much as cognitive faculties of his hearers, without, indeed, considering those two aspects as unconnected or hermetically sealed.

Even when not on campus, Drummond continued to hone his versatility as a speaker. The long college breaks over the summer allowed for full involvement in his local Free Church of Scotland congregation as a "missioner." His efforts to establish a congregation in Possilpark, his regular lectures in support of student work at the University of Edinburgh and his more periodic involvement in Moody's campaigns and in delivering public lectures meant his life was structured according to a hectic schedule of talks. This routine continued, interrupted by some years for travel, until Drummond's death in 1897. His position at the college was made permanent when a bequest given in 1883 provided the means to translate his lectureship into a chair. Part of the process of promotion meant formal ordination to the Free Church of Scotland ministry. This took place in November 1884 and marked Drummond's permanent appointment as a professor of natural science. The chair was important not just because it validated his authority in a general sense—it also reinforced his reputation as a scientific evangelist, or as someone who could speak with authority on science as much as on religious matters. This was a dimension of Drummond's public persona that was given forcible and embodied expression when he stood on a platform to speak.

Drummond also built on his composite reputation as someone worth listening to by courting connections with a more socially elevated audience and set of admirers. This social mobility was pursued through a series of talks given in 1885 at Grosvenor House in London. Repeated again in 1888, they emerged from Drummond's relationship with Lord and Lady Aberdeen, a friendship that strongly influenced both parties.[13] It was Lord Aberdeen who invited Drummond to speak to an elite audience in the ballroom of Grosvenor House, the home of the Duke of Westminster, Hugh Grosvenor. Drummond was at first reluctant, in part because he felt ill-prepared and because he now was arguing that "the lecture, as a weapon, always seemed to me a poor influence in religion." This did not mean a rejection of the power of public speaking. When he learned that what was wanted were "addresses of a simple kind," he changed his mind and consented.[14] His talks attracted a large and highbrow audience and provided a new platform from which to voice his message about evolution and Christianity.

As Drummond's experience as a speaker, national profile and popularity all increased, his style of address became a matter of public interest and comment. Contemporary descriptions picked out his chaste approach to delivery, the lack of gestures, the controlled but not especially strong or distinctive voice and the impeccable attire. Other common observations included Drummond's handsome face and features and, as noted already, his striking eyes and penetrating gaze. As one contemporary put it, Drummond was "a tall, well-built, handsome man—almost a king among men—and no one who has looked into those eyes can ever hold any other opinion that they were attractive. He was rather particularly well and neatly dressed. These things, combined with his skill of style, all had their influence upon his hearers."[15]

Drummond's decorous style and avoidance of dramatic gesture aligned with the general trend that marked the lectures of the other celebrity speakers on science who had toured the United States before him. It is not surprising that Drummond had noticed and kept in his large collection of newspaper clippings of Thomas Henry Huxley's advice to aspiring orators.[16] Published in 1888 in the *Pall Mall Gazette*, Huxley's recommendations concentrated on the importance of controlling nerves and feelings. As we have seen, Huxley prioritized determining an argument in advance and refusing to depart from it in delivery. There should be no unconstrained show of emotion, playing to the crowd or other forms of flamboyance. It is interesting, then, to note the parallels between the two speakers apparent in contemporary descriptions of Drummond's style. One depiction of a talk Drummond gave during his American tour of 1893, for example, commented that the speaker "makes no attempt at oratory, but speaks clearly in a low voice. He seems to have the faculty for emphasizing nearly every word, and every sentence spoken is emphatic in meaning without any use of large words. He makes no use of gestures, but assumes very graceful postures in speaking."[17] Another account from the same year reported that he was "a quiet speaker and [...] makes emphatic by pauses rather than intonation or gesture."[18] Both of these echoed an earlier report, published while Drummond was in America in 1887, which noted that "his speech was slow, deliberate, exact, weighty as to every word, full of thought and force and logic, devoid of any show of embellishments."[19] All of these features followed closely Huxley's own beau ideal of the effective public speaker.[20] Both Huxley and Drummond were influenced by common styles of speech in British platform culture, not least in church and parliament. If Drummond's evangelical style was more pronounced (and more finely honed) than Huxley's, it was also tempered by the latter's "scientific" commitment to studied and controlled delivery. The political platform, too, though more marginal to Drummond's life than the pulpit and lecture hall, had an influence. Drummond, who was known to and admired by William Gladstone, spoke in support of the Liberal Party during election campaigns in the 1880s and early 1890s.[21]

Interestingly, Drummond was also convinced, like Huxley, that dramatic gestures associated with certain styles of oratory were the residua of evolution. In his Lowell lecture on the evolution of language, Drummond argued that gesture and intonation were the earliest form and foundation of language. In this he was following Darwin in *The Descent of Man* and *The Expression of the Emotions*, as well as those like George Romanes and William Dwight Whitney who built on Darwin's insights.[22] It followed from this understanding of language evolution that gestures and intonation, while essential and foundational, were primitive and "child-like." The eloquent orator, though a product of "voices of the forest and the activities of the anthill," avoided gesture and other "auxiliaries" when dealing with more abstract subjects.[23] As Drummond expressed it, "gestures are mainly called in to supplement expression when the subject matter of discourse does not belong to the highest ranges of thought, or the speaker to the loftiest type of oratory;" furthermore, when a speaker "allows his mind to grapple intensely and absorbingly with an exalted theme, he becomes more and more motionless."[24] When that stillness occurred, the orator embodied the ascent of man from primitive language user to sophisticated thinker and talker.

Drummond's own commitment to exemplify "the loftiest type of oratory" was picked up by others who heard and read his Lowell lectures. In 1894 the prominent liberal Congregationalist Washington Gladden described Drummond's *Ascent of Man* as "the very perfection of popular scientific instruction . . . One cannot read without seeing before him the erect but slender frame, and the clear-cut physiognomy; the quiet manner, with no trace of affectation, and the unstrained distinctness of articulation, are reproduced in our memory as we follow these crisp sentences." Quoting Drummond's statement on the gestureless character of speakers dealing with the "highest ranges of thought," Gladden commented that there was "nothing" in Drummond's arguments "which requires the sawing of the air or the pounding of the cushion; but there is in every sentence a freshness, a vitality, and a conviction which enchain thought."[25] None of this meant that Drummond's lectures left audiences instructed but unmoved. His aim was to convert hearts and transform minds in tandem. He had cultivated a way of speaking that combined the scientific oratory of a Huxley with the evangelical appeal of a Moody.

This hybrid form of speech not only echoed Drummond's convictions about the unity of science and spirit but also paved the way for success in America arguably beyond that achieved by Huxley's lecture performances. If Drummond had mastered the art of understatement in service of serious science and serious religion, he was also able to avoid the charge, made by some against Huxley while he was in the United States, of delivering discourses that bored rather than inspired the listener. Whatever criticism might have been leveled against Drummond's manner or message, there is little evidence that he delivered lectures that were thought drab or tedious. His commitment to the "simple message," and to using understatement to powerful emotional effect, helped to avoid accusations of dryness. It was not just that Drummond interlaced his talks with provocative aphorisms about evolution and religion. Reports of his performances repeatedly suggest that he was also able to use his disciplined deportment, glances, dress and demeanor to reinforce the gravity and emotional and spiritual relevance of his spoken discourse.

Washington Gladden's commentary also points to another vital feature of Drummond's career as a popular lecturer and author. The intimate ecology that bound together success in print and on the platform was, if anything, more apparent and crucial in Drummond's career than in those of his contemporaries. His many publications were firmly based—with the partial exception of *Tropical Africa*—on lectures, sermons and addresses delivered on multiple occasions to very different audiences.[26] This reflected Drummond's varied and extensive experience as a preacher, evangelist and lecturer. The genesis of his publications in speech very likely contributed to their astonishing success, being marked as they were by the virtues of carefully composed and rehearsed verbal performances. Drummond here was following a pattern. Many, if not most, of the widely read texts on science and religion in the period had their origins in public lectures. The American Spencerian and theist John Fiske provides one prominent example. As Bernard Lightman has shown, Fiske's *Outlines of Cosmic Philosophy* (first published in 1874) and his *Destiny of Man, Viewed in the Light of His Origin* (first published in 1884) were among the best-selling texts on evolution in late nineteenth-century America.[27] Both began as a series of popular lectures that attracted significant public attention.[28] The oral origins of best-selling books such as Fiske's were not incidental to their success, even if they were not always a necessary condition for it. A set of widely reported popular lectures not only provided publicity for authors—they also helped to fire public debate and interest in the subject matter. Getting a hold of an authorized version of lectures that had stirred controversy and piqued public interest was one good reason to buy a book. But it also forced authors to write for the ear of a listening audience in ways that could make published versions accessible and engaging.

There are many reasons to suppose that Drummond's success as an author was strongly tied in multiple ways to his reputation as a speaker. He took carefully crafting, revising and representing words spoken and then written with earnest seriousness. The influence between oral and printed text went, of course, both ways, and Drummond's spoken deliverances were shaped in part by the rhetorical and compositional techniques associated with particular genres of printed text. Drummond, for example, was intimately connected through his childhood and family ties to a venture that championed a form of evangelical publishing designed to reach and interest a truly mass readership. The Stirling Tract Enterprise, a firm run by his uncle, Peter Drummond, had, by the 1870s, produced and distributed millions of tracts designed to promote evangelical ideas and attitudes. The tract as a print medium demanded concise, accessible and arresting articulation of theological and moral ideas easily printed in a cheap and readily reproducible material format. Drummond grew up close to the heart of a leading manufacturer of this archetypal mode of evangelical print. It is perhaps for this reason that Drummond went to the trouble later in his career of personally selecting the binding, typeface and paper for his many publications.[29]

It was his book *Natural Law in the Spiritual World* that arguably secured Drummond's international reputation as a leading evangelical commentator on evolution, providing further evidence of the mutual influence of speech and written and printed text. This bestseller cannot be understood apart from the intricate mesh of connections between platform and print culture. Much of the content was first aired in a series of Sunday evening talks delivered as part of a church extension project in Possilpark, a suburb just beyond Glasgow's municipal boundary. Drummond had been charged by the Reverend Marcus Dods, the minister of the congregation where he worshipped and was an elder, with establishing a church in a community devastated by the banking crisis of 1878. His talks to "well-to-do" people now facing the hardship of unexpected poverty were designed as practical apologetics and borrowed analogies from the natural world to ground and illustrate spiritual truths and moral injunctions.[30] As Anne Scott has argued, Drummond reworked and redeployed the traditional typological interpretation of the Christian Bible that found in the Old Testament events and characters that prefigured and foreshadowed the life and works of Christ, a commonplace rhetorical trope employed by evangelical preachers in the Presbyterian tradition. Taking his cue from his theological teachers, Drummond read the book of nature typologically, finding precursors of higher spiritual principles. The continuity that he traced between laws of growth and form operating in the material and the spiritual world was only later made explicit. His original talks, while informed by the natural science he was teaching during the week to students at the Free Church College, shared much in common with a conventional kind of practical natural theology and scriptural natural history. In 1881 and 1882, they were published serially in a small and failing periodical, the *Clerical World*.

The book appeared in full a year later, published by the London firm Hodder and Stoughton. The manuscript had been rejected by two other publishers and was significantly revised by Drummond before it went to press. But the content, format and atmosphere of the book had been sub-stantially set by the exigencies of oral delivery. Within five years, seventy thousand copies had sold in Britain alone and the book received sustained and widespread critical attention.[31] While some put its success down to a favorable review in the *Spectator*, a well-respected and widely read London periodical, it was also propelled by the book's mix of evangelical tropes and evolutionary illustrations, pithily and piously expressed. Drummond was keen to insist that his book did much more than "freshen . . . the theolog-ical air with natural facts and illustrations."[32] To Drummond the novelty was in his claim that the same laws governed both the spiritual and the natural world. But he was equally insistent that readers would lose little by skipping his elaboration of this argument in the introduction and moving directly to the more applied chapters that followed. It was in these that the directness, freshness and less systematic character of his original composi-tions was most in evidence.

Drummond, then, was well acquainted with what Tom F. Wright has described as the proliferation of "new textual forms of the talking mind" that accompanied a resurgent culture of public speech in the nineteenth century.[33] This meant he was alert, too, to the problems that came with revising popular lectures into book form. He was particularly sensitive to the proliferation of unauthorized copies of his spoken and printed address-es. Throughout his career as celebrity lecturer and author, his books were pirated freely in America. To counter this, Drummond worked closely with his publishers in both Britain and the United States to ensure authorized versions were clearly marked and marketed as such. Toward the end of his career, shortly before his final illness, he won a pioneering legal challenge against an unauthorized reproduction of his Lowell lectures. Before deliv-ering the lectures, he had taken out a copyright protecting whatever book might emerge from them. The injunction also limited the extent to which American newspapers could report the lectures. The legal challenge, howev-er, was made not on copyright grounds but on the basis of the "servile" and "fraudulent" nature of the unauthorized account, based as it was on an ed-ited version of abridged reports of the lectures that had been printed in the *British Weekly*. That Drummond ended up embroiled in this court case was a natural consequence of his skepticism about the accuracy of press reports. It was also forced by the "great pecuniary injury" caused to Drummond by this form of opportunistic publishing.[34] One consequence of this dispute was that Drummond rushed his Lowell lectures to print in a way that ran against his commitment to the careful crafting of his lectures in book form. It was no wonder, then, that Drummond developed a reputation early on as someone shy of newspaper attention and wary of reporters even though, as will become clear, this was itself a way to lend popular intrigue to his public persona and reputation. This reputation masks, too, the central importance of the newspaper as a medium that popularized and promoted his spoken words, something Drummond was well aware of and, when the time was right, exploited in full.

This, of course, was not a straightforward task. Drummond's strenuous attempts to control how his spoken words appeared in newspaper accounts of them could, on occasion, compound the problems caused by the translation of vocal performance into printed text. In 1890, for example, Drummond gave several addresses designed for students at Oddfellows' Hall, Edinburgh. In them he explored themes that were later more fully developed in his Lowell lectures and other American addresses given during his 1893 tour. But a condition of their delivery was that no reporters would be present. This was Drummond's general preference for occasions he regarded as more sacred than secular. Denying access to reporters, however, brought its own difficulties. Reports of his talks ended up being based, as Drummond's biographer put it, on "bits torn from their context by a young prig or two in his audience and flung to the rapacity of certain of the lower class of religious papers."[35] These then became ammunition for his religious critics, who cited the same reports as evidence of his heterodoxy. Drummond, like Huxley, Tyndall, Proctor and Wallace, could not entirely prevail in his struggle to benefit from, but at the same time control, press attention. Allowing full access risked piracy, plagiarism and pecuniary loss. Restricting access could lead to damaging misrepresentations. It could also mean more, rather than less, unwelcome press interest in his talks or, on other occasions, missing out on a major source of publicity.

In some respects, Drummond's career as celebrity speaker was unique. On the platform, he performed the part of itinerant evangelist, explorer, scientist, theologian and sportsman (fishing, hunting and cricket especially) and gained popular applause for his prowess in all of these roles. He also played the private gentleman in an intensely public way. The fascination all this generated helped propel him to celebrity status in Britain, America and the other parts of the world where he sermonized about science. In contrast to Huxley, Tyndall, Proctor and Wallace, he did not emphasize one of his many social roles above others. He saw all as beneficial to his mission—to evangelize the "outsider" and skeptic, drawing them to Christianity by translating and transforming that faith into the language and cultural forms of their world, not least by transmogrifying it into scientific and evolutionary terms. Yet at the same time, as outlined above, Drummond shared much in common with the celebrity British science lecturers who had toured the United States before him. He aimed not so much to impart information as to redirect the affections and the beliefs of his audiences toward a religiously meaningful scientific outlook. Inspiration was at least as important as instruction. Drummond also adopted a similar stance and style as his predecessors, even if modified and made to suit his own predilections and public persona. He wrestled, too, with dilemmas and demands that others before him had faced. That he did this with arguably greater success than any apart from perhaps John Tyndall some two decades previously both suggests the ongoing importance, vitality and influence of speech on the cultural relevance and authority of science and highlights some of the changes that had occurred in American lecture culture by the 1890s.

By the time Drummond toured the United States in 1893, his worldwide reputation as a scientific evangelist and an evangelist for science had reached its highest pitch. The lecture platform was now a crucial technology for achieving this overriding mission. His reputation as a must-have speaker had been building for at least a decade and he had long been the recipient of constant invitations to lecture at home and much further afield. This was, as we shall see, Drummond's second tour of the United States and came on the back of other international lecture tours. It is important, then, to briefly review these earlier tours as an important component of Drummond's global reputation as a celebrity lecturer.

DRUMMOND ON THE WORLD STAGE

The publication, and publishing success, of *Natural Law in the Spiritual World* undoubtedly helped propel Drummond to fame. His name, and his presentation of Christianity using the language of evolution, became widely known in the United Kingdom, North America, Australia and New Zealand and in several European countries. As a speaker, however, Drummond's geographical range had remained relatively local. He had become a well-known figure on the platforms of public halls across Britain and Ireland, not least through his intense involvement with Moody's campaigns in 1873–1876 and again in the early 1880s. His two series of lectures delivered at Grosvenor House consolidated his reputation and made his voice familiar to a social elite. It was, however, Drummond's participation in a movement among university students in Edinburgh that provided the main impetus for lecture tours first in the United States and Canada and then in Australia (with stops in Hong Kong, Singapore and Japan).

Moody's invitation to come and speak in the United States was a long-standing one. He was anxious on several occasions to involve Drummond with his campaigns there. When Moody embarked on his work with university students at Northfield in upstate New York, Drummond was again approached to assist. It was Drummond's commitment to the student movement work in Edinburgh that, at first, kept him at home. That initiative, begun in 1884, had been inspired in part by the visit to the city of

Charles T. Studd and Stanley P. Smith, two Cambridge University students dedicated to foreign missions. Drummond was among the first invited to support the movement's efforts to evangelize and Christianize students and quickly became a key supporter. His devotion to the cause was among the chief reasons why Drummond decided in 1887 to travel to the United States.

Drummond sailed on June 11 that year and his initial speaking engagements were at Moody's Northfield student convention and at two Chautauqua summer camps. Later in the summer he delivered an evening lecture on the botany and zoology of Central Africa during the annual meeting of the American Association for the Advancement of Science. He saw this and other occasional science lectures as a way to financially and intellectually support his chief aim, launching the student movement in North America. In late summer that work began in earnest. Four coworkers from Edinburgh—Alexander Russell Simpson, professor of midwifery at the University of Edinburgh; William Smith Greenfield, professor of pathology at the University of Edinburgh; Dr. G. P. Smith, secretary of the movement in Edinburgh; and two medical students, J. C. Webster and George Purves Smith—joined Drummond on a tour of campuses in New York, Boston, Toronto and elsewhere. Among the colleges, Drummond spent time at Amherst, Columbia, Dartmouth, Princeton, Smith and Williams. His university stop offs included Harvard, Pennsylvania and Yale. He also visited Hartford Theological School, Union College and various medical schools in New York.[36] On nearly every occasion one or more of his fellow student movement enthusiasts joined him.

This first American lecture tour certainly attracted significant press attention and exposed Drummond to America's fast-paced and diverse lecture culture. His Chautauqua addresses were heard by thousands of summer campers who gathered to listen to the author of *Natural Law*. These talks were important not only in themselves. They also allowed Drummond to become a star speaker at the summer camp meetings at Chautauqua, one of the latest institutional expressions of the American love affair with the public lecture. In many respects, Drummond was made for this manifestation of the middle-class appetite for engaging formal talk on the issues of the day. His manner and message harmonized well with the rhetorical situation created by the progressive Protestant ethic that so strongly informed the Chautauqua movement.[37] With origins in Methodist summer camps in the 1870s and its popularity with vacationists looking for a mix of nature study, social gospelism, entertainment and intellectual stimulation, the movement found in Drummond an exemplary embodiment of its general appeal. Indeed, *Natural Law in the Spiritual World* was among the important influences on the movement's cofounder, John Heyl Vincent.[38]

In his first exposure at Chautauqua in 1887, and when speaking to students at Moody's Northfield student convention on the same tour, Drummond explicitly appealed to natural law and evolution to illustrate and ground the character of Christian life.[39]

During his first American tour, Drummond's overall strategic aim meant that, apart from his talks at Chautauqua, his audiences were largely composed of students. To keep those occasions as private as possible given their religious purpose, he worked hard to restrict press coverage. The reports that did leak out demonstrate that his intent was to reach students skeptical of Christianity and alienated from its more traditional forms. When he spoke at the colleges, his message was more directly evangelistic in tone and import. Evolutionary tropes were less conspicuous, though Drummond did argue for a Christianity repurposed for an age of science.[40]

Drummond's many addresses in and around Boston, New York and Toronto undoubtedly left an impact, evidenced by a large volume of personal correspondence, follow-up meetings and the forming of student organizations that borrowed from the Edinburgh model. It also provided the opportunity for Drummond to deliver at Northfield a version of his "The Greatest Thing in the World," an address that Moody would continue to promote before and after its publication in 1889. In these ways, it laid the groundwork for the unprecedented attention Drummond received when he returned to the United States in 1893. Before then, the student movement work opened further opportunities for foreign tours, the most notable of which took Drummond to Australia.

The Australian tour had its immediate origins in an invitation to Drummond from over two hundred Melbourne University students. This time Drummond conducted the tour alone and restricted his energies to semiprivate talks delivered to student audiences in colleges around Australia.[41] His efforts to prevent reports of the meetings appearing in local newspapers were even more effective than during his American tour. In a letter written while in Australia, Drummond noted that he could "send no reports, as I have discovered how to circumvent the Press and have succeeded most effectually everywhere."[42] His message was broadly the same as the one he had brought to the United States. He promoted a work among students that would bring many to faith in a Christianity free from an overweening concern with doctrine and couched in scientific modes of thought and practice. He aimed again at those more distant from Christian commitment, and at students who had succeeded either academically or in sport. He was after the "right kind of men" and appealed for a muscular Christianity open to change, speculation and social action.[43]

In the years leading up to his second lecture tour of the United States, in addition to becoming familiar with the mechanics of speaking tours in parts of the world that had cultivated vital and elaborate machinery for supporting public lectures, Drummond talked about and wrote much of the content of what would become his book *The Ascent of Man*. This provided him with a more fully formed argument about the unity that existed between the doctrine of evolution and Christianity. More than before, Drummond was committed to lead individual hearers and readers to a sympathy for evolution as much as to inspire them with the religious vision that he weaved into his presentation of the evolutionary science of his day. With his rough script in hand, Drummond set off once again for America to proclaim the gospel, now fully reformulated in the language and logic of evolution.

AN EVANGEL FOR EVOLUTION AT HUNTINGTON HALL

In 1890, with Drummond's reputation riding high, Augustus Lowell took the opportunity to invite him to deliver a series of Lowell Institute lectures. When he received a reply to say Drummond was touring Australia, he bided his time, writing again in March 1891 to reissue the invitation. With all the interest in the journalist-explorer Henry Stanley's imminent visit to the continent, Drummond might "make Africa his subject." But Lowell gave him free range, should he prefer to "select some religious theme."[44] The ideal number of lectures would be twelve, with a payment of $200 for each one delivered. Drummond was in no doubt of his theme. He had been working on it for several years. His title, provided to Lowell, was to be "The Evolution of Man."

Once advertised, the rush for tickets to hear Drummond was unprecedented. Despite the long history of the venerable and popular lecture series, Drummond's evidently broke all records. According to one source, for every ticket purchased, ten hopefuls went away disappointed.[45] It was also reported that a "ring of speculators" had bought up tickets that were then sold at "fabulous prices."[46] Interest in the lectures resounded across a broad swathe of Boston society. One report recounted how "clerks and shop-girls, labouring people of both sexes, went without their support in order to secure a favourable place in the long line." The audience that pressed into Huntington Hall cut across at least some class lines, drawing together "a mixed assemblage of the best elements of the rank and file of our population."[47]

Facing such interest and such a mixed audience, Drummond immediately set about rewriting the script he had brought with him. Holed up in Hotel Brunswick, Drummond reworked his lectures to accommodate the clerks and shop girls.[48] The strategy seemed to work, and Drummond found himself addressing a capacity audience (the hall could seat nine hundred), uncomfortably squeezed into a venue too small to meet demand.[49] Halfway through the course, Drummond was asked to offer a matinée performance, explaining to his mother that "the crowds have been so great and so many turned away that I have been asked to re-deliver the remaining six lectures—repeating the one of the previous evening at 3 in the afternoon of the following day."[50] Even this was not enough to satisfy the Boston public. Writing to his mother a week later, he observed that "my afternoon audience is just as big as the evening ones and I fear a good many were unable to get tickets for the second edition also."[51]

Drummond's opening lecture began with a celebration of the truth of evolution, "the story of creation." The fact of evolution, or that "vision which is revolutionizing the world," was fully established. Its "strength, its value, and its universality" was fully confirmed by science. It was yet more remarkable that the "doctrine of development" had gripped the minds of the leading thinkers of the age. Drummond's task in his Lowell lectures was to give, he claimed for the first time, an outline sketch of the entire drama of evolution.

It was also, however, "the age of the evolution of Evolution."[52] The world of science was "hot with controversies" over the causes of evolution. Wallace stood against Darwin on the subjects of sexual selection and the origins of the human intellect. Weismann opposed "Darwinian assumptions of the inheritability of acquired characteristics," offering his "Germ-Plasm" in the place of "gemmules" or "primordial units." In Drummond's assessment of these debates, Darwinism, kept within certain limits, may yet turn out to be true. But he was anxious to warn his audience of a significant threat. Darwinism applied to the "great moral facts" of human history would prove "false and pernicious."[53] The problem lay in trying to account for all of nature "from beneath." The solution was to begin with the "final product" and approach nature through "man drawn to scale."[54] Already Drummond was addressing in arresting words and phrases a topic—what it means to be human in light of evolution—that was a matter of widespread public fascination.

The second half of Drummond's lecture moved from these more dramatic claims to a detailed account of human embryology. The developing embryo retold the story of evolution in miniature, wandering as it did "among the ghosts of departed types."[55] Drummond qualified his otherwise exuberant embrace of ontogeny as an accelerated version of evolutionary history by noting that human embryonic development did not recapitulate all evolution and that its precise features had not yet been fully described. It was clear, however, that the long history of evolution, organic and inorganic, was swept up into the production of its final end. The origins of humanity could be traced as far back as the breaking up of suns and the cooling of planets. These ancient and primal origins, far from degrading humanity, exalted it and gave it a meaning more sublime than any other possible method of creation. All of nature pointed toward the "evolution of man." It was Wallace's message all over again, but in a more distinctive and compelling key. Using many of his techniques of effective communication, Drummond was staging himself as the evangelist of evolution as much as he was using evolution to recuperate and reenvisage a broad Christian creed.

In his second lecture, again delivered to a packed Huntington Hall, Drummond began with an account of Louis Robinson's experiments demonstrating the persistence in human infants of the reflex grip found among other young primates. An image projected on the screen showing a baby hanging from a cane "had all the appearance of a monkey in its aspect."[56] The image was supplemented by descriptions, borrowed from Alfred Russel Wallace, of the powerful grip of a young orangutan. Drummond's primary focus, however, was not shared characteristics between primates but rather the limits and "terminal points" of evolution. Natural history "was not always a becoming. Sometimes things arrive."[57] Evolution had its ends or organizational "fixtures" that can be seen throughout the living world. The most dramatic was the human body, "the highest branch of the tree of life."[58] Here Drummond explicitly echoed the philosopher John Fiske's claim that humans were the most exalted living form. And it was not only the highest but had made possible a new phase of evolution directed not by nature but by human intelligence. The result, paradoxically, was the atrophy of the human body—of which weaker sight, hearing and musculature all bore witness—and the body's subservience to mental operations. This, Drummond declared, opened the way for "a new and inconceivably loftier order of being."[59]

As well as garnering significant press attention and a packed hall, Drummond's opening talks also drew the eye of one of America's most celebrated and shrewd lecture agents, James Burton Pond. Writing to Drummond on April 11, Pond offered Drummond $5,000 for a lecture tour in the autumn on the condition that he give no free lectures over the summer months. Pond noted that he had worked with Matthew Arnold, Charles Kingsley and Charles Dickens among others, thus placing Drummond among some of the most celebrated British speakers on America's lecture circuit.[60] The autumn tour never materialized but the offer must have been a boost to Drummond and provides another indication of his popularity.

Public fascination with Drummond was in evidence once again when audiences gathered for his third Lowell lecture. His auditors listened with "riveted . . . attention" as he circled back to the vestigial structures in the human body that provided clear evidence of an evolutionary history.[61] Ears, arm hair, the appendix and remains of a tail were vividly described to show that without evolution there was "no possible explanation" for the redundant features of the human body.[62] While Drummond worked hard to correct mistaken assumptions and put forward a strong scientific case for evolution, this was not, finally, his ultimate concern. As he put it to his hearers, the physical evidence of the evolution of the human body was "of importance only in its higher implications." It was the "dissection of the soul" more than the body that mattered. At this point, Drummond brought before his hearers an issue more often associated with theology than biology: the problem of evil. If the physical traces of animal ancestry could be found in the human body, so too could traces of animal nature be found in the moral nature of humans. These were "infinitely more real, because their muscles are not yet atrophied." In typically aphoristic mode, Drummond declared that "the problem really is, not how sin came into the world, but how to get it out." A precise science of sin was needed to overcome the evil within, premised on the assumption that "man's present amphibious life cannot be final."[63] The discussion of the problem of sin was omitted from the authorized published version of the lecture, but this was classic Drummondism and injected spiritual concerns into a grand evolutionary history of humankind.

Drummond's dramatic topics and style of address immediately give inspiration to headline writers and caricaturists. Although Drummond carefully distinguished the claim that humans had descended from apelike creatures from the mistaken account of evolution that pictured humans emerging from existing primates, Boston's dailies, predictably perhaps, made much of the monkey-man connection. *Boston Daily Globe*'s headline for Drummond's third lecture proclaimed, "Man, Child of the Monkey," while the *Boston Evening Record* published a cartoon of his audience with simian-like ears, eyes and tails.[64] These now hackneyed journalistic images of evolution received new life, coming from the mouth (it was supposed) of a celebrity evangelist.

In part spurred by all the press attention, the appeal of Drummond's lectures showed no sign of diminishing. When he stood up to deliver his fourth lecture on "the struggle for life," folding chairs and stools filled the aisles. Perhaps sensitive to observations by journalists that he could not be heard in all parts of the hall, he spoke "more loudly and distinctly than on any previous occasion."[65] This struggle to adapt to the immediate environment was redolent of the message he delivered. The struggle for life was necessary for progressive evolution and was an essential force for achieving nature's final end. If it was only one factor in evolution, it was nevertheless an essential one. But it had been misunderstood as a "vast murderous machine for the annihilation of the majority."[66] The undeniable price of the survival of the fittest could be justified when the history of evolution was viewed as a whole. Struggle was a necessary stimulus for progress and was the source of moral character as much as technology. War had been the manufactory of "all the arts and highest virtues of man."[67] But the time for war had now passed and in the future a more refined struggle would continue in the moral sphere. As Drummond later expressed it, the incessant struggle for life had laid a foundation for a "nobler superstructure" and hinted at a "profounder chemistry," that of the "Struggle for the Life of Others."[68] But in the lecture as delivered, he held back from announcing what became in final print form the central argument of the whole series. These decidely ethical enunciations, superadded to grand visions of evolutionary history, were clearly playing well in Boston and worked together with Drummond's unique delivery, designed as it was to speak directly to the moral and affective "centers" of each auditor.

Having delivered four lectures of his Lowell series, Drummond devoted two intervening evenings to addressing students at Harvard College. Here he largely left off his sweeping account of evolution and appeared as the darling of students dissatisfied with the stagnant and outmoded Christian faith of older generations. The first talk, delivered to an overflowing audience in Appleton Chapel composed as much of curious members of the public as of staff and students, considered "the outsider." Drummond appealed to his hearers to avoid "cant," "exaggerated feeling" and religious superciliousness in order to win back those who had turned away from Christianity. Christianity, he argued, did not consist first in creeds but in a willingness to be a student of Christ. Drummond also batted off a couple of common objections to the Christian life: first, that it was dull, and second that it clashed with scientific truth. In dealing with the latter, Drummond argued that Thomas Henry Huxley "had never said one word against Christ or against the words of Christ" and that the celebrated physicist William Thomson opened his lectures in prayer.[69] Speaking the following evening in Boylston Hall—"crowded to suffocation with a representative college audience"—Drummond explored other reasons why young men in particular drifted from a commitment to the Christian life.[70] The tedium of poor sermons and an apparently irrelevant Bible were chief among them. Against these, Drummond defended churchgoing and Bible reading, pointing out the good in both. Finding the right parts of the library of books that made up the Bible was essential. The Gospel of John, for example, was "better than Emerson."[71]

The interest in Drummond's Lowell lectures continued to increase as the series moved toward its halfway point. When he delivered his fifth lecture, on "the evolution of mind," Drummond had to share the platform with members of the audience who could find nowhere else to sit or stand. Some of them "swung [their] heels from seating postures all around the edge of the platforms," while others filled up the stage itself.[72] After a burst of loud and enthusiastic applause, Drummond addressed the packed hall on the subject of mental evolution. Drawing on the work of the zoologist George Romanes, Drummond, even while admitting the unique capacities of human cognition, declared mind a product of gradual evolution. For the famous coworker of Moody, this was a bold move. Even placed alongside the efforts of coreligionists making peace with evolution, this was striking. Many, following the Catholic anatomist and zoologist St. George Mivart, treated the human intellect an exception, as something infused directly by God.

Drummond made his controversial case by drawing on four areas of inquiry—child development, mind in lower animals, the artefacts of "primeval" humans and the contemporary lives of isolated "savages." A child's mind, that "subtler embryo," recapitulated the order of mental evolution.[73] Traces of complex emotion and intelligence in animals demonstrated not only continuity with humans but also the order of mental evolution. The most primitive animals experienced fear, while "further up" emotions such as jealousy, sympathy, grief and shame could be found. A similar order could be discerned as the traveler passed from one Pacific island to another— from ones occupied by the most "degraded" savage to others demonstrating increasing levels of "civilization." As Drummond spoke of the evolution of emotions, he stirred his audience's feelings, keeping them entranced by the manner and message of his lectures. In this, he was again acting as an ardent and mesmerizing advocate for evolution (including some noxious racial views, though these were not immediately noticed), fully embracing the discoveries of leading theorists while also clearing space for a spiritual rendition of those scientific truths.

Drummond concluded this pivotal lecture with an encomium later incorporated into the final chapter of *The Ascent of Man*. In it he warned his hearers to resist the inclination to argue for a "special interposition of a creative hand" to account for what seemed to be beyond the reach of an evolutionary explanation. This habit of thought risked denying the truth of the continual "sustaining of all living things" by God. It was, Drummond announced, a "mistaken policy to glory in what [evolution] could not account for." God must not be "left to the gaps in our knowledge."[74] Evolution was the method of creation and God upheld all its operations.

One of the most dramatic "gaps" between humans and the rest of nature formed the starting point of Drummond's next lecture. It was clear that there was an extraordinary difference between the "brain" and the "intelligence" of humans and that of their "nearest ally" among animals. There was no need, however, to appeal to divine interposition. There was, after all, "a wonderful unanimity among men of science as to the answer."[75] This lay in the gradual emergence of ever more complex language, a form of communication that provided humans, uniquely, with the ability to acquire skills and knowledge with unprecedented rapidity. The rudiments of language could, Drummond suggested, be found among other animals. Gestures, noises and intonation provided a system of communication among cooperating animals that had obvious survival value. To this was added the imitation of sounds, the basis of the first words. Combined with the necessary physical developments of body, humans were able to develop intelligent speech, a combination of kinesics and utterance embodied by the lecturer before them. This composite technology of communication not only provided a necessary condition for rapid mental evolution but also pointed to the unexplored potential of human intelligence. Drummond closed with the thought that evolution had by no means ceased. The instruments of mind and language would bring about "far greater results" than what had been so far produced.[76] It was conceivable that after the telegraph and telephone, "the next stage in Evolution is telepathy."[77] The potential for this future evolution could be heard and seen in Drummond's own performance, as medium and subject matter merged.

With such sensational claims left hanging in the air, Drummond's popularity continued unabated. The *Boston Daily Advertiser* inserted a note to readers that Drummond's lectures would be repeated the following afternoon. On Sunday, many hundreds turned away disappointed from another student talk at Harvard's Appleton Chapel on discovering a notice announcing that his evening address was not open to the public. Instead, the chapel was jammed with students who had been unable to hear him the previous weekend because "every inch of space was taken by eager outsiders." Drummond's subject was temptation. Echoing his Lowell lectures, Drummond reminded his audience that they contained "remnants of animal bodies [and] the residuum of animal nature." Just as "the struggle for life" was key to organic evolution, so the fight against temptation was crucial for developing moral muscle. Drummond closed his address by stressing the importance of moving beyond the merely animal. The material body, the finished product of evolution, was a dead thing and "every man" was to "rise above himself, and be a man."[78] All this was evolution in action and the best kind of practical and manly Christianity.

Drummond's seventh Lowell lecture started on a similar "higher plane." The body and the mind, the evolution of which he had discussed in previous lectures, hardly accounted for the whole human constitution. Physical desires, and the pursuit of learning, did not finally satisfy. Instead, "man reaches his full height only when love becomes breath and life." That did not mean that love had not evolved. It was, in fact, "distilled on the earth" and "its roots began to grow with the first cell of life which budded on this earth." It was the evolution of sex that, for Drummond, provided the opening chapter in the emergence of love. It had, Drummond noted, "forced association, combination and fellowship upon the world."[79] This was not love but was nevertheless "indirectly necessary to it." Love, sex, evolution: Drummond showed once again that he knew how to orchestrate a talk to create a symphonic sermon harmonizing science, humanity and religion in a way that evoked the full spectrum of human emotion.

The later parts of Drummond's lectures dealt with the differences between the sexes. If the two forces in the evolution of life were the quest for nutrition and for reproduction, the male sex was orientated toward the first and the female toward the second. The first was individualistic, and the second "other-regarding." Summarizing recent breeding experiments with tadpoles and aphides, Drummond suggested that nutrient-poor environments encouraged a greater proportion of male offspring. When access to nutrients was easy, the less outwardly active females thrived. This, in part, lay behind the tendency of men toward warmongering and the "ardour of the chase" and reflected the orientation of women toward the domestic sphere.[80] At the close, Drummond offered a "eulogy of the civilizing effects which woman exerts upon restless man."[81]

Normative assumptions about gender strongly colored Drummond's characterization of the sexes in nature, largely borrowed from the eclectic evolutionist Patrick Geddes and biologist J. Arthur Thomson. At the same time, the priority Drummond gave to the role of maternal sacrifice in evolution and the argument he made about motherhood as the pinnacle of evolution contrasted to a more familiar reading of Genesis that emphasized the subordination of the first woman to Adam and the role of Eve in leading Adam into sin. Drummond did not mean to restrict self-sacrifice on behalf of others to women alone, as if men too were not called to this higher form of living. Sacrificial love reached its fullest natural expression in motherhood but that was but a foretaste of the destiny of all: a higher life directed by altruism. The appearance in nature of the mother made possible, and pointed toward, the greatest of all forces, that of love. Drummond made no mention of female subordination or seduction. Instead, he repositioned Eve, the mother of all the living, as the prefigurement of the final end of humanity. Maternity, as he later expressed it, "was the mother of Love itself."[82]

Drummond's message was welcomed by at least one prominent Bostonian. After hearing him talk on the evolution of sex, Lilian Whiting, a freelance journalist and connoisseur of Boston high society, produced a glowing profile of the Lowell lecturer. Rather than critique Drummond's characterization of the sexes, Whiting hailed him as "a great spiritual leader" whose performances at the Lowell Institute had "rivaled those of Paderewski [a Polish pianist] and Mme. Duse [a celebrated Italian actress] in point of enthusiasm and crowds and intensity." Whiting anticipated that

Drummond's lectures would "grow more and more intensely interesting," confirming his position as "appointed prophet" for reconciling science and religion. While Whiting stressed the appeal of Drummond's writings, she noted that "his charm of manner can hardly be translated into words." No wonder, then, that "the great masses of the people [were] pouring forth" to hear him.[83] This recommendation captured well Drummond's abilities and ambitions as a lecturer and confirmed his status as a celebrity as well as prophet. Drummond was playing the game of famed lecturer well, performing effectively alongside other enthralling sounds and voices that drew Boston's citizens to the city's entertainment quarter.

Drummond's eighth lecture, on the "evolution of a mother," brought further confirmation of his popularity. In his meditations on the theme, Drummond's arguments reached another high point. The evolution of the mother was, he declared, the "chief thing" nature had done. Evolution had aimed toward motherhood and "the goal of animal nature was the creation of a family centred about a mother."[84] Motherhood was born of an earlier maternity that provided the material resources for independent offspring but did not involve prolonged care of young. Four developments were necessary before motherhood proper could appear. The number of offspring had to reduce to one or two, the embryo had to develop to a point of resemblance to the parent before appearing, the young had to be dependent on the mother for survival and the mother had to be equipped to nurture and protect her progeny. The latter two factors in particular marked off humans from other animals, with children requiring a long period of nurture, protection and education. All was based on physiology but "in all these preparations nature had an ethical end."[85] What scientists called "lactation . . . moralists call[ed]love."[86] With this claim, Drummond gave another specific instance of his general principle: nature and spirit were one and love infused all.

Of course, in turning to the evolution of sex and of motherhood, Drummond had trespassed on a vigorous public debate about gender. As a comment in the New York *Sun* put it several days later, "a large proportion of the great intellects of Boston belong to women." It was not surprising that, having "laid down the law that the difference between the sexes is fundamental and eternal: that man is all activity, energy and enterprise and women is all for peace and repose," Drummond "brought down on his head a storm of reproach and logic."[87] When asked by a Boston journalist to respond to this claim, Drummond laughed it off as a "jest." There had, he suggested, only been one letter published in a newspaper about his lecture. He had merely argued that "certain physiological characteristics" distinguished the sexes in nature. Had he talked about "the brain" he would have argued the case more strongly "but with obvious qualifications."[88]

This was not likely to satisfy the letter writer that Drummond may well have been referring to. On the day after his lecture on the evolution of the mother, the suffragist and journalist Ellen Battelle Dietrick wrote to the *Boston Evening Transcript* complaining that Drummond had offered an idealized vision of motherhood that was not, in fact, rooted in the natural world. Citing female animals that were anything but altruistic or stereotypically maternal, Dietrick argued that Drummond "wanders far from discernment in teaching that it is natural and excellent for the human race to divide itself either in its ideals or in its travel toward those ideals."[89] Dietrick had heard Drummond's lecture as a false naturalization of the gender stereotypes that undergirded the deadening and exclusionary effects of a patriarchal public culture.

This was not, however, the only way to interpret Drummond's message about the evolution of women. When Drummond looked back over evolutionary history, he saw not fully formed and fixed laws by which human relations might be governed, but prototypes of ethical life. Evolution in physiological terms may have arrived at its main end point—the evolution of the mother—but spiritual evolution had by no means run its course. The ethical lessons of motherhood, and of the child, provided a universal tutelage and supplied a basis for a future world marked by a higher life of sympathy and self-sacrifice that was not necessarily gendered in the way that Dietrick abhorred. That Drummond could be heard in these different ways, as a preserver of religiously sanctioned gender roles or as a bold prophet of a more egalitarian future, was one consequence of the line he walked as evangelist of evolution.

In the two lectures that followed this celebration of motherhood, arresting themes and arguments were less in evidence. The lectures dealt with the early appearance in nature of family life. The first was subsumed into the chapter in *The Ascent of Man*, entitled "The Evolution of a Father" and the second remained unpublished. Boston's newspapers did not give them the same attention as the preceding discourses, with only shorter summaries provided for readers interested in keeping up with the whole series.

Sandwiched between his two lectures on the evolution of the human family, Drummond visited Wellesley College to address its female students about the "difficulties of the Christian life." It was, he confessed, "out of his beat" to speak to "young ladies," and he admitted he had not spoken to an all-female audience since his last visit to the college nearly six years previously. His message to the college was delivered in a conversational style and repeated some of what he had said at Appleton Chapel weeks before. Among the subjects discussed was that of doubt, a condition Drummond described as a "prelude to truth." In the closing passages of the lecture, Drummond dealt with the "application of the theory of evolution to spiritual life." It was this that "formed such an important part of his message to the world." It was true, after all, that "Christ's whole aim in life was to evolve it," and the way to a "high life" involved "yielding" to the prompting of His Spirit.[90] As with his other lectures to students, this talk, among other things, provided a way of hearing and reading his Lowell lectures that brought them into alignment with more explicitly spiritual aims and longings. It also advertised what would be the parting message of his Lowell series.

The following evening Drummond returned to Huntington Hall and delivered his tenth Lowell talk. Both the *Boston Globe* and the *Boston Evening Transcript* noted the presence of Joseph Cook, the celebrated Boston preacher, in the audience. Cook, the *Globe* reported, sat for much of the lecture with "his eyes closed, having the appearance of a man drinking in beautiful thoughts." Whether or not the lecture had soporific effects, its subject matter was not without interest. Beginning with "a primitive tribe of 1000 men," Drummond sketched the evolution of human variation, emphasizing the role of "geography and geology."[91] The emphasis, according to the reports, was on divisions of labor—between fishermen and shepherds for example—rather than racial difference. Perhaps borrowing from his Scottish compatriot Patrick Geddes's famous valley section (the similarity was not likely to have been accidental), Drummond's focus on human occupations largely avoided the subject of race. The keynote was spiritual evolution. In closing, echoing the thesis for which he was famous, Drummond reminded his audience that "the spiritual fluid and electrical fluid are under the same law." The way "man gets bread is the practice by which he gets soul."[92] Through such epigrams, Drummond impressed on his audience the truth that nature, not "religion," was the origin of human goodness.

It was this theme that provided Drummond with the basis for the sixth and final week of his Lowell lectures. It was now approaching the middle of May and Boston was experiencing a spell of hot weather. Yet the hall remained crowded, as Drummond declared evolution and Christianity to be "one and the same."[93] The lecture on the Tuesday evening investigated the principle of cooperation that gave organisms such advantage in the struggle for life. Mutual aid in plant and animals led on to human progress and civilization. At his most overtly racist, Drummond remarked that Africa "was a waste and a wilderness . . . because the Africans have not acquired the principle of mutual aid." In the "higher nations" cooperation had become a necessity that betokened a universal peace. There was, at the same time, a darker hue to Drummond's prognostications. A "spirit of competition" continued to exert powerful influence and "the final struggle between self-ishness and unselfishness" was the "most significant episode in modern history."[94]

In his final lecture Drummond stepped forward again as evangelist and proclaimed Christianity to be that "further evolution" that was the future of the world. The grand vision of evolution heard by so many of Boston's lecture-loving public was now heralded as a harbinger of Christianity. It was impossible for anyone to "follow up evolution and not come to Christianity."[95] This was because Christianity was itself born of nature and all that it claimed had its origins there. It was not religion or dogma but "self-sacrificing love" that was the ultimate end of evolution. None of this was ethereal. The "Kingdom of God" was simply "society's best men working for the best end with the best methods."[96] The "worship of God" consisted of love and service of others.

These closing thoughts drew admiration. An editorial published by the *Boston Daily Advertiser* several days later applauded Drummond for his approach to evolution and Christianity. Unlike other attempts at reconciliation, Drummond had boldly preached an essential unity that made harmonizing schemes unnecessary. He had not spent time trying to find room for divine agency, assuming instead that evolution was itself infused with divine intent. In securing Drummond, the Lowell Institute had "done a splendid public service." The "multitudes" that had not been able to acquire tickets now waited eagerly for the publication of the lectures.[97]

On May 14, two days after Drummond had concluded his Lowell lectures, he delivered two addresses, one in the morning at the church at Amherst College and one in the evening at nearby Smith College. Reports praised Drummond's discourses on temptation (at Amherst) and on obstacles to the Christian life (at Smith). The lecture on temptation was a repeat of the one given at Appleton Chapel and the talk at Smith dealt again with a subject covered at Harvard and Wellesley. Drummond, perhaps, had decided that the difficulties of religious doubt and of a "dull" Bible were suitable topics for women's colleges and that "temptation" and the more decisive rejection of Christianity by agnostic students better served audiences largely composed of young men. In any case, his college talks continued to act as adjuncts and informal guides to his more serious and substantial addresses at the Lowell Institute.

By any measure, Drummond's talks in the Boston area were an undeniable success. This brought, among other things, financial reward. The college talks did pay, but the Lowell lectures proved especially lucrative. Doubling up his lectures had guaranteed an even bigger fee than expected. Drummond, showing himself a shrewd operator on the lecture circuit, took the opportunity to invest the money in a magazine. Samuel McClure, founded of the McClure Newspaper Syndicate, had come to Boston to reacquaint himself with Drummond after an earlier meeting in Columbia in 1887. On learning of McClure's desire to establish a new illustrated month-

FIGURE 5.1. Photographic portrait of Henry Drummond, *McClure's Magazine*, 1 (1893), frontispiece. Reproduced by permission of the National Library of Scotland.

ly, Drummond bought $2000 worth of stock and loaned McClure a further $1000.[98] The first issue, released on May 29, included Drummond's photograph as the frontispiece and an article by him entitled "Where Man Got His Ears," a piece lifted from a section of his third Lowell lecture.[99] While the circulation of the inaugural issue was not particularly large, Drummond's contribution was widely advertised in newspapers and contributed still further to his ever-rising public profile and celebrity status.

This rising profile did come at something of a cost. In the pause between Drummond's speaking engagements, one Sunday newspaper positioned him among the leaders of the "new theology" and defended him from those that dismissed him as a rank heretic. Yet placing him among such "heretics of today" as the biblical critic Charles Briggs, the theologian Frederic Farrar and the preacher David Swing did nothing to endear Drummond to his conservative critics in America.[100] As Drummond turned from his secular Boston audiences to speak to a more religious constituency about evolution, the challenges of persuading evangelicals that evolution was true and no threat to their most deeply held convictions began to loom on the horizon. His calm and chaste style of delivering daring conjectures about evolution, which served him so well in Boston, would face significant resistance by others more sensitive to the apparent implications for religious doctrine and dogma. This, however, was only part of the story, and on many occasions his appeal and popularity, if anything, experienced a surge, drawing crowds of ten thousand or more for a single talk.

EVOLUTION FOR EVANGELICALS

While he continued to be discussed in newspapers and periodicals, Drummond, although inundated with invitations, kept his diary clear of speaking engagements. Toward the end of May he traveled to Chicago to visit the World's Fair and to meet his friends, Lord and Lady Aberdeen (John Campbell Hamilton-Gordon, lord Aberdeen, had just been appointed governor general of Canada). It was during his visit to the fair that Drummond received an invitation from a group of Methodists to come and lecture in Minneapolis and St. Paul.[101] Perhaps because it fit into his travel plans (which included, from the middle of June, a fishing trip), Drummond accepted and delivered four lectures on the evolution of man to audiences in the Twin Cities during the week beginning June 5.

An introductory lecture on human evolution was followed by lectures on the evolution of the mother, the father and of Christianity (that "further evolution" of human society). The First Baptist Church of Minneapolis and the Ford Hall in St. Paul filled with attentive listeners to hear "one of the most remarkable men of the age."[102] Securing Drummond was a matter of civic pride for Minneapolitans.[103] After an impromptu talk on June 11 at the YMCA rooms in Duluth, the large port city situated at the westernmost point of the Great Lakes, Drummond left to pursue his favorite hobby.[104]

Drummond's departure from Minneapolis and St. Paul did not signal the end of his influence there. In the fortnight following his visit, local divines pronounced on Drummond's lectures and used them as an opportunity to ensure their various congregations were not led astray. The Reverend F. O. Holman, minister of Hennepin Avenue Methodist Episcopal Church, used his Sunday morning sermon to correct possible misunderstanding of Drummond's arguments.[105] While Holman praised Drummond's "poetic genius" and ability to speak on scientific subjects with "perfect clarity" (and in this he captured well Drummond's reputation as a speaker able to speak "scientifically" and "poetically"), he worried that those who had heard Drummond's final lecture had gained the impression that Christianity itself was a product of evolution. Holman himself had "listened as carefully I think as most people and I gained that impression myself from what he said. And yet I know that it was impossible that he meant to convey that impression: for Mr. Drummond is a Christian and he believes (and I am speaking authoritatively), that Christianity is a direct introduction on the part of Almighty God of a new impulse which is for the development of the human race." So much, it would seem, for Drummond's "perfect clarity." It was up to Holman to say what Drummond would and should have said had he had more time to develop his otherwise ambiguous claims.

Holman's sermon, fully reported in the *Minneapolis Daily Tribune*, provoked a long response by James Bradun Alexander, a local engineer and freethinker, who used the occasion to pour scorn on Christians' misapprehensions of evolution and their belief in immortality and eternal punishment.[106] Cutting across this dispute, the Reverend S. W. Sample, minister of the Peoples' Meeting, an outgrowth of the city's All Souls Universalist Church, lectured at the Lyceum Theater on "Darwin, Drummond and Jesus Christ."[107] In his discourse Sample "embraced many of the points made by Prof. Drummond" but suggested that "the Professor should have designated the religion of Jesus Christ and not the religion of all ages" when discussing the relationship between evolution and Christianity.[108] This rather obscure chide from Sample, presumably criticizing a conception of true religion as exclusively Christian, was a point Drummond would meet again when he faced the prospect of communicating his message to the World's Parliament of Religions in Chicago.

Drummond's lectures in Minneapolis, St. Paul and Duluth had marked something of a change in venue and audience. No longer speaking at the elite Lowell Institute or in leading American colleges, Drummond had moved into a more distinctly religious sphere. This foreshadowed his much-anticipated appearance at the Chautauqua Assembly in upstate New York and at the Student's Convention in Northfield, Massachusetts. Drummond was a familiar figure at both gatherings, having made an arresting impression when he spoke there six years previously. At Northfield, he confined himself to religious topics, but at Chautauqua he spoke on both science and religion. It was at Chautauqua that he attracted the largest audiences of his tour (and of any of the tours discussed in this book) and at Northfield where the most intense and sharp-edged controversy arose.

Drummond traveled to the western edge of New York State on Sunday, July 2, to give six lectures on the ascent of man to the twentieth Chautauqua Assembly. His first talk that day, independent of his series on evolution and repeated later at Northfield, took Jean-François Millet's painting *The Angelus* as his subject. The audience gathered in the Hall of Philosophy to hear him fell "under the spell of [his] remarkable personality," a description that echoed the ongoing rise of "charisma" and personal magnetism as the mark of a speaker worth hearing.[109] The following afternoon Drummond started his series in earnest, describing for his audience the evidence of human evolution. His listeners lingered in the Hall of Philosophy afterward and hurled "a fusillade of questions" at the speaker.[110] Drummond waited, however, until the following day to address them. Reinforcing the truth of evolution as an aspect of God's creation, Drummond noted that he was "not responsible for the evolution theory" or for "making man in the way he is made." Humans were indeed "made in God's image" and it was "done by the processes of evolution." Setting such anxieties aside, Drummond then launched into the "origins of mind," countering anticipated objections by denying that an evolved mind supported materialism.[111] Instead, it pointed to an ever-active God and corrected mistaken notions of divine agency as an occasional interruption of the natural order of things.

The third lecture was again at the Hall of Philosophy but the fourth brought a change of venue. The hall had quickly filled to capacity and then overflowed. As a result, the crowds were directed to the assembly's new amphitheater, with its trussed roof, steel columns and open-air atmosphere.[112] With the fifty-five hundred seats filled and more than four thousand standing to listen, Drummond gave an abridged version of his Lowell lecture on the evolution of the mother.[113] His fourth lecture then dealt with the natural emergence of "paternal virtues." Evolution had bequeathed the father's "gift of righteousness" as well as the mother's "gift of mercy," and both were held together by the "mediary" of immortal love, the child. While he was on the subject, Drummond took the opportunity to pay "hearty tribute" to American families, "guarantees of the progress and security of this nation."[114] It was a message that was guaranteed a warm reception from his ten-thousand-strong audience.

The Chautauqua amphitheatre, which had replaced a smaller wooden structure just months before Drummond spoke in it, was among the largest and best auditoria in America. It was set to become a national symbol of charismatic speech and artistic performance. When Drummond walked its boards he filled it to overflowing. Skilled at finding an appropriate register, he brilliantly embodied and expressed the ethos of Chautauqua eloquence. Drummond used the symbolism of the space in which he spoke—a platform for progressive Protestantism and American democracy—to his advantage and made his message about evolution congruent with national progress and civic optimism.

The success at Chautauqua was in many ways predictable and Drummond used that to his advantage. But when he completed his talks there and traveled nearly three hundred miles east to Northfield on the morning of July 9, he was moving on to a different and more challenging event and cultural situation. The arrival at Northfield was, for Drummond, a return to a familiar setting. It was here that Drummond's longstanding relationship with D. L. Moody became a vital element in the talks he delivered and the way they were received. Even though Moody was absent—the evangelist was in Chicago at the World's Fair—his influence and commanding presence were deeply felt.

Drummond and Moody shared much in common, both theologically and in other ways. Their partnership was now nearly twenty years in the making and Moody had no hesitation approving of Drummond's name being on the list of speakers for the Students' Convention. Drummond, like Moody, downplayed doctrinal differences, rarely spoke of divine judgement and subscribed wholeheartedly to an emphasis on the love of God.[115] For all his criticisms of the distorting effects of newspaper reporting and his own self-styled "shyness," Drummond also shared Moody's ability to court the attention of journalists.[116] He repeatedly pointed to the practical bearing of Christian belief in a way that echoed Moody's own lifelong commitment to the economic and social effects of Christian conversion.[117] Both evangelists embraced aspects of "modernism" to further the cause of a down-to-earth and "simple" Christianity, stripped, as they saw it, of hoary and contentious dogmatic accretions.

At the same time, the differences between the two men were also plain to see. The dapper and credentialed Drummond presented a very different figure on the lecture platform from Moody. Drummond's tall and athletic frame contrasted to Moody's thickset and paunchy physique. Their speaking styles, too, were markedly different. Moody spoke with famous rapidity and his diction was frequently poor.[118] These differences did not prevent Drummond from describing Moody's sermons as "eloquence of the highest order" and marked with a "pathos of a quality which few orators have ever reached . . . which not only wholly redeems it, but raises it, not unseldom almost to sublimity."[119] Drummond, for all his difference in posture, pace and physique, borrowed heavily from Moody's techniques. When preaching, Drummond adopted the same unadorned style, used short sentences and uncluttered syntax and marked his sermons with a penchant for anecdotes and aphorisms.

The friendship and formative influence help to account for Drummond's appeal at Northfield in particular. He was beloved not only by

Moody himself (whose feelings on Drummond's death were "akin to those of David on the death of Jonathan") but also by the students who crowded into Northfield's Stone Hall to hear him.[120] For all that, Drummond's arrival was already clouded by controversy. A "delegation," reportedly, had already approached Moody asking him not to include Drummond among the speakers that year.[121] Even before he spoke, a reporter from the *Christian*, a British weekly central to Moody's support base in Britain, informed Drummond that he had been asked to leave him off the list of speakers and not to report anything he said for fear of associating Moody with Drummond's views.[122]

On the day of his arrival Drummond gave the first of three evening talks, tackling on this occasion the subject of overcoming temptation and living a higher life through the careful study of the "anatomy of the soul and the physiology of temptation." Repeating a point made in earlier talks on the subject, Drummond boldly declared that "man's body is developed from an animal creation." To this Drummond added that he had "a suspicion that the evolutionists may be on the right track" and that the legacies of animal ancestries made up the "lower parts" of human nature.[123] Temptation, then, was "the appeal of the animal to the man." [124] Using New Testament language of spirit and flesh, Drummond encouraged his listeners to "live continuously in the spirit." This, Drummond declared, was not "magic" but merely "living along the line of the laws . . . of our human nature." Sermons and prayers were not the answer for those laboring under temptation. Instead a "new environment" was necessary to allow a "new nature [to] grow." To conclude, Drummond urged his listeners "by will power and prayer power and the power of the Spirit of God to walk in the spiritual regions and help others walk there."[125]

In making these claims, and combining elements of evolution with evangelical vocabulary, Drummond was doing what he had long done from platforms and in his publications. There were elements of the argument that had helped make him famous, first aired in Possilpark. The natural and the spiritual world were not two distinct and separated spheres. Rather the latter was simply a "province" of the former.[126] The laws of human nature extended back into natural history and forward into a spiritual future without any discontinuity or break. What was new was the emphasis on human evolution. Here Drummond was tracking Darwin and moving from general claims about natural history and evolution to specific claims about the evolutionary history of humanity. It was not that his talk was especially new in content or touched on areas of controversy that he had remained silent about before. But rather, because he was using the medium of speech and because of where he was employing that medium, Drummond reached a larger audience and significantly raised the theological stakes. His talk was a distinct event around which accusations of heresy or heterodoxy could gather.

The following evening Drummond spoke again at Stone Hall. His audience included the college students gathered for the conference and other day excursionists, like the large group from Brattleboro in Vermont that had traveled on "two tally-hos" to hear Drummond speak.[127] The meeting began with hymn singing, prayer and a solo rendition of "The Ninety and Nine" by Moody's famous musician-accompanist Ira Sankey. When Drummond stood up to speak he told his audience that "for the last three or four years of my life I have had very little to do with the ninety and nine, I have been after the one sheep that was lost, and I have got into the way of talking to that one and trying to make things plain to him."[128] This pursuit of the lost sheep was an exercise in translation, rendering Christianity into a language that agnostics might comprehend. The remaining part of Drummond's talk was an object lesson in doing just that, but the words he used were in keeping with the language of Northfield, translating his Lowell lectures into terms evangelicals might comprehend. He urged his audience not to think of Christianity as fitting the odd individual for heaven but as a "society of the best men" working to transform the world. This, Drummond, declared was what Christ meant by the coming "Kingdom of God." The task of acting as an "anti-septic" to the contagion of sin was urgent and vital. Whatever the superficial appearance, the world was "sunken" and "rotten." The sin of the world's cities presented a "ghastly spectre" to the imagination. It was the "Lamb of God acting through you and through men" that would "take away the sins of the world." There was little here that Drummond had not said many times before. But he did, in closing, allude to the central theme of his Lowell lectures, reminding his audience that "all activities of life are centered in the functions of nutrition and reproduction." If they, at Northfield, had been well fed it was now time to commit to the business of selflessly reproducing the "Christianity of Christ" in the world of civic affairs. In his closing call for action, Drummond appealed to his audience to discern for themselves the "will of God" and give "life for life."[129] This was Christianity as further evolution, working through other-regarding love.

Drummond delivered his final Northfield address the following day, again at Stone Hall. Once again reports observed the rapt attention of his large audience. This is important to note, because of the fuss later generated by the intensely antagonistic reaction to his Northfield talks by a particularly vocal group of critics. On this occasion, Drummond once again used Millet's *The Angelus* as an "illuminated text" to hang his sermon on. The painting, depicting two companions in a potato field stopping work to pray, the Angelus, pointed to the three elements of a "complete life": work, God and love. Labor, Drummond declared, had taken new meaning when God became man and worked as a carpenter. It was through work that "the great Christian graces are communicated to our souls." The presence of God, the second element, suffused the picture just as it suffused the world. Drummond, correcting the false image of God as "up there" in heaven or "back there" six thousand years ago at creation, called for a conception of God as immanent and ever present.[130] This more biblical God, found within, was always at work, now creating the "perfect man." Humans were not yet what they would become, "the buds of our nature are not all developed yet." These thoughts led finally to the third element, that of love. The two peasants, the man and the woman, represented the idea of friendship and the "love of humanity."[131]

Drummond undoubtedly adjusted his message and his language for the audience of students and evangelical excursionists in front of him. But there were clear lines of influence and continuity between what he said at Northfield and his lectures on the "evolution of man." Not surprisingly, then, the reaction of some among his audiences was strongly negative. In a letter to Lady Aberdeen written several weeks later, Drummond reported that at Northfield "many fell upon and rent me." At the same time, a telegraph to Moody from "many of his generous supporters" asked the absent evangelist to denounce Drummond or face being abandoned by them.[132] It is not clear exactly when these denunciations happened, but Drummond left the impression that his opening talk was the one that caused offense, recounting that "before the close of the Conference, I struck an orthodox vein and retrieved myself a little." Whatever the case, his stay at Northfield "was not a happy time."[133]

After Northfield, Drummond largely withdrew from public view. The occasional appearance was noted in the newspapers, but Drummond spent most of his time with friends or on solitary excursions. Among other things he went on an extended salmon fishing trip, traveled to Newfoundland and visited the evangelical financier and philanthropist Morris Ketchum Jesup in Bar Harbor, Maine.[134] This visit was not just about maintaining a friendship. It also aided Drummond's passage through evangelical America and helped secure the backing and support of an influential figure in that movement. Drummond's apparent withdrawal into the northern wilderness did not mean his abandonment of the lecture circuit. It might be read instead as an exercise in regrouping against his detractors and a fitting preparation for relaunching his lecturing ministry. Nevertheless, as August drew to a close, it was one of Drummond's most vocal critics, the transatlantic preacher, premillennialist and missionary enthusiast Arthur Tappan Pierson, who, raking the embers of the Northfield controversy, brought Drummond back into public view.[135]

Writing in the *Christian at Work*, Pierson, recently returned from occupying Charles Spurgeon's old pulpit in London, took the unease surrounding Drummond's talks as a sign of the times. Noting Drummond's "heresy" at Northfield, Pierson warned that a "battle rages" over the Bible and Christ's divinity. In the heat of conflict, "when a man's foes may be they of his own household," the tragic necessity of sacrificing "the dearest friends" for the sake of truth lay on the path ahead.[136] Pierson, with the mantle of Spurgeon on his shoulders, was targeting Drummond and appealing to Moody to make the sacrifice.

Moody ignored appeals to excommunicate Drummond, inviting the latter to take part in his campaign at the World's Fair. Yet while Moody remained loyal, Drummond's reputation among many of Moody's more conservative associates suffered. The following summer, the *Christian Work* reported that no further invitations to Northfield or Chautauqua would be made to the evangelical evolutionist.[137] The same legacy was evident even when Moody was directly involved. Drummond's inaugural address at Northfield was not included in a collection of his talks to the assembly published in 1897 with a tribute by Moody under the title *A Life for a Life*. The collection was part of a series of devotional books "daintily bound in ivory and coloured buckram" published by Moody's brother-in-law's firm, Fleming H. Revell.[138] The first two chapters of the book corresponded, more or less, to versions that appeared under different titles in another volume of Drummond's Northfield addresses published two years later by the New York publishers James Pott and Co. But the rather sketchy final address included in *A Life for a Life* appears to have been delivered at Northfield on Drummond's final evening there in August 1887. Whatever the case, his most controversial talk was silently dropped in the publication that carried Moody's imprimatur.

EVOLUTIONISM AND ESCHATOLOGY IN CHICAGO AND NEW YORK

More immediately, Drummond, about to return to public view, had to face the fallout of Pierson's attack. He had been invited to give a lecture series in early October under the auspices of the University of Chicago and had invitations to speak beforehand at the World's Parliament of Religions and, during the Chicago series, at the Congress of the Evangelical Alliance. The first of these presented a sharp dilemma. To his detractors, the setting and the subject—evolution and Christianity—were guaranteed to confirm his dangerous heterodoxy. Held at the great Hall of Columbus (part of the World's Fair complex), the parliament was widely viewed as a platform for an inclusive religious pluralism. To Arthur Pierson, as to many other conservative churchmen, it was an expression of growing opposition to world evangelization.[139]

Drummond was due to speak at the great hall on the morning of September 27, the final day of the parliament. The other topics for that morning's session included "religious unification," "universal elements in religion" and "religion and music."[140] The hall, with a capacity of four thousand, was full. Drummond, however, had not yet arrived. When the Unitarian minister, the Reverend William Alger got up to speak, he was still not present. Alger closed his address by declaring to the parliament that the idea of Christ went beyond the "mere historic person" that was Jesus of Nazareth. Christ was, instead, ideal or "divine humanity" in all its multiplicity and difference.[141] As the applause died away it became apparent that Drummond's paper, "held until the last moment," was to be read by a Chicago minister, the Reverend Dr. Frank Bristol.[142]

Local press reports gave a full account of Drummond's script, which later appeared in the official proceedings of the parliament. No indication is given as to why Drummond failed to attend in person, but his paper pushed hard at issues that his critics feared would be pursued on the back of his openness to evolution and Biblical criticism. Drummond concentrated on the very issues that Pierson had identified in his jeremiad against Northfield heterodoxy in the *Christian Worker*. He laid out a provisional science of sin and sin's origin in humanity's animal and "savage" past, announcing "fresh developments" in understanding incarnation in its "profound relations to the whole scheme of nature." Evolution had set the religious world ablaze with new insights and higher conceptions of the divine. The combination of science, evolution and Christianity meant that "never before have the attributes of eternity and immensity and infinity clothed themselves in language so majestic in its sublimity." Drummond touched, too, on the question of scripture. Knowledge of evolution had given the world a "new Bible," a book that had "not been made but had grown." Drummond danced around the infallibility of Christian scripture, reassuring listeners and readers that this "transformed ... Bible" remained the record of inspired deeds, words and facts bound "in the matrix of human history." Not "one important word" was lost. Even so, the Bible was a product of evolution, "a nursery of growing truths" that point to "higher things." The line between scriptural and natural revelation was blurred and evolution shone a light that revealed divine truths only now being grasped. In the end theology and science, evolution and Christianity were one. The shared object was "higher and better men in a higher and better world."[143]

It was ironic that in the middle of his paper Drummond lamented the fact that there was "no one to announce in the name of theology that the controversy between science and religion is at an end." In the longer version published in the official proceedings, Drummond noted the "want of a suitable platform" to authoritatively expound the gains of theological science and prevent misplaced attacks on religious views long exploded by theology itself.[144] The Parliament of Religions was apparently not that platform. Whatever his reasons for absenting himself, it had become difficult for Drummond to speak there without being roundly condemned by self-appointed watchdogs of Christian orthodoxy. Whether it would be different at Immanuel Baptist Church, the venue for his University of Chicago lectures, remained to be seen.

Chicago's newspapers announced Drummond's lecture series on "the evolution of man" several weeks in advance. As one notice proclaimed, "In this year of great privileges for Chicago the presence of Prof. Drummond as lecturer in the city will be regarded as one of the greatest."[145] In the year of the World's Fair, this was a high compliment indeed. But it was also controversial. In the lead up to his arrival in Chicago, William Rainey Harper, the university's president, privately admitted that Drummond's views would provoke a "fire of criticism." That, however, did not prevent him from inviting Drummond, in addition to giving his lecture series, to speak at the first formal gathering of the new academic year and to deliver the convocation address the following day. Drummond and Harper shared an enthusiasm for a new vision of Christianity that embraced higher criticism and scientific developments as crucial instruments for the refinement and improvement of Christian belief and as the proper basis for Christian progress.[146] The Presbyterian evolutionist and the progressive Baptist could unite in their commitment to higher education as the agent of Christian renewal.

Drummond's two addresses marking the opening of the university's second year were occasions of considerable public excitement. In both cases, the new Walker Museum provided the venue but was too small to accommodate all those "attracted by the great reputation of the speaker." The first address saw the museum hall fill with "teachers, litterateurs, professional men, businessmen and World's Fair visitors from all parts of the country." Drummond and Harper could not get to the platform until a passageway had been created through the densely packed hall. After scripture reading and prayer, Harper introduced Drummond to the crowd as "one whom we all know, one whom we all love."[147] Drummond's subject was once again *The Angelus*, modified in places to appeal to his Chicago audience. He began with an allusion to the World's Fair before offering his now well-rehearsed reflections on Millet's painting.

The second address proved to be as popular but was framed by more university pomp and circumstance. Before it began, faculty and students with gowns and mortarboard caps walked through the Oxbridge-styled university groves. The traditional architecture and atmosphere were appropriately juxtaposed with the nearby "towering and luminous Ferris wheel," a reminder of the forward press of modernity. The financier George Clarke Walker was there to receive the thanks of the university for a gift of $100,000 to fund the erection of the natural science museum that now bore his name.[148] After processing in with faculty members—accompanied by "an appropriate march" played by an orchestra—Walker sat down on President Harper's right, with Drummond on his left. Again after prayer (and one might note here the contrast with the lack of such public piety at Huxley's address at Johns Hopkins University two decades previously), Harper introduced Drummond, who this time spoke on "some aspects of evolution." His address was a sermon in praise of evolution, the great unifier of human knowledge and all-encompassing reality of life. His plea was that no student would pass through the university without acquainting themselves with evolution "the key to general knowledge."[149] Each should be gripped by a vision of a world in motion, a world that has been and is being evolved by those willing to change along with it. The new museum of natural science could hardly have been a more appropriate place to offer such a gushing encomium to evolution.

Both Harper and Walker privately subscribed to Drummond's views but publicly made no mention of evolution. Instead, in the speeches they delivered in the wake of Drummond's address they stuck closely to a chronicle of practical developments that had led to the new museum and to additional bequests given to the university. There was an opportunity to place the museum's mission, and the university's mission, within Drummond's evolutionary framework. But there is no evidence that either Harper or Walker made any allusion to it. Surrounded by the donors, students and supporters of a pioneer Baptist university, this was, perhaps, politic.

It was not, however, until Drummond delivered the first of six lectures on evolution at Chicago's Immanuel Baptist Church the following evening that an expression of concern was registered in the press. Under the false impression that Drummond was now employed as a University of Chicago lecturer, "leading Baptists" in the city made their discontent known, condemning the official promulgation of, so it was supposed, unqualified Darwinism.[150] What was particularly concerning was Drummond's purported claim that the distinction between some humans and "monkeys" was less clear than the differences observable within humankind. That Drummond spoke for the university in making such claims was very quickly denied. A day later, during a meeting of the State Baptist Convention, Professor Charles Hewitt reassured those present that Drummond was merely a "chance lecturer" (a significant downgrade to put it mildly) and that his views were categorically not endorsed by the university. There was, he declared, "no fear that anything but the pure Baptist teachings would be taught."[151] As further guarantee, he announced that the convention's directors were now inviting state societies to appoint visitors to annually inspect the university's curricula.

Now suitably distanced from the university, Drummond's remaining five lectures passed off unremarked. An abridgement of his Lowell lectures, they dealt with the evolution of the human body and mind, the emergence of language, the evolution of the mother and the father and finally the "evolution of Christianity." An opportunity was given for his audience to submit questions after each lecture, but no further controversy was reported.[152]

At the same time as delivering these more academic discourses, Drummond spoke to the city's YMCA and participated in the Evangelical Alliance's International Conference, held at the Hall of Columbus between October 8 and 14. To the former he gave repeat performances of his Northfield talks, including his lecture on temptation.[153] The latter was arguably of more moment. Organized by the North American branch of the Evangelical Alliance, it was one of many conferences held under the auspices of the World's Congress Auxiliary of the World's Columbian Exposition. The mission of the Evangelical Alliance, founded in England in 1846, aimed to unite Protestant churches around a common doctrinal statement and encourage joint evangelistic and social action. The American branch, after faltering over the slavery question in the late 1840s, reemerged in 1866, propelled by common concerns over religious skepticism, Roman Catholicism, moral licentiousness and other perceived threats to Protestant Christianity. By the 1890s it leaned strongly in the direction of social action, emphasizing the "leavening" effect of Christianity on worldly affairs. The alliance's president, the copper mine magnate William E. Dodge Jr., and its general secretary, Josiah Strong, were proponents of a social gospel orientated toward the alleviation of various societal ills. It was thus not surprising that Dodge and others welcomed the opportunity to associate the alliance's conference with the other congresses held at the Hall of Columbus. It symbolically underlined the progressive form of evangelical cooperation that at least some of the alliance's most influential supporters were now committed to.[154]

The clear emphasis of the alliance's conference on "practical Christianity" provided a hospitable environment for Henry Drummond to speak into. It had doubtless helped that he had been salmon fishing with the alliance's president a few weeks before, but when he rose to bring a greeting to the conference at its opening session, his own high standing within the Evangelical Alliance was secured.[155] Addressing the two thousand delegates gathered in the hall, Drummond delighted in the conference's emphasis on the "social side of Christianity." This, he proclaimed, was "Christ's side of Christianity." In taking up social questions, the alliance had all but rescued itself from extinction and irrelevance. It looked set to rescue the very term "evangelical," a descriptor that had become a byword among some for platitudinous and hypocritical religion.[156]

His second address was a longer, more formal affair delivered five days later. His subject was "Christianity and the Evolution of Society," a title provided by the organizers. Drummond had expected to deal with "the evolution of Christianity," but the change made little difference. The evolution of society *was* the evolution of Christianity and vice versa. While Drummond, deploying the well-worn rhetorical ploy of apophasis, reassured his audience that he would make no mention of "simian ancestry" or "primordial germs," he pushed evolution to the fore and made it his theme. Against the pessimism about the world's future that marked Christianity past, Drummond proclaimed present existence as the stage on which God was working out his creative and redemptive purposes. Drummond surely knew that the premillennial eschatology he referred to was, in fact, on the rise within American evangelicalism. His critic, Arthur Pierson—a reluctant conference participant and speaker—was among its chief proponents.[157] Drummond took care to couch his more arresting claims in softening phrases like "it will not be misunderstood . . ." or "I shall mislead no-one if I remark." Even so, he unashamedly savaged what he took to be a malformed version of Christianity. What he gave with one hand he took away with the other, noting that "next to losing the sense of a personal Christ, the first evil that can befall a man is to have no sense of anything else." Evolution provided the necessary corrective. Sketching a grand epic of an evolving creation, from the "arranging of atoms" to the human mind, Drummond pointed to God's ongoing activity, "carrying on the evolution to still further bounds, evolving men into higher men." The coming "kingdom of God" of which Christ so often spoke was "the evolution of the world." The signs of its coming were everywhere: "the social progress of humanity, the spread of righteousness, the gradual amelioration of life, the freeing of slaves, the elevation of women, the purification of religion." The Christian was to be, above all else, a "practical evolutionist," working to improve the "health," "work" and "wages" of all.[158] If Christians failed to join in this task, the world would advance regardless, driven on by morally earnest people who, though outside the church, deserved to be called Christian.

Drummond's address, coming as it did after a talk on the "science of charity," was thoroughly in tune with the overriding ethos of the Evangelical Alliance's International Conference. It contrasted sharply, however, with Arthur Pierson's account of Christian cooperation and social action, delivered to the conference several days earlier. It was even more starkly different to another talk Pierson gave during the same week to the Congress of Missions, running at the same time. On that occasion, Pierson vehemently attacked any idea that Christ's kingdom could be established

AS A TRAVELLER IN CENTRAL AFRICA. AGE 35 OR 36.

FIGURE 5.2. Henry Drummond in Africa. A model of Christian manliness. *McClure's Magazine* 2 (1893–1894): 436. Reproduced by permission of the National Library of Scotland.

through human effort and based his own impassioned call for social justice for the poor and frontal assault on Christian captivity to wealth and status on an altogether different eschatology.[159] Drummond was chief among those that Pierson aimed to put down.

Despite the vitriol, Pierson's voice was largely drowned out by those much closer to Drummond's optimistic and this-worldly Christian evolutionism. It was a vision Drummond more informally reinforced in two additional talks given as part of the conference's sectional proceedings. In one, he praised sport of all kinds as the best of moral educators and in the other he promoted the cause of the Boys' Brigade. In both cases, he emphasized the "material instruments" that were crucial to real spiritual growth and the development of Christian character. Learning to live for others was cultivated not in church (that provided only a "stimulus") but on the sports field.[160] Team sports were the locus for moral evolution.

With his Evangelical Alliance talks delivered, and his lecture series at Immanuel Baptist Church complete, Drummond made one more appearance in Chicago. On October 16, he shared a platform with Lady Aber-

IN 1893.   FROM A SNAP SHOT IN QUEBEC.

FIGURE 5.3. A "snap-shot" of a suave Henry Drummond relaxing in Quebec away from the lecture circuit. *McClure's Magazine* 2 (1893–1894): 437. Reproduced by permission of the National Library of Scotland.

deen, then visiting Chicago in her capacity as organizer of the World Fair's exhibition of Irish industries.[161] Drummond then traveled to New York to prepare for his departure to England and to give a final lecture at Chickering Hall on the evening of Friday October 20.

The New York address was organized by the Students' Club, a society founded in the wake of Drummond's visit to New York in October 1887.[162] It would be Drummond's farewell to the United States and represented his leading passion—inspiring college students to organize for the sake of the Christianization of higher education. On the afternoon of the lecture a reception was held at the club's rooms on Lexington Avenue and Chickering Hall then filled with students to hear Drummond's final words from an American platform. He spoke on "the ideal man" and touched on the origins of the student movement in Scotland. Some way into his talk he stopped and requested that "no stenographic report of his address be taken."[163] There was no need. His American tour had attracted huge interest and his message about evolution and Christianity had reverberated through packed halls and been printed countless times in American newspapers. He could revert to his practice of making his talk feel private, intimate, and directly addressed to each individual heart. The next day he boarded the RMS *Umbria* and sailed for England.

In the wake of Drummond's visit, an advertisement appeared in the popular religious periodical, the *Outlook*, alerting readers to an upcoming article by Drummond on the Boys' Brigade. The catch line announced that "we all listen now when Professor Henry Drummond talks."[164] It captured something of the success of Drummond's tour. He had been heard by tens of thousands of Americans, and reports of his lectures had been read by many more.

At least part of Drummond's mass appeal to American audiences was due to his composite public image. This was well represented several months later in the April issue of *McClure's Magazine*. As part of a standard feature, "Human Documents," the magazine included several portraits of Drummond. Among them was one of Drummond as a geologist-traveler, photographed in Africa. The image displayed an ideal specimen of the "higher man," impeccably dressed in a waistcoat, and with a shirt collar and silk neckwear, but nevertheless "practical" and scientific, the well-formed Christian scientist. A second portrait, taken while Drummond was holidaying in Quebec during his 1893 tour, reproduced the image of the suave and fashionable lecturer who had courted evangelical moguls, mesmerized audiences and attracted huge crowds.

It was clear, too, that how Drummond talked—without pronounced gestures, with a measured tempo and a clear articulation—vocally and bodily enacted the type of humanity that anticipated evolution toward a yet-more Christlike future race. It also reproduced a way of speaking that had become associated with the most celebrated of the British scientists who had toured America in search of large and appreciative audiences. But it also did more than this. Drummond managed to superadd to this scientific style of speaking a phraseology, prosody and poise that skilfully adapted a "straight-to-the-heart" way of talking most evident in his good friend D. L. Moody. Drummond, measured by levels of positive public attention and by audience figures, had managed to attractively communicate what others had failed to—a sense that a profound and familiar religiosity could be united with a full embrace of a scientific life. Controversies over the heterodoxy of his message notwithstanding, he bound together religious and scientific themes through artful argument and vocal performances that, if the flood of enthusiastic reporting is to be believed, entranced his listeners and drew them back for more. It might be reasonably argued that he had succeeded where Huxley had failed and become in America a consummate evangelist for evolution.

Henry Drummond's success in speaking about the science of evolution in a way that infused it with spiritual and emotive appeal provides the clearest example among all five of our intrepid lecturers of the power of speech—spoken and reported—to fuel and shape the cultural "energies" of science. James Moore, in his otherwise superb analysis of Drummond's thought, argued that the "chief predisposing cause" of the fame of the evangelical evolutionist was the success of *Natural Law in the Spiritual World*. This chapter has marshalled evidence that suggests such an assessment must be qualified and supplemented. Moore notes in passing that "only a small proportion of [Drummond's] readers were ever exposed . . . to his renowned magnetic personality." When all of Drummond's talks and sermons to mass audiences are gathered together, that proportion was surely not so small after all. It is also likely that only a fraction of those who did directly encounter Drummond's "magnetic personality" read his books (though they may, with untold numbers of others, have read reports of Drummond's lectures and sermons in America's newspapers). It was never, of course, about a zero-sum game between numbers of readers and hearers. For his admirers, reading Drummond made one desire to hear him, and hearing him, to read him, again and again. For all that, it is worth pausing to reflect on the fact that Drummond's voice, his reported power over an audience and his arresting presence on the stage could not be reproduced on the pages of books. Nor could books generate the extensive newspaper reporting or the intense local controversies that so often followed in the wake of Drummond's lectures in America. Drummond's American tours, perhaps more than those of any of the other lecturers tracked by this book, demonstrate the underappreciated power of speech to generate and transform the cultural meanings invested in public science.

CONCLUSION

SCIENCE, HISTORICALLY SPEAKING

In 1885 the Oxford theologian Aubrey Moore noted a distinction "of absolute importance" between science and religion. It was, Moore declared, "a difference of tone and temper rather than of anything else" that divided the two great human enterprises. Science, Moore noted, had its "thinkers." Religion "its prophets and its priests."[1] The comments appeared in Moore's review of the Bampton lectures delivered the year before by the bishop of Exeter, the Reverend Frederick Temple. Temple's lectures, Moore noted, had been given by a celebrated champion of "the great facts of morals and religion." But they did not quite conform to "tone and temper" of religious discourse as Moore had defined them. The style of Temple's "sermons" was "clear and clean-cut," characterized "at every turn" by "the analytical tendencies of the mathematician."[2] This blending of tendencies undermined Moore's crisp distinction between scientific and religious discourse.

The speakers who have been closely followed in this book also operated with their own preferred demarcations between different styles of address. They were united in their conviction, whether explicitly or implicitly expressed, that speaking publicly about science demanded careful attention to the embodied nature of the act. All considered overly dramatic, premeditated gestures and exaggerated vocal inflections as inappropriate for delivering a science lecture (or for any lecture—each aimed to be consummately scientific in speech as in life). They generally stood still and used controlled diction and a limited variation in pitch to enhance clarity and, perhaps more importantly, visibly and vocally embody the virtues of a scientific life.

There were, in the wider world of public speech, models that they could and did borrow and adapt to achieve their aims. This borrowing, of course, brought risks. It could threaten the distinctiveness of a style believed appropriate for communicating science. Nevertheless, it was a risk they all took. For example, all five more or less consciously followed a dominant trend in British parliamentary speech. At Westminster, a commitment to perspicuity and restrained emotion was designed to draw attention to the message rather than the man (parliamentary politics like public science was predominantly a man's world). Among other effects, this made science lectures delivered in the same manner a help to democracy, promoting a form of speech appropriate for an age of enfranchisement. At the same time, this chaste style did not always travel well. When taken to America, it did not always appeal to audiences that increasingly favored a more charismatic and colorful delivery.

Perhaps more significantly, if paradoxically, all five lecturers were connoisseurs of the evangelical sermon. This form of direct and extempore address that appealed to the heart could provide lessons for the advocate of converting society to an all-embracing scientific outlook. Toward the end of his life, Alfred Russel Wallace fondly recalled the charming cadences and evocative language he had heard from the pulpits of Methodist chapels as a young man. Thomas Henry Huxley, even if he avoided expressing admiration for evangelical preachers, billed at least some of his own lectures as lay sermons. This was more than a superficial appropriation. Huxley used extempore address to not only instruct but also, and from his point of view more importantly, to move (and convert) his hearers. If many of his secular congregants in America felt he had failed to master the art of the evangelical sermon, his efforts did not entirely fail. Henry Drummond, however, was altogether more successful at adopting and adapting evangelical style to communicate and embody a Lebenswelt that was at once scientific and religious. He had been schooled by one of the most popular preachers of his age and became a compelling evangelist for evolution and an ardent evolutionist doing the work of an evangelist. It proved to be a winning, if controversial, combination on the American lecture circuit.

THE MANNER AS THE MESSAGE

If all five speakers had shared influences on their speaking styles, each also developed certain prosodic or postural distinctives that embodied sometimes contrasting messages about science and its cultural ramifications. John Tyndall may have exemplified a commitment to the parsimonious use of studied gesture, but, according to contemporary descriptions of his lectures, his lithe, athletic body frequently shimmered with energy, enthusiasm and visceral absorption in his subject matter. This was, indeed, part of his appeal and added to the frisson and fascination produced by the metaphysical provocations that injected interest and intrigue into his platform performances. In contrast to this, Alfred Russel Wallace struggled to disassociate himself from another well-known English naturalist who had, at least by some accounts, failed dismally to engage his American audiences.

The Reverend John G. Wood, even with his famous blackboard sketches, delivered lectures that were altogether too safe and colorless. Wallace, despite his fame as the codiscoverer of the theory of natural selection, lectured too much like Wood. He was certainly clear enough in diction and in his use of language. But the aging naturalist apparently failed to ignite excitement in his hearers by adopting a rather staid and static style of delivery that lacked musicality and motion (by his own admission, Wallace could not hold a tune).

Both Thomas Henry Huxley and Richard Proctor, at least in one respect, managed to merge more effectively their message and mode of delivery. They did this, however, in completely contrasting ways. Huxley was well known for speaking at a tempo somewhere between andante and adagio. This was appreciated, at the very least, by stenographers. It also reinforced a sense of measure and control that Huxley valued as marks of scientific thinking. The slow tempo gave an impression of precision and of words that had been carefully weighed. Of course, Huxley was capable of varying his pace and altering other aspects of his delivery, but by all accounts, he rarely spoke in haste.

Richard Proctor, on the other hand, seemed incapable of speaking at a tempo slower than allegro. His lectures, as a result, were much harder to record with the kind accuracy that at least some journalists aimed for. But the fast pace was part of the appeal. It gave a dramatic sense of the speed of scientific change and of an astronomer and lecturer able to quickly digest and disseminate the latest results of scientific discovery. Proctor not only spoke quickly but, in a darkened hall, swiftly moved through hundreds of lantern slides, overwhelming his audience with images and information. The lanternist, like the stenographer, struggled to keep up. This rapid-fire approach did not only demonstrate the pace of scientific change. It also underlined the mind-numbing immensity and variability of the physical universe and reinforced Proctor's oft-stated conviction that whatever the ultimate meaning of the universe and of humanity might be, it was far beyond the grasp of even the most agile mind.

Proctor was not alone in his efforts to reinforce a particular metaphysical message through prosodic strategies (conscious or not). All five speakers communicated strong convictions about the metaphysical implications of modern science. For all five, evolution in particular provided a concentrated theme to speak about an adequate metaphysics or religious outlook in a world of dramatic scientific developments. Each shared a conviction that religion was a realm of deep feelings before it was a set of cognitive claims or doctrinal commitments. It followed that their living voices could be used to evoke what otherwise could not be articulated or aroused. This was, I want to argue, one reason why the lecture tour as much (or more than) the printed word was considered a worthwhile enterprise. Whatever other motives were in play, all five shared a widespread assumption that speech was a peculiarly effective mode through which to engender cultural change consonant with science. Each speaker brought to America a religious vision as much as a scientific one and used the versatility of voice and embodied action to recommend it to a mass audience.

FALTERING SPEECH

The medium, of course, was not always the message. There was, in many cases, an instability in the styles of address adopted by individual lecturers. Thomas Henry Huxley's career-long struggle to speak loudly enough to be heard provides just one example. On several occasions while he was in America, audiences who could not hear him gave up and left his lecture long before it had finished. It was a common complaint made in press reports. But the reputational damage was more severe than losing listeners. As an index of character, a weak voice indicated a lack of conviction and authentic masculinity. His poorly projected voice, fairly or not, provided Huxley's critics with ammunition to shoot down an infamous agnostic who was threatening the creeds and convictions of Christian America.

Richard Proctor provides another instance of this disconnect between the style and intentions of the lecturer. For Proctor, the problem was volubility rather than volume. On the one hand, it confirmed his capacity to keep in step with astronomical science. On the other, it risked reinforcing a reputation for inconstancy. During his second lecture tour, Proctor had to campaign against widespread accusations of being overly mercurial. It seemed to some that his religious views changed as quickly as his scientific ones. As he took advantage of American railroads and traveled from town to town, news of his radical change of opinion on a variety of science-religion issues circulated with equal speed via syndicated newspaper reports. Proctor's religious outlook *had* changed but not, he was forced to argue, in ways that cast doubt on the stability and consistency of his intellectual judgements.

Henry Drummond, while succeeding where others failed in uniting the style and substance of his lectures, nevertheless faced acute challenges precisely on account of that success. His composite or hybrid mode of address—at once scientific and spiritual in tone and temper—threatened to be undone in contexts where severe crosscutting pressures were in play. At Northfield, for example, Drummond struggled to pitch his lectures in a way that avoided accusations of theological heterodoxy. He spoke and sounded too much like a heretic. It was difficult, after that, for Drummond to entirely recover his poise. His otherwise unexplained absence from the World's Parliament of Religions in Chicago two months later certainly suggests this. Speaking there would surely have aroused further suspicion that Drummond was indeed diluting or discarding core doctrinal convictions in his efforts to unite evolution and Christianity.

The issue for Alfred Russel Wallace was not so much the volatility of vocal performance. The difficulty was more fundamental. Wallace simply took too long to find his voice, or to speak about his chosen subjects in ways that drew in an audience and attracted the kind of intense press attention required for a profitable and consequential tour. The inauspicious beginning at Chickering Hall may have put Wallace off balance. His solid but rather unexciting set of Lowell lectures largely failed to generate invitations to speak elsewhere. He just about managed to deliver a sufficient number of lectures to pay his way to California. It was only there that he began to exploit, with the help of his brother, the technologies of celebrity and speak on a subject controversial enough to stir up at least local excitement. In earlier lectures Wallace *did* make arguments that were, for many, heterodox and unconventional. One of his most repeated talks finished with the claim that the human spirit was not a product of natural selection and that material evolution, which is to say the universe itself, ultimately subserved the emergence of this indestructible reality. But this was not enough to produce the kind of public attention required to make his tour more than the sum of the forty-odd lectures he gave while he was in America.[3]

VOICES IN PLACE AND IN PRINT

The tensions, vulnerabilities and varying interpretations of the lecturers discussed (in all their dimensions) were a consequence of a range of factors operating at different scales. There was, for example, a shared, if dynamic, set of expectations about what constituted an eloquent and compelling platform performance. It would be wrong to claim the American lecture culture was in any sense homogeneous, but by the late nineteenth century it was a nationwide enterprise connected by agents, newsprint, networks of local lyceums and other civic organizations and fueled by a mass appetite for hearing, seeing and judging celebrity speakers. At the same time, local circumstances could profoundly influence the performance and impact of a lecture or lecture series. Under certain conditions, lectures could function as a lightning rod, electrifying preexisting debates. It was an experience shared by all five lecturers. In various places, they witnessed their lectures being consumed by local controversy.

The fraught politics of higher education was one common cause of lectures being overwhelmed by intense local debates. Thomas Henry Huxley's address at Johns Hopkins University, then in its first year, is a case in point. His vision for university education was drowned out, at least in the press reporting, by controversy over accusations of ceremonial impropriety. Behind this lay an intense contest between those who, like the university's president Daniel Gilman, regarded modern American universities as secular spaces neutral with respect to religious distinctives, and his critics, who wanted a designedly Christian institution of higher learning.

Alfred Russel Wallace, while lecturing in Toronto (one of the few times any of the five scientists strayed north of the border to speak), found his lectures being roughly thrust into the throes of a similar debate over the control and direction of college education. His lectures were treated as evidence of secularizing and antireligious trends by defenders of local independent denominational colleges threatened by the centralization of educational provision. What these and other examples show is that the lecturers were never in control of the immediate and longer-lasting effects of their lectures. More often than not, they were tossed about by turbulent debates caused by long-standing tensions between different accounts of the proper foundations and governance of educational and public culture. If this at times increased ticket sales, it also meant that local agendas frequently rode roughshod over the hopes and intentions of all five speakers.

Journalists and publishers were among the chief agents fanning the flames of debate and controversy and were also critical players in making or breaking a lecture tour. All five lecturers were again united in believing the press and publishers to be necessary if untrustworthy allies. All five made strenuous efforts to make use of contacts in the world of print to shape the public profile and public face of their tour. Such exertions to control the message and to profit from book sales and media exposure certainly made some difference. Huxley's personal ties to the *New York Tribune* and the Nashville-based *Daily American* were used to good effect. And Drummond's success in court to ban a pirated and partial version of his Lowell lectures is further evidence of how seriously printed records and reports of lectures were taken. Arguably, the lecturers were more at the mercy of pressmen and publishers than anything else. Devastating attacks on their character and performances were far from rare. Misquotations and deliberate distortions kept them busy writing letters to editors throughout their tours. At the same time, newspapers in particular were a crucial part of making and maintaining their celebrity status. When the reports were generous and long, it helped generate invitations and audiences. To be interviewed by a major regional or national newspaper translated directly into ticket and book sales. And, perhaps above all, the often-careful transcriptions of lectures translated platform performances into print for a truly mass audience. It was the appeal and fascination of living scientific voices that attracted stenographers, journalists and editors to record and print lectures across several columns on the front pages of American newspapers. Science that had been spoken was considered eminently newsworthy. It sold newspapers as well as tickets. And if scientific talk, or talk inserted in a scientific lecture, stoked religious controversy, the geographical reach of the spoken word, rendered into print, only increased.

SCIENTIFIC TALK OF THE FUTURE

Much evidence has been amassed here to demonstrate the vibrancy and visibility of science lectures delivered by scientific celebrities in the Gilded Age. But what of the fate of the unique and powerful kinds of scientific speech heard on the American lecture circuit? At least into the early twentieth century, the American lecture tour remained an appealing option for popularizers of science, even if lecture culture itself was losing some of its prominence in an age of rapid university expansion and profound changes in cultures of public entertainment and communication. As Thomas Augst has shown, by the 1890s an increasing number of nostalgic elegies were written for the lecture platform, "as a site . . . for shared experience of communication that had been so intimately tied to the human body."[4] Lecturers and their audiences also increasingly encountered, and competed with, new ways to verbally and visually communicate science in a fast-changing media environment. The advent of the phonograph, the microphone and of radio and motion pictures all helped to alter market demand, performance techniques and cultural expectations around spoken public discourse. From the 1920s, the radio created new and potentially larger audiences for science.[5] The visual was of course lost in this medium and the body of the speaker, so often commented upon in reports of lectures, disappeared from direct view. This may have accomplished what the cast of speakers considered in this book seemed at times to desire. They all consciously held their bodies still, hoping their physical presence would not distract from the intrinsic appeal of scientific truths and a scientific life. But these attempts to control their movements on stage were part of their message. To see the still form of the lecturer was to encounter an embodiment of scientific virtue. At the same time, such composure was never perfectly achieved or necessarily always desired. This was not invariably a threat to success, at least by some important measures. The appearance and motion of a speaker's body, whether contrived or not, was considered a critical technology for effective and emotive verbal communication and for generating the kind of public fascination that fueled celebrity status.

The parallel rise of film, and then later television, offered further opportunities to connect image and word in the service of mass science communication. A fascinating and growing literature on science in film and TV documentaries points to trends that may have deeper roots in the lecture culture of the Gilded Age, even if the contrasts and novel developments are also critical to understand science on screen.[6] Tim Boon, for example, has argued that, in the 1960s, there was "a distinct shift to a form of television science that emphasised the personality of scientists, and which required them to deliver a new kind of performance of self." This, Boon argues, meant that "the skills required of scientists here are televisual—essentially kinds of communicative performance of self—rather than those they would necessarily need to undertake the science that forms the ostensible subject of the programme."[7] This was, in many if not in all respects, a return to an issue that celebrity scientist lecturing in Gilded Age America were acutely aware of. The cultivation of a scientific persona was of critical importance and the effective use of oratorical skills that helped constitute and communicate that persona was paramount to the success of a lecture tour.

In closing, we might ask whether, in the very different media environments of today, the arguments made here continue to have resonance. It is, of course, a different project to attempt to answer this. Nevertheless, as the lectures of the scientists followed here fall silent, we might recall the living voices of celebrity scientist-broadcasters in our own time. We may hear the polite Oxford intonation that mediates (if not mellows) Richard Dawkins's hawkish scientific atheism or summon up the soothing timbre of David Attenborough evoking the variety and vulnerability of the natural world. Perhaps some readers have encountered the conversational, open-shirted evangelical science of Francis Collins (sometimes accompanied with guitar) easing the passage of biocultural change. Or maybe others are reminded of Brian Cox's vocals, with their earnest "imagine there's no heaven" tone, declaring the wonders of the material cosmos a glorious substitute for ancient superstitions.[8] These voices supply prima facie evidence that the transformative but unstable cultural power of scientific speech persists, even if heard and seen in media unimaginable to Victorian scientist-lecturers and their audiences.

NOTES

PREFACE AND ACKNOWLEDGMENTS

1. Arthur Conan Doyle, "The Voice of Science," Strand Magazine 1 (1891): 312–17. For two recent interpretations, see John M. Picker, Victorian Soundscapes (Oxford: Oxford University Press, 2003): 127–29 and Jonathan Cranfield, Twentieth-Century Victorian: Arthur Conan Doyle and the Strand Magazine, 1891–1930 (Edinburgh: Edinburgh University Press, 2016): 24–25.

2. Doyle, "The Voice of Science," 313.

3. Doyle, "The Voice of Science," 314.

4. Charles Lyell to Charles Darwin, September 8, 1838, Darwin Correspondence Project, "Letter no. 425," accessed on December 3, 2020, https://www.darwinproject.ac.uk/letter/DCP-LETT-425.xml.

5. John Tyndall to Bence Jones, September 9, 1872, Royal Institution of Great Britain, MS JT/1/TYP/3/804.

INTRODUCTION: SCIENCE LECTURES IN AN AGE OF ORATORY

1. Thomas Henry Huxley, Discourses Biological and Geological (London: Macmillan, 1894), vi–vii.

2. Huxley, Discourses, vii.

3. Thomas Henry Huxley, "The School Boards: What They Can Do, and What They May Do," Contemporary Review 16 (1870): 11.

4. The literature on science and print culture is now extensive. Notable work published within the last two decades includes: Geoffrey Cantor et al., eds., Science in the Nineteenth-Century Periodical: Reading the Magazine of Nature (Cambridge, UK: Cambridge University Press, 2004); Geoffrey Cantor and Sally Shuttleworth, eds., Science Serialized: Representations of the Sciences in Nineteenth-Century Periodicals (Cambridge, MA: MIT Press, 2004); Gowan Dawson et al., eds., Science Periodicals in Nineteenth-Century Britain: Constructing Scientific Communities (Chicago: University of Chicago Press, 2020); Louise Henson et al., eds., Culture and Science in the Nineteenth-Century

Media (Farnham: Ashgate, 2004); and Aileen Fyfe and Bernard Lightman, eds., *Science in the Marketplace: Nineteenth-Century Sites and Experiences* (Chicago: University of Chicago Press, 2007).

5. See, for example, Ted Curtis Smythe, *The Gilded Age Press, 1865–1900* (Westport, CT: Praeger, 2003) and Jean Chalaby, *The Invention of Journalism* (London: Macmillan, 1998). For popular science in nineteenth-century America in particular, see Katherine Pandora, "Popular Science in National and Transnational Perspective: Suggestions from the American Context," *Isis* 100, no. 2 (June 2009): 346–58.

6. Donald Scott, "The Popular Lecture and the Creation of a Public in Mid-Nineteenth-Century America," *Journal of American History* 66, no. 4 (March 1980): 797; 805–6.

7. Joseph S. Meisel, *Public Speech and the Culture of Public Life in the Age of Gladstone* (New York: Columbia University Press, 2001).

8. Moses Coit Tyler, "How They Manage Their Lectures in England," *Putnam's Magazine* 13 (1869): 99.

9. Angela G. Ray, *The Lyceum and Public Culture in the Nineteenth-Century United States* (East Lansing: Michigan State University Press: 2005).

10. Carolyn Eastman, "Placing Platform Culture in Nineteenth-Century American Life," in *Thinking Together: Lecturing, Learning and Difference in the Long Nineteenth Century*, ed. Angela G. Ray and Paul Stob (Philadelphia: Pennsylvania University Press, 2018), 191.

11. Thomas Wentworth Higginson, "The American Lecture-System," *Macmillan Magazine* 18 (1868): 53.

12. Thomas F. Wright, *Lecturing the Atlantic: Speech, Print and an Anglo-American Commons, 1830–1870* (Oxford: Oxford University Press, 2017).

13. Sandra M. Gustafson, *Eloquence Is Power: Oratory and Performance in Early America* (Chapel Hill: University of North Carolina Press, 2000), 118.

14. See also William Clarke, *Academic Charisma and the Origins of the Research University* (Chicago: University of Chicago Press, 2006).

15. Jeremy C. Young, *The Age of Charisma: Leaders, Followers and Emotions in American Society, 1870–1940* (Cambridge, UK: Cambridge University Press, 2017), xiii.

16. Young, *Age of Charisma*, xviii.

17. Josephine Hoegarts, "Speaking Like Intelligent Men: Vocal Articulations of Authority in the House of Commons in the Nineteenth Century," *Radical History Review* 121, no. 1 (January 2015): 125.

18. Hoegarts, "Speaking Like Intelligent Men," 130.

19. Amanda Adams, *Performing Authorship in the Nineteenth-Century Transatlantic Lecture Tour* (Farnham: Ashgate, 2014), 10.

20. Thomas Augst, *The Clerk's Tale: Young Men and Moral Life in Nineteenth-Century America* (Chicago: University of Chicago Press, 2003), 138.

21. The literature on science lecturing in the nineteenth century, at least for the British case, is large. For a selection of more recent work, see Sophie Forgan, "Lis-

tening and Learning: Audiences and Their Roles in Nineteenth-Century Britain," in *Participating in the Knowledge Society: Researchers beyond the University Walls*, ed. Ruth Finnegan (Basingstoke: Palgrave Macmillan, 2005), 65–78; Jill Howard, "'Physics and Fashion': John Tyndall and His Audiences in Mid-Victorian Britain," *Studies in History and Philosophy of Science* 35, no. 4 (2004): 729–58; David Knight, "Scientific Lectures: A History of Performance," *Interdisciplinary Science Reviews* 27, no. 3 (2002): 217–24; Bernard Lightman, "Lecturing in the Spatial Economy of Science," in Fyfe and Lightman, *Science in the Marketplace*, 97–132; and Iwan R. Morus, "Seeing and Believing Science," *Isis* 97, no. 1 (March 2006): 101–10.

22. Earlier versions of this argument are made in Diarmid A. Finnegan, "Placing Science in an Age of Oratory: Spaces of Scientific Speech in Mid-Victorian Edinburgh," in *Geographies of Nineteenth-Century Science*, ed. David N. Livingstone and Charles W. J. Withers (Chicago: University of Chicago Press, 2011), 153–77, and Diarmid A. Finnegan, "Finding a Scientific Voice: Performing Science, Space and Speech in the Nineteenth Century," *Transactions of the Institute of British Geographers* 42, no. 2 (June 2017): 192–205. Sarah Zimmerman has done something broadly comparable in her study of literary lectures in Georgian Britain. See Sarah Zimmerman, *The Romantic Literary Lecture in Britain* (Oxford: Oxford University Press, 2019).

23. Juliet Pollock, "Michael Faraday," *Saint Paul's Magazine* 6 (1870): 294.

24. For more on Faraday's vocal performances and its place in oratorical culture more generally, see Finnegan, "Finding a Scientific Voice."

25. James Secord, "How Scientific Conversation Became Shop Talk," *Transactions of the Royal Historical Society* 17 (2007): 129–56.

26. Martin Hewitt, "Beyond Scientific Spectacle: Image and Word in Nineteenth-Century Popular Lecturing," in *Popular Exhibitions, Science and Showmanship*, ed. Joe Kember, John Plunkett and Jill A. Sullivan (London: Pickering and Chatto, 2011), 79–96.

27. David N. Livingstone, "Science, Site and Speech: Scientific Knowledge and the Spaces of Rhetoric," *History of the Human Sciences* 20, no. 2 (May 2007): 71–98. I have developed this argument in a number of places. See, for example, Finnegan, "Placing Science in an Age of Oratory" and idem., "Daniel William Cahill (1796–1864) and the Rhetorical Geography of Science and Religion," in Kember, Plunkett and Sullivan, *Popular Exhibitions*, 97–114.

28. Erving Goffman, *Forms of Talk* (Philadelphia: University of Pennsylvania Press, 1981).

29. I am grateful to David Livingstone for first pointing this out to me in one of his memorable research seminars (which is to say, coffee shop conversations).

30. Goffman, *Forms of Talk*, 188.

31. The concept of the communication circuit can be found in: Richard Darnton, "What Is the History of Books," *Daedalus* 111, no. 3 (Summer 1982): 65–83.

32. On the general theme of celebrity in this period, see Páiric Finnerty and Rod Rosenquist, "Transatlantic Celebrity: European Fame in Nineteenth-Century Ameri-

ca," *Comparative American Studies: An International Journal* 14, no. 1 (2016): 1–6 and the papers that follow. On press and publicity, see Lori Lyn Bogle, "Pandering to the Crowd: The American Governing Elite's Changing Views on Mass Media and Publicity during the Nineteenth Century," *Journalism History* 43, no. 2 (2017): 62–74.

33. James B. Pond, *Eccentricities of Genius* (London: Chatto and Windus, 1901), 8.

34. Michael Schudson, "Questioning Authority: A History of the News Interview in American Journalism, 1860s–1930s," *Media, Culture and Society* 16, no. 4 (October 1994): 565–87.

35. David Friedman, *Wilde in America: Oscar Wilde and the Invention of Modern Celebrity* (New York: W. W. Norton, 2015), 132–40.

36. Wright, *Lecturing the Atlantic*, 23–24.

37. On the Belfast example, see Finnegan, "Finding a Scientific Voice," 202.

38. Joshua Nall, *News from Mars: Mass Media and the Forging of a New Astronomy, 1860–1910* (Pittsburgh: University of Pittsburgh Press, 2019), 21–68.

39. Wright, *Lecturing the Atlantic*, 23.

40. Harriette K. Smith, *History of the Lowell Institute* (Boston: Lamson, Wolffe, 1898); Margaret W. Rossiter, "Benjamin Silliman and the Lowell Institute: The Popularization of Science in Nineteenth-Century America," *New England Quarterly* 44, no. 4 (December 1971): 602–26. The Wagner Free Institute of Science in Philadelphia also offered free lectures from its inception in 1855. See Erin Lorraine Anderson, "Nature on Display: The Wagner Free Institute of Science, 1855–1900," PhD diss., University of Delaware, 2020.

41. Robert Dott Jr., "Lyell in America: His Lectures, Field Work and Mutual Influences, 1841–1853," *Earth Sciences History* 15, no. 2 (1996): 113.

42. Robert H. Silliman, "The Hamlet Affair: Charles Lyell and the North Americans," *Isis* 86, no. 4 (December 1995): 541–61.

43. See Paul H. Theerman, "Dionysius Lardner's American Tour," in *Experiencing Nature*, ed. Paul H. Theerman and Karen Hunger Parshall (Dordrecht: Kluwer Academic Publishers, 1997), 211–36.

44. Jo N. Hays, "The Rise and Fall of Dionysius Lardner," *Annals of Science* 38, no. 5 (1981): 527–42.

45. See Kentwood D. Wells, "Dionysius Lardner: Popular Science Showman of the 1840s," *Magic Lantern Gazette* 29, no. 1 (Spring 2017): 3–18.

46. John P. Nichol, *Views of Astronomy* (New York: Greeley and McElrath, 1848), iii.

47. Trudy E. Bell, "Mitchel, Ormsby MacKnight," in *Biographical Dictionary of Astronomers*, ed. Thomas Hockey, vol. 2 (New York: Springer, 2007), 790. My thanks to an anonymous reviewer who drew my attention to the understudied Mitchel and to the (regrettably) unpublished work of the late Craig B. Waff on Mitchel's lectures.

48. Rossiter, "Benjamin Silliman and the Lowell Institute."

49. Quoted in Diarmid A. Finnegan, "Lectures," in *A Companion to the History of Science*, ed. Bernard Lightman (Chichester: John Wiley and Sons, 2016), 422.

50. See Anderson, "Nature on Display."

51. G. A. Bremner and Jonathan Conlin, "Consolations of Amero-Teutonism: E. A. Freeman's Tour of the United States, 1881–82," in *Making History: Edward Augustus Freeman and Victorian Cultural Politics*, ed. G. A. Bremner and Jonathan Conlin (Oxford: Oxford University Press, 2015), 103–4.

52. Amanda Adams, *Performing Authorship*, 9, 12.

53. On Collins, see Susan R. Hanes, *Wilkie Collins's American Tour, 1873–74* (London: Pickering and Chatto, 2014).

54. For example, "A Day among the Newspapers," *Cincinnati Daily Gazette*, October 16, 1872, 2.

55. On the importance of female "vocality" in British public culture in this period, see Janice Schroeder, "Speaking Volumes: Victorian Feminism and the Appeal of Public Discussion," *Nineteenth-Century Contexts* 25, no. 2 (2003): 97–117.

56. Friedman, *Wilde in America*.

57. Michèle Mendelssohn, *Making Oscar Wilde* (Oxford: Oxford University Press, 2018).

58. See chapter 4.

59. Lightman, "Lecturing in the Spatial Economy of Science."

60. W. Valentine Ball, ed., *Letters and Reminiscences of Sir Robert Ball* (London: Cassell, 1915), 216.

61. See, for example, Granville Ganter, "Women's Entrepreneurial Lecturing in the Early National Period," in Ray and Stob, *Thinking Together*, 41–55.

62. Phebe Mitchell Kendall, *Maria Mitchell: Life, Letters and Journals* (Boston: Lee and Shepard, 1896), 261.

63. On female suffrage and lecturing in the postbellum period, see Angela G. Ray, "What Hath She Wrought: Women's Rights and the Nineteenth-Century Lyceum," *Rhetoric and Public Affairs* 9, no. 2 (Summer 2006): 183–213 and Lisa Tetrault, "The Incorporation of American Feminism: Suffragists and the Postbellum Lyceum," *Journal of American History* 96, no. 4 (March 2010): 1027–56. On African American involvement in a mostly parallel lecture culture, see Shirley Wilson Logan, *Liberating Language: Sites of Rhetorical Education in Nineteenth-Century Black America* (Carbondale: Southern Illinois University Press, 2008). On Frederick Douglass's postbellum lecturing, see John R. McKivigan, "'A New Vocation before Me': Frederick Douglass's Post-Civil War Lyceum Career," *Howard Journal of Communication* 29, no. 3 (2018): 268–81; and Wright, *Lecturing the Atlantic*, chapter 2.

64. See Bonnie Carr O'Neill, *Literary Celebrity and Public Life in Nineteenth-Century United States* (Athens: University of Georgia Press, 2017).

65. For fascinating and important examples of prominent female speakers who talked extensively about evolution in the context of women's rights, see Kimberly A. Hamlin, *From Eve to Evolution: Darwin, Science and Women's Rights in Gilded Age America* (Chicago: University of Chicago Press, 2014).

66. "Out and About," *Worcester Daily Spy*, March 14, 1895, 4.

67. Recorded in Orison Swett Marden, *How They Succeeded* (Boston: Lothrop, 1901), 94.

68. David Cahan, "Helmholtz in Gilded Age America: The International Electrical Congress of 1893 and the Relations of Science and Technology," *Annals of Science* 67, no. 1 (2010): 1–38. See also David Cahan, *Helmholtz: A Life in Science* (Chicago: University of Chicago Press, 2018).

69. See Paul Lucier, "The Professional and the Scientist in Nineteenth-Century America," *Isis* 100, no. 4 (December 2009): 699–732. For the classic account of the invention and fate of the word, see Sydney Ross, "Scientist: The Story of a Word," *Annals of Science* 18, no. 2 (1962): 65–85.

70. Bernard Lightman, "The Theology of the Victorian Scientific Naturalists," in Peter Harrison and Jon H. Roberts, *Science without God: Rethinking the History of Scientific Naturalism* (Oxford: Oxford University Press, 2019), 234–54. See also Bernard Lightman and Michael Reidy, eds., *The Age of Naturalism: Tyndall and his Contemporaries* (Pittsburgh: University of Pittsburgh Press, 2014) and Gowan Dawson and Bernard Lightman, eds., *Victorian Scientific Naturalism: Community, Identity, Continuity* (Chicago: University of Chicago Press, 2014).

71. Tyler, "How They Manage Their Lectures in England," 106.

72. Quoted in John R. McKivigan, *Forgotten Firebrand: James Redpath and the Making of Nineteenth-Century America* (Ithaca: Cornell University Press, 2008), 135.

73. Augst, *Clerk's Tale*, 143.

74. David N. Livingstone, *Dealing with Darwin: Place, Politics and Rhetoric in Religious Engagements with Evolution* (Baltimore: Johns Hopkins University Press, 2014).

75. Morus, "Seeing and Believing Science"; idem., "Worlds of Wonder: Sensation and the Victorian Scientific Performance," *Isis* 101, no. 4 (December 2010): 806–16; Simon Schaffer, "Transport Phenomenon: Space and Visibility in Victorian Physics," *Early Popular Visual Culture* 10, no. 1 (2012): 71–91.

CHAPTER 1: SCIENCE, SPEECH AND CHARACTER

1. John Tyndall, *Fragments of Science for Unscientific People* (New York: D. Appleton, 1871), 103.

2. Tyndall, *Fragments*, 104.

3. John Tyndall, *New Fragments* (New York: D. Appleton, 1892), 47–77.

4. For further exploration of the influence of Emerson, Carlyle and Fichte on Tyndall, see Ian Hesketh, "Technologies of the Scientific Self: John Tyndall and His Journals," *Isis* 110, no. 3 (September 2019): 460–82.

5. On this, see Stefan Collini, *Public Moralists: Political Thought and Intellectual Life in Britain, 1860–1930* (Oxford: Clarendon, 1991).

6. Tyndall, *New Fragments*, 77.

7. Tyndall, *Fragments*, 103.

8. Ralph Waldo Emerson, "Character," in *Collected Works of Ralph Waldo Emerson*, vol. 3 (Cambridge, MA: Harvard University Press, 1983), 53–67. For further reflections on Emerson, character and lecture culture, see Thomas Augst, "Composing the Moral Senses: Emerson and the Politics of Character in Nineteenth-Century America," *Political Theory* 27, no. 1 (February 1999): 85–120.

9. Tyndall, *Fragments*, 103.

10. Tyndall, *New Fragments*, 37.

11. Emerson, "Character," 58, 60.

12. See, for example, Tyndall, *Fragments*, 51–2.

13. John Tyndall, *The Forms of Water in Clouds and Rivers, Glaciers and Ice*, 4th ed. (New York: D. Appleton, 1877), xvii.

14. John Tyndall, "My Schools and Schoolmasters," *Popular Science Monthly* 26 (1885): 333–35.

15. Tyndall, "My Schools and Schoolmasters," 335.

16. Patrick Joyce, *Democratic Subjects: The Self and the Social in Nineteenth Century England* (Cambridge, UK: Cambridge University Press, 1994), 103.

17. Tom F. Wright, "Listening to Emerson's 'England' at Clinton Hall," *Journal of American Studies* 46, no. 3 (August 2012): 646.

18. Bonnie Carr O'Neill, "'The Best of Me Is There': Emerson as Lecturer and Celebrity," *American Literature* 80, no. 4 (December 2008): 756.

19. See, for example, Tyndall, *Fragments*, 105.

20. Thomas H. Huxley, "Professor Tyndall," *Nineteenth Century* 35 (1894): 5.

21. Tyndall Journal, February 11, 1853, Royal Institution of Great Britain (RI hereafter) MS JT2/V, 189.

22. Tyndall Journal, May 18, 1855, RI MS JT2/VIa, 59.

23. "Prof. Tyndall on Light," *New York Tribune*, December 18, 1876, 1.

24. Quoted in Ronald A. Bosco and Joel Myerson, eds. *The Selected Lectures of Ralph Waldo Emerson* (Athens: University of Georgia Press, 2005), xvii.

25. John Fiske, *Edward Livingston Youmans: Interpreter of Science for the People* (New York: D. Appleton, 1894), 140.

26. See, for example, "The Voice and Character," *Bow Bells* 5 (1866): 89; "The Voice as an Index of Character," *Musical Standard* 29 (1885): 192; "How Voice Reveals Character," *Review of Reviews* 9 (1893): 156.

27. Ralph Waldo Emerson, "Eloquence," in *Collected Works of Ralph Waldo Emerson*, vol. 8 (Cambridge, MA: Harvard University Press, 2010), 64.

28. Joseph Dalton Hooker to Charles Darwin, September 4, 1866, Darwin Correspondence Project, "Letter no. 5206," accessed on November 16, 2019, https://www.darwinproject.ac.uk/letter/DCP-LETT-5206.xml.

29. P. Chalmers Mitchell, "Professor Tyndall," *New Review* 10 (1894): 79.

30. Tyndall to Huxley, September 12, 1867, RI MS JT/1/TYP/9/2912.

31. Fiske, *Youmans*, 140.

32. Huxley to Tyndall, January 1, 1873, Imperial College London, Records and Archives, Thomas Henry Huxley Collection, 8.131.

33. See Jane Smeal Thompson and Helen Thompson, *Silvanus Phillips Thompson: His Life and Letters* (London: T. Fisher Unwin, 1920), 157.

34. Tyndall Journal, February 17, 1866, RI MS JT2/VIIIa/p. 117.

35. For evidence of a significant shift in Tyndall's public reputation after his Belfast address, see Bernard Lightman, "Scientists as Materialists in the Periodical Press: Tyndall's Belfast Address," in *Science Serialized: Representations of Science in Nineteenth-Century Periodicals*, ed. Geoffrey Cantor and Sally Shuttleworth (Cambridge, MA: MIT Press, 2004), 199–237.

36. Stefan Collini, "The Idea of 'Character' in Victorian Political Thought," *Transactions of the Royal Historical Society* 35 (1985): 45–46.

37. For a cogent exploration of this connection, see Michael S. Reidy, "Evolutionary Naturalism on High: The Victorians Sequester the Alps," in *Victorian Scientific Naturalism: Community, Identity, Continuity*, ed. Bernard Lightman and Gowan Dawson (Chicago: University of Chicago Press, 2014), 55–78.

38. John Tyndall to William Spottiswoode, November 11, 1877, RI MS HT/1/T/1313.

39. In particular, Iwan Rhys Morus, "Worlds of Wonder: Sensation and the Victorian Scientific Performance," *Isis* 101, no. 4 (December 2010): 806–16; and Jill Howard, "'Physics and Fashion': John Tyndall and His Audiences in Mid-Victorian Britain," *Studies in History and Philosophy of Science* 35, no. 4 (2004): 729–58.

40. John Tyndall, "*Life and Letters of Faraday*," *Academy* 1, no. 8 (1870): 205.

41. Martin Hewitt, "Beyond Scientific Spectacle: Image and Word in Nineteenth-Century Popular Lecturing," in *Popular Exhibitions, Science and Showmanship, 1840–1910*, ed. Joe Kember, John Plunkett and Jill A. Sullivan (London: Pickering and Chatto, 2012), 79–96.

42. Tyndall, *Fragments*, 105.

43. Ralph Waldo Emerson, "Eloquence," in *Collected Works of Ralph Waldo Emerson*, vol. 7 (Cambridge, MA: Harvard University Press, 2007), 49.

44. Tyndall, *Fragments*, 103.

45. Tyndall, *New Fragments*, 232.

46. Tyndall Journal, January 17, 1854, RI MS JT2/V, 309.

47. Howard, "'Physics and Fashion.'"

48. Jeremiah Rankin and Ruth Barton, "Tyndall, Lewes and Popular Representations of Scientific Authority in Victorian Britain," in *The Age of Scientific Naturalism*, ed. Bernard Lightman and Michael S. Reidy (London: Pickering and Chatto, 2014), 51–70.

49. Emerson, "Eloquence," 40.

50. For a good overview of Tyndall's North British critics, see Howard, "'Physics and Fashion.'"

51. See the account in Cargill Gilston Knott, *The Life and Scientific Work of Peter Guthrie Tait* (Cambridge, UK: Cambridge University Press, 1911), 171–75.

52. For a close contextual reading of Maxwell's poem, see Daniel Brown, *The Poetry of Victorian Scientists: Style, Science and Nonsense* (Cambridge, UK: Cambridge University Press, 2013).

53. Gowan Dawson, *Darwin, Literature and Victorian Respectability* (Cambridge, UK: Cambridge University Press, 2007), 106.

54. John Tyndall, "On a New Series of Chemical Reactions Produced by Light," *Proceedings of the Royal Society* 17 (1868): 102.

55. James Clerk Maxwell, "To the Chief Musician on Nubla [*sic*]: A Tyndallic Ode," *Nature* 4 (1871): 291.

56. James Clerk Maxwell, "Section A: President's Address," *Nature* 2 (1870): 420.

57. Tyndall, *Fragments*, 3.

58. John Tyndall to Thomas Hirst, July 1, 1872, British Library (BL hereafter) Manuscript Collections, Add MS 63092, ff. 118–21.

59. "Biographical Sketch of Professor Tyndall," *Appleton's Journal* 2 (1869): 339.

60. John Tyndall to Thomas Hirst, July 1, 1872, BL Add MS 63092, ff. 118–21.

61. "Biographical Sketch," 340; 341.

62. Joseph Henry et al. to John Tyndall, September 1, 1871, RI MS JT/1/H/76.

63. Hector Tyndale to John Tyndall, August 5, 1872, RI MS HT/1/T/67.

64. On Dickens's American tour, see Malcolm Andrews, *Charles Dickens and His Performing Selves* (Oxford: Oxford University Press, 2006).

65. "Spirit and Spice of Leading Religious Periodicals," *New York Herald*, October 29, 1872, 6.

66. Edward L. Youmans, "Professor Tyndall," *Popular Science Monthly* 1 (1872): 751.

67. "Professor Tyndall's Lectures," *Galaxy* 14 (1872): 561.

68. On Morton, and his involvement with Tyndall and his American tour, see Jeremy Brooker, "A Lecture on Locust Street: Morton, Tyndall, Pepper, and the Construction of Scientific Identity," in *Science Museums in Transition: Cultures of Display in Nineteenth-Century Britain and America*, ed. Carin Berkowitz and Bernard Lightman (Pittsburgh: University of Pittsburgh Press, 2017), 111–38.

69. For a detailed comparison of Pepper and Tyndall, see Bernard Lightman, *Victorian Popularizers of Science: Designing Nature for New Audiences* (Chicago: University of Chicago Press, 2007).

70. John Tyndall to Hector Tyndale, August 21, 1872, RI MS HT/1/T/454.

71. "Professor Tyndall's Lectures," *Galaxy* 14 (1872): 561.

72. On the *Christian Union*, see Frank Luther Mott, *A History of American Magazines, 1865–1885* (Cambridge, MA: Harvard University Press, 1938), 66–7.

73. John Yates, "Professor Tyndall," *Christian Union* 6 (1872): 343–44.

74. "Table Talk," *Appletons' Journal* 8 (1872): 248.

75. "English Lecturers in America," *Pall Mall Gazette*, October 30, 1872.

76. John Tyndall and Henry Thompson, "The 'Prayer for the Sick.' Hints Towards a Serious Attempt to Estimate Its Value," *Contemporary Review* 20 (1872): 205–10. On the "prayer gauge debate" see Frank M. Turner, "Rainfall, Plagues and the Prince of Wales: A Chapter in the Conflict of Religion and Science," *Journal of British Studies* 13, no. 2 (May 1974): 46–65 and Robert Bruce Mullin, "Science, Miracles and the Prayer-Gauge Debate," in *When Science and Christianity Meet*, ed. David C. Lindberg and Ron Numbers (Chicago: University of Chicago Press, 2003), 203–24.

77. John Tyndall, "The Constitution of the Universe," *Fortnightly Review* 3 (1865): 144.

78. John Tyndall to John Clausius, June 27, 1873, RI MS JT/1/T/205.

79. John Tyndall, "Science and Religion," *Popular Science Monthly* 2 (1872): 81. Note that the article was also reprinted in the *New York Tribune* (October 28, 1872), thus reaching a readership that likely overlapped with the audience that heard Tyndall in the city.

80. Emerson, "Eloquence," 49.

81. John Tyndall, *Lectures on Light* (New York: D. Appleton, 1873), 4.

82. John Tyndall, *Notes of a Course of Nine Lectures on Light* (London: Longmans, Green, 1870).

83. Brooker, "Lecture on Locust Street."

84. Harriette Knight Smith, *History of the Lowell Institute* (Boston: Lamson, Wolffe, 1898).

85. "Professor Tyndall at Boston," *New York Tribune*, October 16, 1872, 1.

86. "Popular Science," *Boston Daily Advertiser*, October 25, 1872, 2.

87. "Professor Tyndall at Boston," 1.

88. Bence Jones to John Tyndall, November 18, 1872, RI MS JT/1/J/90.

89. "Professor Tyndall's Plea for Original Science," *New York Herald*, December 16, 1872, 7.

90. "Professor Tyndall's Exertions in New York," *Boston Evening Journal*, January 2, 1873, 4.

91. John Tyndall to Thomas Hirst, October 20, 1872, RI MS JT/1/T/999.

92. John Tyndall to Thomas Hirst, November 23, 1872, RI MS JT/1/T/696.

93. John Tyndall to Joseph Henry, October 26, 1872, William Jones Rhees Papers, The Huntington Library, San Marino, California, Box 43, mssRH 3970.

94. E.g. John Tyndall to Edward Youmans, December 1, 1872; Eliza A. Youmans, "Tyndall's American Visit," *Popular Science Monthly* 44 (1894): 511.

95. *Annual Report of the Provost to the Trustees of the Peabody Institute* 10 (1877): 15.

96. Joseph Henry to John Tyndall, October 6, 1872. Smithsonian Institution Archives, Joseph Henry Collection, RU 7001, Box 4A, Folder 1, 651.

97. For confirmation of the president's attendance, see "Notes," *Nature* 7 (1872): 150. For his being "delighted" see Joseph Henry to John Tyndall, January 7, 1873, RI MS JT/1/H/78. For "thirty congressmen," see "Washington Letters," *Cincinnati Commercial*, December 16, 1872, 2.

98. "Prof. Tyndall: His Sixth and Last Lecture," *National Republican,* December 13, 1872, 4.

99. John Tyndall to Thomas Hirst, January 3, 1873, RI MS JT/1/T/698.

100. John Tyndall to Thomas Hirst, December 22, 1872, RI MS JT/1/T/697.

101. "Cheap Literature," *New York Tribune,* February 26, 1873, 4.

102. John Tyndall to Bence Jones, January 21, 1873, RI MS JT/1/TYP/3/823.

103. "Tyndall and Froude," *Brooklyn Daily Eagle,* December 9, 1872, 2.

104. John Tyndall to Thomas Hirst, January 11, 1873, RI MS JT/1/T/699.

105. "Prof. Tyndall on Light," 1.

106. Emerson, "Character," 61.

107. "Washington Letters," 2.

108. "Spectrum Analysis," *New York Tribune,* January 1, 1873, 2.

109. "Spectrum Analysis," 2.

110. Emerson, "Eloquence," vol. 7, 48.

111. "Professor Tyndall's Lectures," *Albany Journal,* December 28, 1872, 2.

112. Cited in "Modern Skepticism," *Scribner's Monthly* 6 (1873): 429.

113. "Christmas Week in the Religious Press," *New York Herald,* December 29, 1872, 6.

114. "The Invisible Rays," *New York Tribune,* December 30, 1872, 5.

115. On Emerson's lecturing style, see R. Jackson Wilson, "Emerson as Lecturer," in *Cambridge Companion to Ralph Waldo Emerson,* ed. Joel Porte (Cambridge, UK: Cambridge University Press, 1999), 76–96.

116. George Ripley, "Ten Days in London," *New York Tribune,* August 17, 1869, 1. That Ripley was the author was revealed in "Biographical Sketch of Professor Tyndall," 340.

117. John Tyndall to Heinrich Debus, December 26, 1872, RI MS JT/1/T/269.

118. See for example, "Discussion of Prof. Tyndall's Test," *Cincinnati Daily Gazette,* October 16, 1872, 2.

119. *Indianapolis Sentinel,* October 12, 1872, 1.

120. John Tyndall to Bence Jones, January 21, 1873, RI MS JT/1/TYP/3/823.

121. As discussed below, John Hall was identified as the minister in correspondence published in the *Popular Science Monthly.* Confirmation can also be found in Thomas C. Hall, *John Hall: Pastor and Preacher* (New York: Fleming H. Revell, 1901), 275.

122. "Polarized Light," *New York Tribune,* December 27, 1872, 2.

123. "Tyndall's Fourth Lecture," *New York Times,* December 27, 1872, 5. A letter published in the *New York Tribune* asserted that the quote containing "pray for them" from the Sermon on the Mount had not been used by Tyndall but was added (perhaps mischievously) by a reporter. "*Tribune* Reports of the Tyndall Lectures," *New York Tribune,* December 31, 1872, 2.

124. "Tyndall's Fourth Lecture," 5.

125. "Prof. Tyndall and the Presbyterians," *New York Times,* December 30, 1872, 4.

126. "Letters to the Editor," *New York Times,* December 31, 1872, 2.

127. Albert Jackson to John Tyndall, January 13, 1873, RI MS JT/2/12/a.

128. John Tyndall to Bence Jones, January 3, 1873, RI MS JT/1/TYP/3/821.

129. Cited in Andrew P. Haley, *Restaurants and the Rise of the American Middle Class, 1880–1920* (Chapel Hill: University of North Carolina Press, 2011), 25.

130. John Tyndall to Joseph Henry, January 14, 1873, William Jones Rhees Papers, The Huntington Library, San Marino, California, Box 43, mssRH 3979.

131. "The Tyndall Dinner," *New York Times*, February 5, 1873, 5.

132. "Prof. Tyndall Honored," *New York Tribune*, February 5, 1873, 1.

133. "Professor Tyndall's Deed of Trust," *Popular Science Monthly* 3 (1873): 100.

134. See Debby Applegate, *The Most Famous Man in America: A Biography of Henry Ward Beecher* (New York: Doubleday, 2006).

135. "The Tyndall Dinner," 5.

136. "True and False Science," *New York Evangelist*, February 13, 1873, 1.

137. Oliver Wendell Holmes, "Letter from Dr. Holmes," *Sunday Times*, February 9, 1873, 12. I might suggest that the reference to "authoritative accents" was not at all incidental or merely metaphorical.

138. "Autocratic Impertinence," *Cincinnati Daily Gazette*, February 10, 1873, 4.

139. Edward L. Youmans, "A Correction: Letter from Prof. Tyndall," *Popular Science Monthly* 3 (1873): 241–43.

140. Alfred M. Mayer to John Tyndall, March 1, 1873, RI MS JT/1/M/75.

141. *Proceedings at the Farewell Banquet to Professor Tyndall* (New York: D. Appleton, 1873).

142. Joseph Henry to Benjamin Silliman Jr., February 28, 1873, in *The Papers of Joseph Henry*, vol. 11, *January 1866–May 1878*, ed. Marc Rothenberg et al. (Sagamore Beach, MA: Watson Publishing/Science History Publications, 2007), 449.

143. William C. Brownell, "English Lecturers in America," *Galaxy* 20 (1875): 62–72.

144. "Professor Tyndall's Lectures in New York," *Galaxy* 15 (1873): 273.

145. Graeme Gooday, "Ethnicity, Expertise and Authority: The Cases of Lewis Howard Latimer, William Preece and John Tyndall," in *Scientists' Expertise as Performance*, ed. Joris Vandendriessche, Evert Peeters and Kaat Wils (London: Pickering and Chatto, 2015), 15–30.

146. Emerson, "Eloquence," 49.

147. Emerson, "Character," 59; 61.

148. John Tyndall, "Personal Recollections of Thomas Carlyle," *Fortnightly Review* 48 (1890): 28.

149. Augst, "Composing the Moral Senses," 106.

150. "Professor Huxley's First Lecture," *New York Herald*, September 19, 1876, 6.

CHAPTER 2: REASON'S RHETOR

1. Charles Darwin, *The Descent of Man*, vol. 2 (London: John Murray, 1871), 336–37.

2. "Professor Huxley," *Sphinx* 3 (1870): 367.

3. "Men of the Day, No. 19," *Vanity Fair*, January 28, 1871, 306.

4. Thomas H. Huxley, "Autobiography," in *Collected Essays*, vol. 1 (London: Macmillan, 1893), 3.

5. Thomas H. Huxley, "English Literature and the Universities," *Pall Mall Gazette*, October 22, 1886, 1.

6. Josephine Hoegaerts, "Speaking Like Intelligent Men: Vocal Articulations of Authority and Identity in the House of Commons in the Nineteenth Century," *Radical History Review* 121, no. 1 (January 2015): 123–44. Janette Martin has discerned a similar turn away from classical modes of oratory among speakers championing a range of popular movements in provincial settings. Janette Martin, "Popular Political Oratory and Itinerant Lecturing in Yorkshire and the North East in an Age of Chartism, 1837–60," PhD diss., University of York, 2010.

7. William Robertson, *The Life and Times of Right Hon. John Bright* (London: Cassell, 1883), 243.

8. Huxley is reported to have said that Bright's "was the only oratory which ever really held me." See Wilfrid Ward, "Thomas Henry Huxley: A Reminiscence," *Nineteenth Century* 40 (1896): 286.

9. For Faraday, see introduction.

10. Thomas H. S. Escott, "The House of Commons: Its Personnel and Its Oratory," *Fraser's Magazine* 10 (1874): 516.

11. Thomas H. Huxley, *Collected Essays*, vol. 8 (London: Macmillan, 1894), v–vi.

12. Leonard Huxley, *Life and Letters of Thomas Henry Huxley*, vol. 1 (New York: D. Appleton, 1900), 124.

13. Huxley, *Life and Letters*, vol. 1, 202.

14. Huxley, "Autobiography," 15.

15. Thomas H. Huxley, "How to Become an Orator," *Pall Mall Gazette*, October 24, 1888, 1–2.

16. Reported in Fabian Franklin, *The Life of Daniel Coit Gilman* (New York: Dodd, Mead, 1910), 222–23.

17. Robert H. Ellison, *The Victorian Pulpit: Spoken and Written Sermons in Nineteenth-Century Britain* (Selinsgrove, PA: Susquehanna University Press, 1998).

18. Emily Murphy Cope, "'Inspiration of Delivery': John A. Broadus and the Evangelical Underpinnings of Extemporaneous Orator," *Rhetoric Society Quarterly* 45, no. 4 (2015): 279–99.

19. Ellison, *The Victorian Pulpit*.

20. The nature and extent of Huxley's debt to Nonconformist Christianity is discussed in Bernard Lightman, "Interpreting Agnosticism as a Nonconformist Sect: T. H. Huxley's 'New Reformation,'" in *Science and Dissent in England, 1688–1945*, ed. Paul Wood (Aldershot: Ashgate, 2004), 197–214. See also John H. Brooke, "The History of Science and Religion: Some Evangelical Dimensions," in *Evangelicals and Science in*

Historical Perspective, ed. David N. Livingstone, D. G. Hart and Mark Noll (Oxford: Oxford University Press, 1999), 17–42.

21. Adrian Desmond, *Huxley: From Devil's Disciple to Evolution's High Priest* (Reading, MA: Addison-Wesley, 1997), 209.

22. J. D. Hooker to C. Darwin, July 2, 1860, Darwin Correspondence Project, "Letter no. 2852," accessed on February 26, 2016, http://www.darwinproject.ac.uk/DCP-LETT-2852.

23. "Professor Huxley at Oxford," *Speaker* 7 (1893): 596.

24. J. Vernon Jensen, *Thomas Henry Huxley: Communicating for Science* (Newark: University of Delaware Press, 1991), 174.

25. Edward L. Youmans, "Prof. Huxley," *Popular Science Monthly* 9 (1876): 622.

26. Charles H. Spurgeon, *Lectures to My Students* (London: Passmore and Alabaster, 1875), 122.

27. John. A. Broadus, *A Treatise on the Preparation and Delivery of Sermons* (Philadelphia: Smith, English, 1871), 450.

28. George W. Smalley, "Mr. Huxley," *Scribner's Magazine* 18 (1895): 521.

29. Oliver Anderson, "Hansard's Hazards: An Illustration from Recent Interpretation of Married Woman's Property Law and the 1857 Divorce Act," *English Historical Review* 112, no. 449 (November 1997): 1202–15. See also Hoegaerts, "Speaking Like Intelligent Men" and Martin, "Popular Political Oratory," 130–42.

30. Thomas Allen Reed, a celebrated phonographer with experience recording Huxley's lectures, suggested that the average pace of a speaker was 120 words per minute. Lawyers, he reported, could speak up to two hundred words per minute. Huxley is generally described as adopting a slower pace of delivery. See "How Fast Can People Talk?," *Shorthand Review* 5 (1893): 51.

31. Huxley, *Life and Letters*, vol. 1, 446.

32. On the substantial relations between "location" and "locution," or speech and space, in the context of scientific debates in this and other periods, see David N. Livingstone, "Science, Site and Speech: Scientific Knowledge and the Spaces of Rhetoric," *History of the Human Sciences* 20, no. 2 (May 2007): 71–98.

33. For a detailed account of Huxley's involvement in Sunday lecturing, see Ruth Barton, "Sunday Lecture Societies: Naturalistic Scientists, Unitarians and Secularists Unite against Sabbatarian Legislation," in *Victorian Scientific Naturalism: Community, Identity, Continuity*, ed. Gowan Dawson and Bernard Lightman (Chicago: University of Chicago Press, 2014), 189–219.

34. On the uneasy relationship between Huxley's lectures and some working-class and "freethinking" constituencies, see Desmond, *Huxley*, 292–93, 440; and for a later period, see Paul White, *Thomas Huxley: Making the "Man of Science"* (Cambridge, UK: Cambridge University Press, 2003), 141–48.

35. Thomas H. Huxley, "Professor Huxley and His Critics," *Scotsman*, January 24, 1862.

36. Joseph D. Hooker to Charles Darwin, January 19, 1862, Darwin Correspondence Project, "Letter no. 3395," accessed on March 4, 2016, http://www.darwinproject.ac.uk/DCP-LETT-3395.

37. William Macdonald, "The Origin of Species," *Scotsman*, January 29, 1862.

38. For a fuller account, see Diarmid A. Finnegan, "Placing Science in an Age of Oratory: Spaces of Scientific Speech in Mid-Victorian Edinburgh," in *Geographies of Nineteenth-Century Science*, ed. David N. Livingstone and Charles W. J. Withers (Chicago: University of Chicago Press, 2011), 153–77.

39. Desmond, *Huxley*, 463–82; William Peirce Randel, "Huxley in America," *Proceedings of the American Philosophical Society* 114, no. 2 (April 1970): 73–99; J. Vernon Jensen, "Thomas Henry Huxley's Lecture Tour of the United States, 1876," *Notes and Records of the Royal Society of London* 42, no. 2 (July 1988): 181–95.

40. Randel, "Huxley in America"; Jensen, "Thomas Henry Huxley's Lecture Tour."

41. Youmans, "Prof. Huxley," 622. On Youmans's wider role in American debates about science and religion, see James C. Ungureanu, "Edward L. Youmans and the 'Peacemakers' in the *Popular Science Monthly*," *Fides et Historia* 51, no. 2 (Summer–Fall 2019): 13–32.

42. Smalley, "Mr. Huxley," 522.

43. *New York Tribune*, August 8, 1876, 4.

44. "Huxley's Impressions of America," *New York Tribune*, August 26, 1876, 12.

45. "A Fling at Huxley," *Sunday Times*, September 3, 1876, 6.

46. Cited in James M. Smith, "Thomas Henry Huxley in Nashville: Part II," *Tennessee Historical Quarterly* 33, no. 3 (Fall 1974): 328.

47. James Summerville, "Albert Roberts, Journalist of the New South: Part II," *Tennessee Historical Quarterly* 42, no. 2 (Summer 1983): 179–202.

48. Cited in Smith, "Huxley in Nashville: Part II," 334.

49. Smith, "Huxley in Nashville: Part II," 335.

50. "Professor Huxley in Nashville," *Cincinnati Commercial*, September 10, 1876, 6.

51. "Was It Prof. Huxley or One of Burch's Jokes?" *Commercial and Legal Reporter*, September 13, 1876, 1.

52. Summerville, "Albert Roberts," 182.

53. Erving Goffman, *Forms of Talk* (Philadelphia: University of Pennsylvania Press, 1981), 168.

54. Daniel C. Gilman, *Address at the Inauguration of Daniel C. Gilman* (Baltimore: John Murphy, 1876), 38.

55. Gilman, *Address*, 36.

56. White, *Thomas Huxley*, 114.

57. For a wider look at these maneuvers, see James C. Ungureanu, "Science, Religion and the 'New Reformation' of the Nineteenth Century," *Science & Christian Belief* 31, no. 1 (2019): 41–61.

58. Thomas H. Huxley, "The School Boards: What They Can Do, and What They May Do," *Contemporary Review* (1870): 1–15.

59. On the essentially "moralistic" character of Gilman's faith, see Daryl Hart, "Faith and Learning in the Age of the University: The Academic Ministry of Daniel Coit Gilman," in *The Secularization of the Academy*, ed. George M. Marsden and Bradley J. Longfield (Oxford: Oxford University Press, 1991), 107–45. While I agree with Hart that Gilman was not the lapsed Congregationalist he is sometimes presented as, I differ from Hart in questioning a clear breach between a religious Gilman and a "secularist" Huxley. The unifying ground is, to me, the interesting feature of that relationship.

60. Hart, "Faith and Learning," 125

61. This description is derived from "Musical and Dramatic," *Cincinnati Commercial*, January 4, 1875, 3. Further details can be found in George W. Howard, *The Monumental City* (Baltimore: J. D. Ehler, 1873–1876), 333.

62. The text of Huxley's peroration used here is from "Professor Huxley," *New York Herald*, September 13, 1876, 10. Somewhat different wording appears in the *New York Tribune*: "Huxley at Baltimore," September 13, 1876, 8. The text that appeared in *Nature* differs again. All, however, end with "your success as [his] joy." That was a line Huxley did not forget.

63. Huxley's account of this episode is given in Huxley, "How to Become an Orator," 2. Gilman provides his account in, Daniel C. Gilman, *The Launching of a University* (New York: Dodd, Mead, 1906), 21.

64. On Huxley's interest in Punshon, see W. F. Bynum and Caroline Overy, *Michael Foster and Thomas Henry Huxley Correspondence, Medical History Supplement, No. 28* (London: Welcome Trust Centre for the History of Medicine, 2009), 27. For Punshon's remarkable powers of recollection see, Frederic W. Macdonald, *The Life of William Morley Punshon, LL.D.* (London: Hodder and Stoughton, 1887), 37.

65. The *New York Tribune* recorded Huxley as speaking at a rate of about 120 words per minute, perhaps a little faster than his normal pace. See "Huxley at Baltimore," 8.

66. Cited in Jensen, "Thomas Henry Huxley's Address," 264–65.

67. *Baltimore Sun*, September 13, 1876. Cited in Randel, "Huxley in America," 88–89.

68. Gilman, *Launching of a University*, 21.

69. Cited in Jensen, "Thomas Henry Huxley's Address," 265.

70. "The Religious Press on Huxley," *New York Herald*, September 24, 1876, 8.

71. "Huxley's Visit and Its Consequences," *Boston Evening Journal*, September 25, 1876, 2.

72. Gilman, *Launching of a University*, p. 22.

73. See Jensen, "Thomas Henry Huxley's Address," 264 and "Johns Hopkins University," *Boston Evening Journal*, September 26, 1876, 2.

74. "Prof. Huxley and the Johns Hopkins University," *Boston Evening Journal*, September 26, 1876, 2.

75. Cited in Gilman, *Launching of a University*, 23.

76. "Prof. Huxley as a Lecturer," *New York Tribune*, September 13, 1876, 4; "Professor Huxley," *New York Herald*, September 18, 1876, 4.

77. This advertisement appeared in several places. For example, *New York Tribune*, September 14, 1876, 6.

78. "Private View of Chickering Hall," *Sunday Mercury*, November 14, 1875, 4.

79. *New York Herald*, September 16, 1876, 5.

80. *New York Herald*, October 8, 1876, 11.

81. *New York Herald*, November 7, 1875, 4.

82. "Evidence of Evolution," *New York Tribune*, September 19, 1876, 8.

83. "The Theory of Evolution," *New York Times*, September 19, 1876, 8.

84. Randel, "Huxley in America," 93.

85. "Evidence of Evolution," *New York Tribune*, September 19, 1876, 1.

86. "Prof. Huxley's First Lecture," *New York Herald*, September 19, 1876, 6.

87. "Prof. Huxley's First Lecture," *Sun*, September 19, 1876, 2.

88. "Prof. Huxley's Evasion," *Sun*, September 19, 1876, 2.

89. "Evolution," *Cincinnati Daily Gazette*, September 21, 1876, 6.

90. "Professor Huxley's Second Lecture," *New York Herald*, September 21, 1876, 6.

91. "Prof. Huxley and John Milton," *New York Tribune*, September 23, 1876, 2. James McCosh published a longer response a few weeks later in the *World*, repeating at greater length the arguments made in his *Tribune* letter. The longer piece also appeared in the *Popular Science Monthly* and as one chapter in a short collection of papers. See James McCosh, *The Development Hypothesis: Is It Sufficient?* (New York: Robert Carter and Brothers, 1876).

92. "Professor Huxley's Lectures," *Sacramento Daily Union-Record*, October 7, 1876, 4.

93. "Professor Huxley," *New York Herald*, September 21, 1876, 3.

94. "Professor Huxley," *New York Herald*, September 21, 1876, 3.

95. "Professor Huxley," *Public Ledger*, September 27, 1876, 1.

96. "Evidences of Evolution," *New York Tribune*, September 23, 1876, 1.

97. "Huxley's Closing Lecture," *Sun*, September 23, 1876, 3.

98. "Evidences of Evolution," *New York Tribune*, September 23, 1876, 1.

99. *Portland Daily Press*, September 28, 1876, 2; Bayard Taylor, "Letter," *Cincinnati Commercial*, September 27, 1876, 5.

100. "Physiological Discoveries," *Daily Graphic*, September 27, 1876, 601.

101. A. G. "Letter," *Boston Morning Journal*, September 27, 1876, 2.

102. Reprinted in "Professor Huxley a Disappointment," *Cincinnati Daily Gazette*, September 27, 1876, 6.

103. "The Reportorial Hypothesis," *New York Times*, September 22, 1876, 4.

104. Huxley, *Life and Letters*, 500.

105. Huxley, *Life and Letters*, 502. Emphasis in original.

106. "Professor Huxley a Disappointment," 6.

107. Randel, "Huxley in America," 94.

108. Randel, "Huxley in America," 87. Here also, "greatness had been in their midst."

109. Desmond, *Huxley*, 482.

110. Desmond, *Huxley*, 480.

111. Taylor, "Letter," 5.

112. Martin Hewitt, "Beyond Scientific Spectacle: Image and Word in Nineteenth-Century Popular Lecturing," in *Popular Exhibitions, Science and Showmanship, 1840–1910*, ed. Joe Kember, John Plunkett and Jill A. Sullivan (London: Pickering and Chatto, 2012), 91.

113. See Gough's remarks along these lines in W. Carlos Martyn, *John B. Gough: The Apostle of Cold Water* (New York: Funk and Wagnalls, 1894), 301.

CHAPTER 3: RICHARD PROCTOR AND THE TEMPO OF SCIENCE

1. "Professor Proctor at the Opera," *New York Herald*, January 14, 1880, 10.

2. "Mr. Proctor's Visit," *New York Tribune*, April 8, 1874, 4.

3. "Sketch of R. A. Proctor," *Popular Science Monthly* 4 (1874): 487.

4. On Proctor's astronomical career and dramatic influence on the development of the culture of astronomical science in the late nineteenth century, see Joshua Nall, *News from Mars: Mass Media and the Forging of a New Astronomy, 1860–1910* (Pittsburgh: University of Pittsburgh Press, 2019), 21–68.

5. Richard A. Proctor, "Colours of the Double Stars," *Cornhill Magazine* 8 (1863): 679–87.

6. Richard A. Proctor, "Old and New Astronomy: An Autobiographical Sketch," *Knowledge* 11 (1888): 112–13.

7. On Clodd's correspondence with Proctor about his conversion, see Bernard Lightman, *Victorian Popularizers of Science: Designing Nature for New Audiences* (Chicago: University of Chicago Press, 2007), 309–11.

8. E.g., "Personal," *Daily Critic*, December 26, 1873, 1.

9. Job 27:33. King James Version.

10. See, for example, Richard A. Proctor, "Editorial Gossip," *Knowledge* 4 (1883): 318.

11. "Mr. Proctor on Star Drift and Nebulae," *Scientific Opinion* 3 (1870): 417.

12. 1871 Census of England & Wales, RG10/670, folio 134, page 13, accessed July 13, 2018, ancestry.co.uk.

13. Nall, *News from Mars*, 39–44.

14. "Sunday Lecture Society," *Graphic*, October 29, 1870, 430. On the Sunday Lecture Society, see Ruth Barton, "Sunday Lecture Societies: Naturalistic Scientists, Unitarians, and Secularists Unite against Sabbatarian Legislation," in *Victorian Scientific Naturalism: Community, Identity, Continuity*, ed. Gowan Dawson and Bernard Lightman (Chicago: University of Chicago Press, 2014), 189–219.

15. "Stoppage of Mr. Proctor's Sunday Lecture," *Sydney Morning Herald*, September 6, 1880, 6.

16. Richard A. Proctor, "Sydney and Sunday Lecturing," *Knowledge* 3 (1883): 57. For a full account of this episode, see Martin Bush, "The Proctor-Parkes Incident: Politics, Protestants and Popular Astronomy in Australia in 1880," *Historical Records of Australian Science* 28, no. 1 (2017): 26–36.

17. Richard A. Proctor, "Letters Received and Short Answers," *Knowledge* 8 (1885): 190.

18. Richard A. Proctor, "Editorial Gossip," *Knowledge* 3 (1883): 155.

19. Proctor, "Editorial Gossip," *Knowledge* 4 (1883): 292.

20. Richard A. Proctor, "The Type Writer," *Knowledge* 2 (1882): 402.

21. Proctor, "Old and New Astronomy," 112–3.

22. Both Lightman in *Victorian Popularizers* and Nall in *News from Mars* underline Proctor's commitment to this vision for astronomical science.

23. On this aspect of lecture culture, see Carolyn Eastman, "Placing Platform Culture in Nineteenth-Century American Life," in *Thinking Together: Lecturing, Learning and Difference in the Long Nineteenth Century*, ed. Angela. G. Ray and Paul Stob (Philadelphia: Pennsylvania University Press, 2018): 187–202.

24. Nall, *News from Mars*.

25. *Tribune Popular Science* (Boston: Henry L. Shepard, 1874). For an advert posted on the day of publication, see *New York Tribune*, July 21, 1874, 6. An earlier version appeared as *Tribune Extra no. 9*. See *New York Tribune*, January 24, 1874, 12.

26. On sale figures of the *Tribune Extra no. 9* version, see *New York Tribune*, March 6, 1874, 4.

27. Richard A. Proctor, *Six Lectures on Astronomy* (New York: Truth Seeker, 1876).

28. Richard A. Proctor, "Gossip," *Knowledge* 8 (1885): 228.

29. Richard A. Proctor, "Astronomy in America," *Popular Science Review* 15 (1876): 360.

30. Richard A. Proctor, "Our Two Brains," *Knowledge* 5 (1884): 309–10.

31. Ralph Waldo Emerson, "Eloquence," in *Collected Works of Ralph Waldo Emerson*, vol. 7 (Cambridge, MA: Harvard University Press, 2007), 34.

32. *New York Tribune*, April 4, 1874, 4.

33. Lightman, *Victorian Popularizers*, 295–350.

34. "Proctor-Agassiz," *New York Tribune*, January 24, 1874, 1.

35. Richard A. Proctor, "Letters to the Editors," *English Mechanic and World of Science* 20 (1874–1875): 473.

36. James Rush, *Philosophy of the Human Voice* (Philadelphia: n. p., 1827), 149, 446.

37. James E. Murdoch, *Orthophony, or, the Cultivation of the Voice, in Elocution*, 7th ed. (Boston: Ticknor, Reed and Fields, 1851), 164.

38. Joshua Hall McIlvaine, *Elocution: The Source and Elements of Its Power* (New York: Charles Scribner, 1871), 347.

39. Richard A. Proctor, "Lecturing Notes," *Knowledge* 3 (1883): 40; Proctor, "Editorial Gossip," *Knowledge* 4 (1883): 318.

40. Proctor, "Lecturing Notes," 40.

41. Proctor, "Editorial Gossip," *Knowledge* 4 (1883): 318.

42. Richard A. Proctor, "Gossip," *Knowledge* 8 (1886): 195.

43. "Theatrical and Personal Gossip in the Metropolis," *Cincinnati Daily Gazette*, November 28, 1879, 5.

44. Proctor, "Editorial Gossip," *Knowledge* 4 (1883): 318.

45. *Christian Union*, July 9, 1873, 1.

46. J. L. E. Dryer and H. H. Turner, *History of the Royal Astronomical Society* (London: Royal Astronomical Society, 1923), 173.

47. Richard A. Proctor, "Lectures and the London Papers," *Knowledge* 3 (1883): 217.

48. A full account of the dispute appears in Jessica Ratcliff, *The Transit of Venus Enterprise in Victorian Britain* (London: Pickering and Chatto, 2008). See also Dryer and Turner, *History of the RAS* and Nall, *News from Mars*.

49. Proctor, "Old and New Astronomy," 113.

50. Proctor, "Sydney and Sunday Lecturing," 57.

51. See relevant comments in Proctor, "Astronomy in America," 356.

52. Proctor, "Astronomy in America," 362–63.

53. C. A. Young, "Constitution of the Sun," *New York Tribune*, January 28, 1873, 8.

54. "First Appearance of Prof. Proctor," *Boston Evening Journal*, October 22, 1873, 4.

55. Proctor, "Gossip," *Knowledge* 8 (1885): 228.

56. Proctor announced he had "conclusively traced" the authorship to Holden some five years later. See Richard A. Proctor, "Personal Intelligence," *New York Herald*, January 22, 1880, 6.

57. "Literary News," *Atlantic Monthly* 34 (1874): 363–64.

58. Richard A. Proctor, "To the Editor of the *Atlantic Monthly*," *Atlantic Monthly* 34 (1874): 750–51; Richard A. Proctor, "Mr. R. A. Proctor and the *Atlantic Monthly*," *New York Times*, December 31, 1874, 4.

59. *Christian Union*, July 9, 1873, 1.

60. *Christian Union*, October 8, 1873, 1.

61. *Popular Science Review*, 4 (1873–1874): 116.

62. "Richard Anthony Proctor," *Scribner's Monthly* 7 (1873): 172; 175.

63. The extravagant title for Bellew was used in advertisements of his readings. E.g., *Boston Daily Advertiser*, September 27, 1873, 1. For a full account of Collins's tour, see Susan R. Hanes, *Wilkie Collins's American Tour, 1873–4* (London: Pickering and Chatto, 2008).

64. Proctor, "Gossip," *Knowledge* 8 (1885): 228.

65. *New York Tribune*, April 4, 1874, 4; *Indianapolis Sentinel*, March 4, 1874, 4.

66. "The Greatest Living Astronomer," *Salem Register*, August 28, 1873, 2.

67. "Lowell Institute," *Boston Daily Globe*, October 24, 1873, 4.

68. "First Appearance of Prof. Proctor," 4.

69. "Proctor's Tenth lecture," *Boston Daily Advertiser*, November 22, 1873, 1.

70. "Lowell Institute," *Boston Daily Globe*, October 22, 1873, 5.

71. "The Lecture Room," *Boston Evening Journal*, October 22, 1873, 4.

72. "The Lecture Season," *Boston Daily Globe*, October 25, 1873, 5.

73. "The Lecture Field," *Boston Daily Advertiser*, November 12, 1873, 1; "Professor Proctor's Eighth Lecture," *Boston Evening Journal*, November 17, 1873, 1.

74. On the modern history of the lively and controverted debates about the plurality of worlds, see Michael J. Crowe, *The Extraterrestrial Life Debate 1750–1900: The Idea of a Plurality of Worlds from Kant to Lowell* (Cambridge, UK: Cambridge University Press, 1988). On Proctor's influential and shifting views on the subject, see pp. 367–77.

75. "Lectures," *Boston Evening Journal*, October 28, 1873, 3. For Mitchell's career, see Renée Bergland, *Maria Mitchell and the Sexing of Science: An Astronomer among the American Romantics* (Boston: Beacon, 2008).

76. On the appeal of Gough's "Street life in London," see Tom F. Wright, "The Transatlantic Larynx in Wartime: John Gough's London Voices," in *Transatlantic Traffic and (Mis)Translations*, ed. Robin Peel and Daniel Maudlin (Durham: University of New Hampshire Press, 2013), 45–62.

77. For the hall's capacity see "The New Music Hall," *Boston Evening Transcript*, November 20, 1852, 2.

78. "Proctor's Eighth Lecture."

79. "The Lyceum Platform," *Boston Daily Advertiser*, November 26, 1873, 1.

80. "Mr. Proctor's Lectures," *Boston Daily Advertiser*, November 5, 1873, 2.

81. "Prof. Proctor's Lectures," *New York Tribune*, December 9, 1873, 4.

82. "Proctor's Lecture before the Young Men's Association," *Albany Evening Journal*, December 5, 1873, 3.

83. John Elderkin, *A Brief History of the Lotus Club* (New York: Lotus Club, 1895), 13.

84. "An English Tribute to American Scientists," *New York Tribune*, December 15, 1873, 7.

85. Maria Mitchell, "American Astronomers Abroad," *New York Tribune*, December 19, 1873, 3.

86. Richard A. Proctor, "The Astronomer Royal and American Astronomy," *New York Tribune*, December 24, 1874, 7.

87. "Reception to Richard A. Proctor," *New York Tribune*, January 5, 1874, 5.

88. "A Study of the Sun," *New York Tribune*, January 9, 1874, 10. See also "The Sun," *English Mechanic and World of Science* 18 (1873–1874): 528. On Morton's own lecturing success, see Jeremy Brooker, "A Lecture on Locust Street: Morton, Tyndall, Pepper, and the Construction of Scientific Identity," in *Science Museums in Transition: Cultures of Display in Nineteenth-Century Britain and America*, ed. Carin Berkowitz and Bernard Lightman (Pittsburgh: University of Pittsburgh Press, 2017), 111–38.

89. *New York Tribune*, January 9, 1874, 6.

90. *New York Tribune*, February 11, 1874, 4.

91. "Prof. Proctor's Lectures on Astronomy," *Brooklyn Daily Eagle*, January 5, 1874, 4.

92. "Prof. Proctor's Lectures," *Brooklyn Daily Eagle*, January 14, 1874, 2.

93. "Mr. R. A. Proctor's Lecture Last Night," *Cincinnati Commercial*, March 5, 1874, 4. On Vaughan, see DeLisle Stewart, "Vaughan, Daniel," in *American Dictionary of National Biography*, vol. 19, ed. Dumas Malone (New York: Charles Scribner's Sons, 1936), 235–36.

94. Richard A. Proctor, "A Note of Apology," *New York Herald*, March 19, 1874, 10.

95. On the physical demands of the latter part of his tour, see Richard A. Proctor, "Waking at Will," *English Mechanic and World of Science* 20 (1875): 551.

96. "This Morning's Gazette," *Cincinnati Daily Gazette*, February 14, 1874, 4.

97. "Against Prof. Proctor," *Cincinnati Daily Gazette*, March 16, 1874, 8.

98. Richard A. Proctor, "Letter to the Editors," *English Mechanic and World of Science*, 22 (1875–6): 220.

99. *New York Tribune*, October 23, 1875, 6.

100. First advertised in *Boston Daily Advertiser*, October 30, 1875, 1.

101. Richard A. Proctor, *Albany Evening Journal*, November 5, 1875, 2. For a fuller account of Proctor's changing views about life on other planets, see Lightman, *Victorian Popularizers*, 313.

102. "Prof. Proctor on the Origin of Matter and Life," *Albany Evening Journal*, November 8, 1875, 2.

103. "Prof. Proctor in Boston," *New York Tribune*, November 8, 1875, 1.

104. "A Convert of Science," *New York Tribune*, November 8, 1875, 4.

105. "Mr. Proctor's Change of Base," *Cincinnati Daily Gazette*, November 10, 1875, 4.

106. *Portland Daily Press*, November 10, 1875, 1.

107. Henry Ward Beecher, *Christian Union*, November 10, 1875, 386.

108. Richard A. Proctor, "Drift of Scientific Theory," *New York Tribune*, November 11, 1875, 5.

109. *New York Tribune*, November 11, 1875, 4.

110. As reported in "The Church of Rome and Science," *Cincinnati Daily Gazette*, November 19, 1875, 2.

111. See Ernan McMullin, "Darwin and the Other Christian Tradition," *Zygon* 46, no. 2 (June 2011): 291–316.

112. On this, and the yet more problematic case of Eve's creation, see Diarmid A. Finnegan, "Eve and Evolution: Christian Responses to the First Woman Question, 1860–1900," *Journal of the History of Ideas* 75, no. 2 (April 2014): 281–303.

113. Richard A. Proctor, "Note from Professor Proctor," *Daily Graphic*, November 11, 1875, 83.

114. Proctor, "Note from Professor Proctor," 83.

115. Proctor, "Astronomy in America," 363.

116. See, in particular, James Ungureanu, *Science, Religion and the Protestant Tradition: Retracing the Origins of Conflict* (Pittsburgh: University of Pittsburgh Press, 2019).

117. "Prof. Proctor in Boston," *New York Tribune*, November 15, 1875, 1.

118. "Prof. Proctor in Boston," *New York Tribune*, November 22, 1875, supplement, 2.

119. "Prof. Proctor in Boston," *New York Tribune*, November 29, 1875, 10.

120. Proctor, *Six Lectures on Astronomy*, 58–65.

121. Proctor, *Six Lectures on Astronomy*, 4.

122. Proctor, *Six Lectures on Astronomy*, 27.

123. Proctor, *Six Lectures on Astronomy*, 47.

124. Proctor, *Six Lectures on Astronomy*, 56–57.

125. *Columbian Register*, April 1, 1876, 1.

126. For his full itinerary, see Richard A. Proctor, "The Lecture System in America and England," *English Mechanic and the World of Science* 24 (1876): 1.

127. An advertisement for Proctor's lecture in Cincinnati, for example, promised the display of "600 stereopticon views." In "Prof. Proctor's Coming," *Cincinnati Daily Gazette*, February 5, 1876, 8.

128. "New Publications," *New York Herald*, December 22, 1875, 7; Richard A. Proctor, *Our Place among the Infinities* (New York: D. Appleton, 1876). The British edition appeared a month or so earlier and was published by the London publisher H. S. King.

129. "Personal," *New York Tribune*, November 23, 1875, 4.

130. Later rumors that his wife had traveled with him to Australia were false. See "Proctor on Perihelion," *Indianapolis Sentinel*, May 7, 1881, 2.

131. "Professor Proctor to Revisit America," *Boston Evening Journal*, July 11, 1879, 2. On Smyth's evangelical pyramidology, see David Gange, *Dialogues with the Dead: Egyptology in British Culture and Religion, 1822–1922* (Oxford: Oxford University Press, 2013): 131–35.

132. "Chat by the Way," *New York Herald*, October 5, 1879, 6.

133. "Richard A. Proctor," *New York Herald*, October 29, 1879, 5.

134. "Literature," *New York Herald*, December 19, 1879, 5. The book was available for sale the following day.

135. "Poetry of Astronomy," *New York Tribune*, November 11, 1879, 5.

136. "Infinitesimally Small Men," *New York Herald*, November 19, 1879, 6.

137. "Other Worlds and Suns," *New York Tribune*, November 21, 1879, 8.

138. "Other Worlds and Suns," 3. De Quincey's dream vision, which Proctor had been quoting for years, first appeared in Thomas De Quincey, "System of the Heavens, as Revealed by Lord Rosse's Telescope," *Tait's Edinburgh Magazine* 13 (1846): 564–78.

139. "Astronomy and Religion," *New York Times*, November 24, 1879, 8.

140. "Religion in Astronomy," *New York Tribune*, November 24, 1879, 5.

141. W. H. Barlow, "Mr. Proctor and Man," *New York Herald*, November 25, 1879, 3.

142. "Prof. Proctor's Next Lecture," *New York Times*, November 23, 1879, 12.

143. "Astronomy and Religion," 8.

144. "Astronomy and Religion," 8.

145. *Northern Christian Advocate*, December 4, 1879, 4.

146. Barlow, "Mr. Proctor and Man,"; "Professor Proctor Challenged," *New York Herald*, November 26, 3.

147. "Reply to Professor Proctor," *Brooklyn Daily Eagle*, December 2, 1879, 2.

148. Gange, *Dialogues with the Dead*. See also Charles W. J. Withers, *Zero Degrees: Geographies of the Prime Meridian* (Cambridge, MA: Harvard University Press, 2017), 132–36.

149. See R. J. Helmstadter, "Wild, Joseph," in *Dictionary of Canadian Biography*, vol. 13, (Toronto: University of Toronto/Université Laval, 2003), accessed July 18, 2018, http://www.biographi.ca/en/bio/wild_joseph_13E.html.

150. "Professor Proctor Criticized," *New York Herald*, December 12, 1879, 5.

151. "Proctor on Edison," *New York Herald*, December 22, 1879, 3.

152. See Proctor's essay on the phonograph in Proctor, *Pleasant Ways in Science*, 274–95. For a full American history of the development of electric light, see Ernest Freeberg, *The Age of Edison: Electric Light and the Invention of Modern America* (New York: Penguin, 2013).

153. "Prof. Proctor's Lectures," *Cincinnati Daily Gazette*, February 14, 1880, 8.

154. *Indianapolis Daily Sentinel*, February 12, 1880, 4.

155. See, for example, *Cincinnati Daily Tribune*, February 22, 1880, 4. There is also a hint that his lectures in Boston were something of a financial failure. See *New York Times*, May 25, 1880, 4.

156. The spatial fix can, among other things, be thought of as a temporary solution to weakening market demand through expansion into new and unexploited territories. For a fuller discussion, see David Harvey, "Globalization and 'the Spatial Fix,'" *Geographische Revue* 2 (2001): 23–30.

157. "Professor Proctor," *Daily Alta California*, March 28, 1880, 2.

158. "The Sun," *Daily Alta California*, April 6, 1880, 1.

159. "Professor Proctor," *Daily Alta California*, April 13, 1880, 1.

160. "The Pyramids," *Daily Alta California*, April 11, 1880, 1.

161. "The Genesis of Worlds," *Daily Alta California*, April 12, 1880, 1.

162. Richard A. Proctor, "Professor Proctor's Lectures," *New York Herald*, April 21, 1880, 6.

163. *New York Times*, May 25, 1880, 4.

164. "Myths and Marvels of Astronomy," *New York Herald*, March 8, 1880, 8.

165. Proctor was on Williams's roster from 1875 or earlier. See "List of Lectures," *Boston Daily Journal*, June 8, 1875, 2.

166. For example, Richard A. Proctor, "Evolution and Religion," *New York Herald*, January 25, 1880, 15.

167. Proctor, "Evolution and Religion," 15.

168. Richard A. Proctor, "The Past and Future of Our Earth," *Contemporary Review* 25 (1874): 74–92.

CHAPTER 4: ALFRED RUSSEL WALLACE

1. Evan Charteris, *The Life and Letters of Sir Edmund Gosse* (London: William Heinemann, 1931), 169.

2. Alfred R. Wallace (1886), letter to Edmund William Gosse, dated February 16, 1886. In G. W. Beccaloni, ed., Wallace Letters Online, WCP4127.4144, http://www .nhm.ac.uk/research-curation/scientific-resources/collections/library-collections/wal lace-letters-online/4127/4144/T/details.html.

3. Susan Goodman and Carl Dawson, *William Dean Howells: A Writer's Life* (Berkeley: University of California Press, 2005), 261.

4. Charteris, *Life and Letters*, 168; 173–75.

5. "Mr. Gosse in New York," *New York Tribune*, January 14, 1885, 3.

6. Alfred R. Wallace, *My Life: A Record of Events and Opinions*, 2 vols. (London: Chapman and Hall, 1905), vol. 1, 199–200.

7. Wallace, *My Life*, vol. 1, 224–25.

8. Wallace, *My Life*, vol. 2, 209.

9. Alfred R. Wallace, "An Essay, on the Best Method of Conducting the Kingston Mechanics' Institution," in *The History of Kington*, ed. Richard Parry (Kington: Charles Humphreys, 1845), 66–70.

10. Wallace, "An Essay," 66–70.

11. John Rowland, William Llewellyn and Alfred. R. Russell, "Corresponding Societies and Lecture Rooms Bill (in Favour), communication on meeting places," in *Appendix to the Reports of the Select Committee of the House of Commons on Public Petitions. Session 1846* (London: House of Commons, 1846), 319, http://people.wku.edu/charles. smith/wallace/S001B.htm.

12. Wallace, *My Life*, vol. 1, 233.

13. Alison Winters, *Mesmerized: Powers of Mind in Victorian Britain* (Chicago: University of Chicago Press, 1998), 130–35.

14. Wallace suggests 1844 as the year he heard Buckingham (*My Life*, 130). This was probably before he took up his teaching position in Leicester. Buckingham had established the British and Foreign Institute in 1844 and lectured on Palestine at the Hanover Square Rooms on George Street under its auspices in February and March of that year. See, for example, *Morning Post*, February 7, 1844, 1.

15. James Silk Buckingham, *An Improved Syllabus of Mr. Buckingham's Lectures on the Oriental World* (London: Hurst, Chance, 1830), 13.

16. Wallace, *My Life*, vol. 1, 168.

17. Wallace, *My Life*, vol. 1, 240.

18. Wallace, *My Life*, vol. 1, 224.

19. Wallace, *My Life*, vol. 1, 323.

20. "Literary and Philosophical Society," *Newcastle Courant*, November 15, 1867, 8.

21. Alfred Russel Wallace, "Lecture 1. On the Climate & Vegetation of the Tropics," in Beccaloni, Wallace Letters Online, WCP: 6022.6971, https://www.nhm.ac.uk/research-curation/scientific-resources/collections/library-collections/wallace-letters-online/6022/6971/S/details.html#S5. I date this manuscript of Wallace's lecture on tropical vegetation to 1867 and not 1863, as this lecture, and the two associated with it held by the Natural History Museum (reference number WP9/2) correspond to the list reported in the *Newcastle Courant* (see note 15 above) and do not correlate to lectures given in 1863.

22. Wallace, Alfred Russel, list of lectures (1867–[1895]), in Beccaloni, Wallace Letters Online, http://www.nhm.ac.uk/research-curation/scientific-resources/collections/library-collections/wallace-letters-online/5105/5610/S/details.html.

23. "The Claims of Labour," *Glasgow Herald*, June 15, 1886, 1; "The Claims of Labour," *Dundee Courier and Argus*, June 17, 1886, 1.

24. Alfred Russel Wallace, "The Depression of Trade: Its Causes and Its Remedies," 1886, https://people.wku.edu/charles.smith/wallace/S387.htm.

25. Charles H. Smith, "Wallace, Spiritualism and Beyond: 'Change' or 'No Change,'" in *Natural Selection and Beyond: The Intellectual Legacy of Alfred Russel Wallace*, ed. Charles H. Smith and George Beccaloni (Oxford: Oxford University Press, 2008), 391–424.

26. See Richard Milner, "Charles Darwin and Associates, Ghostbusters," *Scientific American* 275, no. 4 (October 1996): 96–101 and Ross A. Slotten, *Heretic in Darwin's Court: The Life of Alfred Russel Wallace* (New York: Columbia University Press, 2004), 337–45.

27. "The Slade Prosecution," *Times*, October 30, 1876, 11.

28. Alfred Russel Wallace, "Modern Spiritualism," *Boston Sunday Herald*, April 26, 1885, 9.

29. Alfred Russel Wallace to A. C. Swinton, December 23, 1885, in *Alfred Russel Wallace Letters and Reminiscences*, ed. James Marchant (London: Cassell, 1916), vol. 2, 155. See also Alfred Russel Wallace, letter to Othniel Charles Marsh, dated January 23, 1886, in Beccaloni, Wallace Letters Online, WCP5360.6081, http://www.nhm.ac.uk/research-curation/scientific-resources/collections/library-collections/wallace-letters-online/5360/6081/T/details.html.

30. Alfred Russel Wallace, letter to Raphael Meldola, dated March 2, 1886, in Beccaloni, Wallace Letters Online, WCP4490.4798, http://www.nhm.ac.uk/research-curation/scientific-resources/collections/library-collections/wallace-letters-online/4490/4798/T/details.html.

31. Carl Wilhelm Ernst, letter to Daniel Coit Gilman, dated February 2, 1886, in Beccaloni, Wallace Letters Online, WCP4853.5252, http://www.nhm.ac.uk/research-curation/scientific-resources/collections/library-collections/wallace-letters-online/4853/5252/T/details.html.

32. Othniel Charles Marsh, letter to Daniel Coit Gilman, dated February 12, 1886, in Beccaloni, Wallace Letters Online, WCP4854.5253, http://www.nhm.ac.uk/

research-curation/scientific-resources/collections/library-collections/wallace-letters
-online/4854/5253/T/details.html.

33. Alfred Russel Wallace to Othniel Charles Marsh, dated January 23, 1886, in Beccaloni, Wallace Letters Online, WCP5360.6081, http://www.nhm.ac.uk/research-curation/scientific-resources/collections/library-collections/wallace-letters -online/5360/6081/T/details.html.

34. Thomas Henry Huxley, letter to H. N. Martin, dated March 4, 1886, in Beccaloni, Wallace Letters Online, WCP4855.5254, http://www.nhm.ac.uk/research-curation/scientific-resources/collections/library-collections/wallace-letters-online/4855/5254/T/details.html.

35. "Personal Intelligence," *New York Herald*, April 13, 1886, 6.

36. For George's suggestion of Redpath, see Henry George, letter to A. R. Wallace, dated January 20, 1886, in Beccaloni, Wallace Letters Online, WCP2665.2555, http://www.nhm.ac.uk/research-curation/scientific-resources/collections/library-collections/wallace-letters-online/2665/2555/S/details.html#S2. On Wood's advice, see Wallace, *My Life*, vol. 2, 105.

37. See Angela Ray, *Lyceum and Public Culture in the Nineteenth-Century United States* (East Lansing: Michigan State University Press, 2005), 39–41.

38. "List of Lectures," *Boston Daily Journal*, June 8, 1875, 2.

39. Alfred Russel Wallace, letter to Francis Galton, dated March 7, 1886, in Beccaloni, Wallace Letters Online, WCP4142.4160, http://www.nhm.ac.uk/research-curation/scientific-resources/collections/library-collections/wallace-letters-online/4142/4160/T/details.html.

40. Wallace, *My Life*, vol. 2, 106.

41. Alfred Russel Wallace, letter to Raphael Meldola, dated August 7, 1886, in Beccaloni, Wallace Letters Online, WCP4494.4802, http://www.nhm.ac.uk/research -curation/scientific-resources/collections/library-collections/wallace-letters-online /4494/4802/T/details.html.

42. On Wallace reading *Progress and Poverty*, see Slotten, *Heretic in Darwin's Court*, 369–70. Quote from Alfred Russel Wallace, letter to Annie Wallace (née Mitten), dated October 23, 1886, in Beccaloni, Wallace Letters Online, WCP422.422, http://www .nhm.ac.uk/research-curation/scientific-resources/collections/library-collections/wal lace-letters-online/422/422/T/details.html.

43. "Henry George's Vigorous Canvass," *New York Tribune*, October 26, 1886, 1.

44. "Not Exactly Harmonious," *New York Times*, October 26, 1886, 2.

45. "George Makes More Speeches," *Sun*, October 26, 1886, 1. A report in the *New York Herald* confirms that George replied not to Wallace but to Miller. "George's Rapid Rounds," *New York Herald*, October 26, 1886, 3.

46. See Alfred Russel Wallace, letter to Annie Wallace (née Mitten), dated January 29, 1887, in Beccaloni, Wallace Letters Online, WCP435.435, http://www.nhm.ac.uk/research-curation/scientific-resources/collections/library-collections/wallace-letters -online/435/435/T/details.html.

47. See Harriette Knight Smith, *History of the Lowell Institute* (Boston: Lamson, Wolffe, 1898), 83.

48. *Boston Evening Transcript*, October 30, 1898, 2.

49. For example, *Boston Evening Journal*, October 29, 1886, 3.

50. "The English Naturalist," *Boston Sunday Herald*, October 31, 1886, 13.

51. Alfred Russel Wallace, "Sir Charles Lyell on Geological Climates and the Origin of Species," *Quarterly Review*, 126 (1869): 394. For a full account of Wallace's reflections on human evolution, see Sherrie Lyons, "The Many Influences Shaping Wallace's Views of Human Evolution," in *An Alfred Russel Wallace Companion*, ed. Charles H. Smith, James T. Costa and David A. Collard (Chicago: University of Chicago Press, 2019): 167–90.

52. "Alfred Russel Wallace, LL.D.," *Banner of Light*, November 6, 1886, 4.

53. "The Darwinian Theory," *Boston Daily Advertiser*, November 2, 1886, 8.

54. "Wallace and Darwin," *Banner of Light*, November 13, 1886, 4.

55. "Application of Darwinism," *Boston Herald*, November 5, 1886, 6; "Dr. Wallace at the Lowell Institute," *Banner of Light*, November 13, 1886, 5.

56. "The Harvard Anniversary," *Boston Evening Transcript*, November 9, 1886, 2.

57. "Sowing and Reaping," *Boston Evening Transcript*, November 16, 1886, 2.

58. See Smith, *History of the Lowell Institute*, 26.

59. "A Group of Lectures," *Boston Evening Transcript*, November 19, 1886, 2.

60. Alfred Russel Wallace, letter to Annie Wallace (née Mitten), dated November 14, 1886, in Beccaloni, Wallace Letters Online, WCP425.425, http://www.nhm.ac.uk/research-curation/scientific-resources/collections/library-collections/wallace-letters-online/425/425/T/details.html.

61. Alfred Russel Wallace, letter to Violet Isabel Wallace, dated December 12, 1886, in Beccaloni, Wallace Letters Online, WCP428.428, http://www.nhm.ac.uk/research-curation/scientific-resources/collections/library-collections/wallace-letters-online/428/428/T/details.html.

62. *Boston Herald*, 26 November 1886, 6.

63. Alfred Russel Wallace, letter to Annie Wallace (née Mitten), dated December 19, 1886, in Beccaloni, Wallace Letters Online, WCP431.431, http://www.nhm.ac.uk/research-curation/scientific-resources/collections/library-collections/wallace-letters-online/431/431/T/details.html.

64. Alfred Russel Wallace, letter to Annie Wallace (née Mitten), dated November 14, 1886, in Beccaloni, Wallace Letters Online, WCP425.425, http://www.nhm.ac.uk/research-curation/scientific-resources/collections/library-collections/wallace-letters-online/425/425/T/details.html.

65. Reproduced in *Banner of Light*, January 8, 1887, 5.

66. Reproduced in *Banner of Light*, January 15, 1887, 4.

67. Cited in Charles H. Smith and Megan Derr, eds., *Alfred Russel Wallace's 1886–1887 Travel Diary: The North American Lecture Tour* (Manchester: SIRI Scientific Press, 2013), 24.

68. Smith and Derr, *Travel Diary*, 26.

69. "The Peabody Institute," *New York Tribune*, October 26, 1866, 1.

70. *Annual Report of the Provost to the Trustees of the Peabody Institute* 2 (1869): 19–21.

71. *Annual Report of the Provost to the Trustees of the Peabody Institute* 20 (1887): 11–12.

72. Cited in Smith and Derr, *Travel Diary*, 27.

73. Daniel Coit Gilman, letter to Alfred Russel Wallace, dated December 8, 1886, in Beccaloni, Wallace Letters Online, WCP1365.1144, http://www.nhm.ac.uk/research-curation/scientific-resources/collections/library-collections/wallace-letters-online/1365/1144/T/details.html.

74. Enclosure by unknown hand, sent to Daniel Coit Gilman, dated December 9, 1886, in Beccaloni, Wallace Letters Online, WCP1365.1144, http://www.nhm.ac.uk/research-curation/scientific-resources/collections/library-collections/wallace-letters-online/4858/5258/S/details.html. Emphasis in the original.

75. Dawson's Lowell lectures were published as J. William Dawson, *The Geological History of Plants* (New York: D. Appleton, 1888). The volume was part of the International Scientific Series. For a fuller account of Dawson's views on evolution, see Nanna Katrine Lüders Kaalund, "Of Rocks and 'Men': The Cosmogony of John William Dawson," in *Historicizing Humans: Deep Time, Evolution and Race in Nineteenth-Century British Sciences*, ed. Efram Sera Shriar (Pittsburgh: University of Pittsburgh Press, 2018), 44–67.

76. "A Spiritualist Exposed," *New York Times*, February 3, 1887, 1.

77. "Ocean Islands and Their Relations," *New York Tribune*, January 12, 1887, 5.

78. Alfred Russel Wallace, letter to Violet Isabel Wallace, dated February 12, 1887, in Beccaloni, Wallace Letters Online, WCP436.436, http://www.nhm.ac.uk/research-curation/scientific-resources/collections/library-collections/wallace-letters-online/436/436/T/details.html.

79. Renamed George Washington University in 1904.

80. See Margaret W. Rossiter, *Women Scientists in America: Struggles and Strategies to 1940* (Baltimore: Johns Hopkins University Press, 1982), 80–82.

81. Wallace to Violet, February 12, 1887.

82. Anita N. McGee, "The Women's Anthropological Society of America," *Science* 13 (1889): 241.

83. "Not Ripe for Socialism," *National Republican*, February 16, 1887, 1.

84. "Not Ripe for Socialism."

85. "Mr. Alfred Russell Wallace," *Washington Critic*, February 16, 1887, 2.

86. Wallace, *My Life*, vol. 2, 129.

87. Kenneth E. Rowe, "Building Monumental Methodist Cathedrals in America's Capital City, 1850–1950," *Methodist History* 50, no. 3 (April 2012): 172.

88. "Grandeur in Living," *National Republican*, February 28, 1887, 1.

89. Smith and Derr, *Travel Diary*, 54–55.

90. Smith and Derr, *Travel Diary*, 61.

91. "Dr. Wallace's Lecture," *Varsity*, March 12, 1887, 205.

92. "Darwinian Theory," *Toronto Mail*, March 11, 1887, 6.

93. "Dr. Wallace's Lecture." For "silvery," see Smith and Derr, *Travel Diary*, 61.

94. "University College and Darwinism," *Toronto Daily Mail*, March 26, 1887, 6.

95. Sir Daniel Wilson's Journal, photocopy of B1965–0014/003(01), p. 112, University of Toronto Discover Archives, https://discoverarchives.library.utoronto.ca/index.php/sir-daniel-wilsons-journal-photocopy-of-b1965-0014-003-01-p-1-107.

96. On Wilson's views on Darwin and human origins, see Livingstone, *Dealing with Darwin: Place, Politics and Rhetoric in Religious Engagements with Evolution* (Baltimore: Johns Hopkins University Press, 2014), 97–102. For a later restatement of his skepticism about the evolutionary origins of mind or "the gift of [human] reason," see Daniel Wilson, "Proceedings at the Dinner," *Proceedings of the American Philosophical Society* 27 (1889): 31–34.

97. These claims can be found in Alfred Russel Wallace, *Darwinism* (London: Macmillan, 1889): 445–78.

98. Smith and Derr, *Travel Diary*, 61–62.

99. Smith and Derr, *Travel Diary*, 71. See also Alfred Russel Wallace, letter to Annie Wallace (née Mitten), April 5, 1887, in Beccaloni, Wallace Letters Online, WCP441.441, http://www.nhm.ac.uk/research-curation/scientific-resources/collections/library-collections/wallace-letters-online/441/441/T/details.html.

100. Alfred Russel Wallace, letter to Annie Wallace (née Mitten), April 22, 1887, in Beccaloni, Wallace Letters Online, WCP443.443, http://www.nhm.ac.uk/research-curation/scientific-resources/collections/library-collections/wallace-letters-online/443/443/T/details.html.

101. Smith and Derr, *Travel Diary*, 93.

102. Smith and Derr. *Travel Diary*, 96.

103. Wallace, *My Life*, vol. 2, 157.

104. For an informative account of this part of Wallace's tour, see Salvatore John Manna, "A Brothers' Reunion: Evolution's Champion Alfred Russel Wallace and Forty-Niner John Wallace," *California History* 85, no. 4 (2008): 4–25, 70–71.

105. "Wallace," *San Francisco Examiner*, May 24, 1887, 3.

106. Alfred Russel Wallace, letter to Annie Wallace (née Mitten), dated May 24, 1887, in Beccaloni, Wallace Letters Online, WCP448.448, http://www.nhm.ac.uk/research-curation/scientific-resources/collections/library-collections/wallace-letters-online/448/448/T/details.html.

107. For a full description, see "Pioneer Hall," *San Francisco Examiner*, May 6, 1886, 6.

108. "Darwinism," *San Francisco Examiner*, May 26, 1887, 2.

109. Respectively, "Man and Monkey," *San Francisco Chronicle*, 26 May 1887, 6; "Darwinism"; "Evolution," *Daily Alta California*, May 26, 1887, 1.

110. "Evolution"; "Man and Monkey."

111. "Darwinism."

112. "Darwinism."

113. "Man and Monkey"; "Darwinism."

114. "Darwinism."

115. First published in John Henry Dell, *The Dawning Grey* (London: Simpkin Marshall, 1885).

116. "Colors in Animals," *San Francisco Examiner*, May 28, 1887, 3.

117. "If a Man Dies, Shall He Live Again," *Golden Gate*, June 11, 1887, 1.

118. Other contributors included Asa Gray, Thomas H. Huxley, Daniel Gilman and William James. See Samuel J. Barrows, ed. *Science and Immortality: The* Christian Register *Symposium* (Boston: George H. Ellis, 1887).

119. *Golden Gate*, June 4, 1887, 4. Emphasis in original.

120. *Golden Gate*, May 28, 1887, 5.

121. "Wallace Comes Out," *San Francisco Examiner*, June 6, 1887, 2.

122. A detailed account is given in "Professor Wallace's Séance with Fred Evans," *Golden Gate*, June 4, 1887, 4.

123. The figure of one thousand was Wallace's estimate. See Smith and Derr, *Travel Diary*, 109.

124. Alfred Russel Wallace, "Lecture on Spiritualism," 1887, WP9/5/1, Natural History Museum Library and Archives.

125. Wallace, "Lecture on Spiritualism," 1.

126. Wallace, "Lecture on Spiritualism," 1–2.

127. Wallace, "Lecture on Spiritualism," 3.

128. Wallace, "Lecture on Spiritualism," 3a. Emphases in original.

129. Smith and Derr, *Travel Diary*, 107.

130. Wallace, "Lecture on Spiritualism," 12.

131. Wallace, "Lecture on Spiritualism," 17. Emphases in original.

132. Allen's poem later appeared in his *A Mortal Coil: A Novel* (London: Chatto and Windus, 1888) and, in full in Grant Allen, *The Lower Slopes* (London: Stone and Kimball, 1894). It is not clear where Wallace had encountered the poem but, being well acquainted with Allen and having spent many evenings with Allen's parents while in Washington, DC, he may have seen a prepublication version of *A Mortal Coil* or a previously published version of the poem itself.

133. Wallace, "Lecture on Spiritualism," 34.

134. For a complete list, see Charles H. Smith, "Bibliography of the Writings of Alfred Russel Wallace," http://people.wku.edu/charles.smith/wallace/bib2.htm, accessed December 5, 2018].

135. Smith and Derr, *Travel Diary*, 107.

136. "Prof. Wallace's Lecture," *Golden Gate*, June 11, 1887, 4.

137. See, for example, "The Life Hereafter," *San Francisco Chronicle*, June 6, 1887, 8; "A Great Problem," *Daily Alta California*, June 6, 1887, 1.

138. "The New Teaching and the Editor," *Golden Gate*, June 11, 1887, 5.

139. "Alfred Russel Wallace," *Mining and Scientific Press*, 54 (1887): 397.

140. Alfred Russel Wallace, letter to Annie Wallace (née Mitten), July 1, 1887, in Beccaloni, Wallace Letters Online, WCP452.452, http://www.nhm.ac.uk/research-curation/scientific-resources/collections/library-collections/wallace-letters-online/452/452/T/details.html.

141. Alfred Russel Wallace, letter to Annie Wallace (née Mitten), dated July 31, 1887, in Beccaloni, Wallace Letters Online, WCP457.457, http://www.nhm.ac.uk/research-curation/scientific-resources/collections/library-collections/wallace-letters-online/457/457/T/details.html.

142. Frank M. Turner, *Between Science and Religion: The Reaction to Scientific Naturalism in Victorian England* (New Haven, CT: Yale University Press, 1974).

143. Robert van Krieken, *Celebrity Society* (London: Routledge, 2012), 10.

CHAPTER 5: EVOLUTION'S EVANGELIST

1. For the Christmas release, see "Professor Drummond's Booklets," *Wesleyan-Methodist Magazine* 114 (1891): 713 and Cuthbert Lennox [John H. Napier], *The Practical Life of Henry Drummond* (New York: James Pott, 1901), 147.

2. Lennox, *Practical Life*, 147.

3. Lennox, *Practical Life*, 147.

4. George Adam Smith, *The Life of Henry Drummond* (New York: Doubleday and McClure, 1898), 34.

5. On the intense and extensive interest in mesmerism in the Victorian period, see Alison Winters, *Mesmerized: Powers of Mind in Victorian Britain* (Chicago: University of Chicago Press, 1998).

6. Smith, *Henry Drummond*, 31.

7. Smith, *Henry Drummond*, 104.

8. Smith, *Henry Drummond*, 35.

9. Henry Drummond, *The New Evangelism and Other Addresses* (New York: Dodd, Mead, 1899), 260.

10. On Moody's career as a celebrity evangelist and media savvy revivalist, see Bruce J. Evensen, *God's Man for the Gilded Age: D. L. Moody and the Rise of Modern Mass Evangelism* (Oxford: Oxford University Press, 2003).

11. Smith, *Henry Drummond*, 93.

12. See Kenneth Ronald Schott, "An Analysis of Henry Drummond and His Rhetoric of Reconciliation" (PhD diss., Ohio State University, 1972), 97–99.

13. On Drummond's influence on the Aberdeens, see Veronica Strong-Boag, *Liberal Hearts and Coronets: The Lives and Times of Ishbel Marjoribanks Gordon and John Campbell Gordon, the Aberdeens* (Toronto: University of Toronto Press, 2015), 83–86.

14. Smith, *Henry Drummond*, 274.

15. Quoted in Smith, *Henry Drummond*, 354.

16. Papers of Professor Henry Drummond, volume of press-cuttings 1885–1896, compiled by Henry Drummond, Acc. 5890/16, National Library of Scotland, Manuscripts Division.

17. "What It Ought to Be," *Duluth Evening Herald*, June 12, 1893, 5. The New York publishers James Pott published an authorized version in 1884.

18. "Best and Last," *Minneapolis Daily Tribune*, June 6, 1893, 6.

19. "The Northfield Convocation," *Springfield Republican*, August 8, 1887, 4.

20. See chapter 2.

21. Smith, *Life of Henry Drummond*, 287–88; Lennox, *Practical Life*, 92–93.

22. For the wider intellectual context, see Gregory Radick, *The Simian Tongue: The Long Debate about Animal Language* (Chicago: University of Chicago Press, 2007).

23. Henry Drummond, *The Ascent of Man* (London: Hodder and Stoughton, 1894), 203.

24. Drummond, *Ascent of Man*, 211.

25. Washington Gladden, "The New Evolution," *McClure's Magazine* 3 (1894): 238.

26. In 1894 Drummond stated that *Tropical Africa* was "the only book I have written" but that even it "was mostly lectured." See Arthur Warren, "Henry Drummond," *Woman at Home*, 2 (1894): 13.

27. Benjamin Kidd's *Social Evolution* (London: Macmillan, 1894) provides at least one exception to this trend. It sold extremely well despite Kidd being a relative unknown. I owe this important counterpoint to one of this book's reviewers. On Kidd's book, see Bernard Lightman, "Darwin and the Popularization of Evolution," *Notes and Records of the Royal Society* 64, no. 1 (March 2009): 5–24.

28. Fiske notes that it was the reports of his lectures in the *New York World* that persuaded him to work them up into book form. John Fiske, *Outlines of Cosmic Philosophy* (London: Macmillan, 1874), vol. 1, vii. For an illuminating report of one of the lectures included in revised form in this book, see "Science from the Rostrum," *New York Tribune*, May 10, 1872, 4. Interestingly, Fiske is described there as one of the few Americans who compared favorably to British science lecturers.

29. Smith, *Henry Drummond*, 316.

30. While Drummond observed that his Possilpark audience was composed of "for the most part working men" (see Henry Drummond, *Natural Law in the Spiritual World* [New York: James Pott, 1884], vi) one biographer noted that they were (also) "well-to-do." Smith, *Henry Drummond*, 134. It is important to look behind the language of "working men" in this period. It was not always a reference to a particular socioeconomic group.

31. James Moore, "Evangelicals and Evolution: Henry Drummond, Herbert Spencer and the Naturalisation of the Spiritual World," *Scottish Journal of Theology* 38, no. 3 (August 1985): 385–86.

32. Drummond, *Natural Law*, viii.

33. Tom F. Wright, *Lecturing the Atlantic: Speech, Print and an Anglo-American Commons, 1830–1870* (Oxford: Oxford University Press, 2017), 6.

34. A full account of the legal case is given in "The American Piracy of 'The Ascent of Man,'" *Bookman* 6 (1894): 109–10.

35. Smith, *Henry Drummond*, 353.

36. This list is taken from William Reed Bigelow, "Harvard's Better Self," *New England Magazine* 9 (1890): 504.

37. See Andrew C. Rieser, *The Chautauqua Moment: Protestants, Progressives and the Culture of Modern Liberalism* (New York: Columbia University Press, 2003).

38. Rieser, *Chautauqua Moment*, 327.

39. "The Northfield Convocation," 4–5; "Sunday at Chautauqua," *New York Times*, August 1, 1887, 1.

40. See, for example, "Edinburgh Professors at Amherst," *Springfield Republican*, September 19, 1887, 4; "Christ in Colleges," *New York Herald*, October 3, 1887, 4; "Extra Christians," *Boston Daily Advertiser*, October 10, 1887, 1.

41. Brief accounts can be found in Smith, *Henry Drummond*, 386–401 and, more recently, Elizabeth Wilson, "Wandering Stars: The Impact of British Evangelists in Australia, 1870–1900" (PhD diss., University of Tasmania, 2011).

42. Cited in Smith, *Henry Drummond*, 395.

43. Smith, *Henry Drummond*, 395.

44. Augustus Lowell to Henry Drummond, March 11, 1891, Papers of Professor Henry Drummond, Acc. 5890/2, National Library of Scotland.

45. Source cited in Smith, *Henry Drummond*, 451.

46. Lennox, *Practical Life Work*, 165.

47. Howard A. Bridgman, "Henry Drummond," *New England Magazine* 14 (1893): 725.

48. Lennox, *Practical Life*, 166.

49. *The Massachusetts Institute of Technology, a Brief Account of Its Foundation, Character, and Equipment* (Boston: MIT Press, 1895), 14.

50. Drummond to Jane Drummond, April 21, 1893, NLS, Papers of Professor Henry Drummond, Acc. 5890/1.

51. Henry Drummond to Jane Drummond, April 28, 1893, NLS, Papers of Professor Henry Drummond, Acc. 5890/1

52. Henry Drummond, *Evolution of Man*, edited by William Templeton (Philadelphia: Henry Altemus, 1893), 51.

53. Drummond, *Evolution of Man*, 39, 43. Drummond removed this indictment of Darwinism from his own authorized version of his Lowell lectures.

54. Drummond, *Evolution of Man*, 39.

55. Drummond, *Evolution of Man*, 88.

56. "Man's Descent," *Boston Daily Globe*, April 8, 1893, 2.

57. Drummond, *Evolution of Man*, 114–15.

58. Drummond, *Evolution of Man*, 117.

59. Drummond, *Evolution of Man*, 128.

60. James B. Pond to Henry Drummond, April 11, 1893, Papers of Professor Henry Drummond, Acc. 5890/2, National Library of Scotland.

61. "A Curiosity Shop," *Boston Daily Advertiser*, April 12, 1893, 4.

62. "Man, Child of the Monkey," *Boston Daily Globe*, April 12, 1893, 2.

63. Drummond, *Evolution of Man*, 157–63.

64. *Boston Evening Record*, April 13, 1893.

65. "Man Grew in Warfare," *Boston Daily Globe*, April 15, 1893, 5.

66. Drummond, *Evolution of Man,* 172.

67. "Man Grew in Warfare," 5.

68. Drummond, *The Ascent of Man*, 273.

69. "Hints to Harvard Men," *Boston Daily Globe*, April 17, 1893, 6.

70. "Drummond at Harvard," *Boston Daily Globe*, April 18, 1893, 8.

71. "In Harvard's Halls," *Boston Daily Journal*, April 18, 1893, 8.

72. "The Evolution of Mind," *Boston Daily Advertiser*, April 19, 1893, 5.

73. Drummond, *Evolution of Man*, 191.

74. Drummond, *Evolution of Man*, 203–6.

75. Drummond, *Evolution of Man*, 213.

76. "Evolution of Mind," *Boston Daily Advertiser*, April 22, 1893, 4.

77. Drummond, *Evolution of Man*, 229.

78. "Drummond's Address," *Boston Daily Globe*, April 24, 1893, 8. This address, along with the two given at Harvard the previous Sunday and Monday, was later published as Henry Drummond, *Sermons Rolled Away and Other Addresses to Young Men Delivered in America* (New York, James Pott, 1899). A fourth talk on the Boys' Brigade, given at Boylston Hall on the previous Thursday, was also published in this collection but was not reported by Boston newspapers. Comparing the summaries published in newspapers with the fuller versions suggests that the former were broadly accurate, if severely truncated.

79. "Man and Woman," *Boston Daily Globe*, April 26, 1893, 6.

80. Drummond, *Ascent of Man*, 330.

81. "Man and Woman," 6.

82. Drummond, *Ascent of Man*, 331.

83. Lilian Whiting, "Life in Boston," *Sunday Inter Ocean*, April 30, 1893, 34.

84. "Mother and Child," *Boston Daily Advertiser*, April 29, 1893, 4.

85. Drummond, *Evolution of Man*, 251.

86. "Prof. Drummond's Lecture," *Boston Daily Globe*, April 29, 1893, 8.

87. "Metaphysical Disturbance in Boston," *Sun*, May 11, 1893, 6.

88. "One of the Sun Jokes," *Boston Daily Advertiser*, May 12, 1893, 8.

89. Elle B. Dietrick, "Professor Drummond on Sex," *Boston Evening Transcript*, April 29, 1893. For further discussion of Dietrick, see Thomas Dixon, *The Invention*

of Altruism: Making Moral Meanings in Victorian Britain (Oxford: Oxford University Press, 2008), 299–300.

90. "Dined the Lecturer," *Boston Daily Globe*, May 6, 1893, 12.

91. "Joseph Cook Was Present," *Boston Daily Globe*, May 6, 1893, 4; "Their Most Noble Monument," *Boston Evening Transcript*, May 6, 1893, 3.

92. "Joseph Cook Was Present," 4.

93. "They Are One and the Same Thing," *Boston Evening Transcript*, May 13, 1893, 3.

94. "Selfishness and Unselfishness," *Boston Evening Transcript*, May 10, 1893, 3.

95. "They Are One and the Same Thing," 3.

96. "Evolution Epigrams," *Boston Daily Globe*, May 13, 1893, 10.

97. "The Drummond Lectures," *Boston Daily Advertiser*, May 15, 1893, 4.

98. Samuel S. McClure (with Willa Cather), "My Autobiography," *McClure's Magazine* 42 (1914): 87–88.

99. "New Publications," *Boston Daily Advertiser*, May 27, 1893, 5.

100. "Heretics of Today," *Sunday Inter Ocean*, May 28, 1893, 29.

101. *St. Paul Daily Globe*, May 23, 1893, 8.

102. "The Famous Scientist," *Minneapolis Daily Tribune*, June 2, 1893, 4.

103. "Prof. Drummond's Visit," *Minneapolis Daily Tribune*, June 4, 1893, 5.

104. "What It Ought to Be," 5.

105. "A Big 'If,'" *Minneapolis Daily Tribune*, June 12, 1893, 2.

106. "Evolution," *Minneapolis Sunday Tribune*, June 18, 1893, 16.

107. Isaac Atwater, *History of the City of Minneapolis* (New York: Munsell, 1893), 238–39.

108. "Theory of Evolution," *Minneapolis Sunday Tribune*, June 19, 1893, 6.

109. "Services by the Lake," *New York Tribune*, July 3, 1893, 4. On the emerging importance in this period of "charisma" in cultures of public speech in the United States, see Jeremy C. Young, *The Age of Charisma: Leaders, Followers and Emotions in American Society, 1870–1940* (Cambridge, UK: Cambridge University Press, 2017).

110. "Opening of Chautauqua," *New York Tribune*, July 10, 1893, 3.

111. "The Origin of Mind," *New York Tribune*, July 5, 1893, 5.

112. Jesse L. Hurlbut, *The Story of Chautauqua* (New York: G. P. Putnam's Sons, 1921), 252.

113. The estimate of ten thousand listeners is derived from "Opening of Chautauqua," 3.

114. "The Evolution of Man," *New York Tribune*, July 8, 1883, 5.

115. For a full account of Moody's theology, see Stanley Gundry, *Love Them In: The Life and Theology of D. L. Moody* (Chicago: Moody Press, 1999).

116. On Moody's relationship with the media, see Evensen, *God's Man for the Gilded Age.*

117. This aspect of Moody's mission is most clearly attested to in Timothy E. W. Gloege, *Guaranteed Pure: The Moody Bible Institute, Business, and the Making of Modern Evangelicalism.* (Chapel Hill: University of North Carolina Press, 2015).

118. For Moody's rapid speaking, see Evensen, *God's Man for the Gilded Age*, 11, 165.

119. Henry Drummond, "Mr. Moody: Some Impressions and Facts," *McClure's Magazine* 4 (1894–1895), 62.

120. Dwight L. Moody, "A Tribute," in Henry Drummond, *A Life for a Life and Other Addresses* (New York: Fleming H. Revell, 1897), 10.

121. Smith, *Life of Henry Drummond*, 282.

122. James Young Simpson, *Henry Drummond* (New York: Charles Scribner's Sons, 1901), 93. On the *Christian*, see James F. Findlay, *Dwight L. Moody: American Evangelist 1837–1899* (Eugene, OR: Wipf and Stock, 2007), 151–52.

123. "Include the Doubting Thomas," *Springfield Daily Republican*, July 11, 1893, 5.

124. "Professor Drummond at Northfield," *New York Tribune*, July 11, 1893, 5.

125. "Include the Doubting Thomas." A full version of the address can be found in Henry Drummond, *Stones Rolled Away*, 125–37. There are changes in emphasis and in word choice, but the substance is captured relatively well by the newspaper reports.

126. Drummond, *Natural Law*, vi.

127. "Moody's College Conference," *Springfield Republican*, July 12, 1893, 4.

128. Henry Drummond, *A Life for a Life*, 14.

129. "Moody's College Conference," 4.

130. Newspaper reports and other published versions of Drummond's address use "imminent." This is corrected to "immanent" in Drummond, *Stones Rolled Away*, 177.

131. "Students' Conference Ends," *Springfield Republican*, July 13, 1893, 4.

132. Charles F. Goss, *Echoes from the Pulpit and Platform* (Hartford, CT: A. D. Worthington, 1900), 86.

133. Quoted in Smith, *Life of Henry Drummond*, 453. See also Moore, "Evangelicals and Evolution," 401–2.

134. Smith, *Life of Henry Drummond*, 454–55; Lennox, *Practical Life*, 124; *New York Tribune*, August 31, 1893, 6.

135. On Pierson, see Dana L. Robert, *Occupy until I Come: Arthur T. Pierson and the Evangelization of the World* (Grand Rapids: William B. Eerdmans, 2003).

136. As reported in the *Outlook*, September 2, 1893, 442. That Drummond was in mind is clear from "What a Pastor Sees and Hears," *New York Tribune*, September 2, 1893, 3.

137. *Christian Work*, July 5, 1894, 9. This periodical is not the same as the *Christian at Work* cited above.

138. Drummond, *A Life for a Life*, front matter. Revell published four books authored by Drummond. The first, *Love: The Supreme Gift*, published in 1890, was Drummond's bestselling book in America. See Allan Fisher, *Fleming H. Revell Company: The First 125 Years, 1870–1995* (Grand Rapids: Revell, 1995).

139. Robert, *Occupy until I Come*, 225–34.

140. "Doings of the Sixteenth Day's Sessions of the Parliament of Religions," *Sunday Inter Ocean*, September 27, 1893, 1.

141. "Rev. William W. Alger," *Daily Inter Ocean*, September 28, 1893, 2.

142. "Henry Drummond's Paper," *Daily Inter Ocean*, September 28, 1893, 2.

143. "Henry Drummond's Paper," 2.

144. Henry Drummond, "Evolution and Christianity," in *The World's Parliament of Religions*, ed. John Henry Barrows (Chicago: Parliament Publishing, 1893), vol. 2, 1319.

145. "Prof. Henry Drummond at Chicago University Early in October," *Chicago Tribune*, September 17, 1893, 6.

146. On Harper and the University of Chicago, see George Marsden, *The Soul of the American University* (Oxford: Oxford University Press, 1994), 236–62; Michael Lee, "Higher Criticism and Higher Education at the University of Chicago: William Rainey Harper's Vision of Religion in the Research University," *History of Education Quarterly* 48, no. 4 (November 2008): 508–33. For a full history of Harper's relationship with more conservative Baptist supporters of the university, see John W. Boyer, *The University of Chicago: A History* (Chicago: University of Chicago Press, 2015).

147. "Elements of a Life," *Chicago Tribune*, October 2, 1893, 3.

148. Boyer, *University of Chicago*, 26.

149. "Convocation of the Great University of Chicago," *Daily Inter Ocean*, October 3, 1893, 1.

150. "Upholds Darwin's Theory," *New York Herald*, October 5, 1893, 8. See also *Northern Christian Advocate*, October 11, 1893, 12.

151. "University in Line with Baptists," *Chicago Tribune*, October 6, 1893, 3.

152. Details taken from official programme booklet, Papers of Professor Henry Drummond, Acc. 5890/17/f.57, National Library of Scotland.

153. *Chicago Tribune*, October 8, 1893, 9.

154. See Philip D. Jordan, *The Evangelical Alliance for the United States of America, 1847–1900* (New York: E. Mellen, 1982).

155. For mention of salmon fishing with William E. Dodge Jr., see The Evangelical Alliance for the United States, *Christianity Practically Applied: The Discussions of the International Christian Conference*, 2 vols (New York: Baker and Taylor, 1894), vol. 2, 117.

156. The Evangelical Alliance for the United States, *Christianity Practically Applied*, vol. 1, 13–15.

157. Robert, *Occupy until I Come*, 227.

158. Henry Drummond, "Christianity and the Evolution of Society," in *Christianity Practically Applied*, vol. 2, 466–71.

159. See Robert, *Occupy until I Come*, 225–34.

160. *Christianity Practically Applied*, vol. 2, 120.

161. "Many Women Meet Lady Aberdeen," *Chicago Tribune*, October 17, 1893, 3.

162. "An Address by Professor Drummond," *New York Tribune*, October 12, 1893, 4.

163. "Professor Drummond to Start for Home," *New York Tribune*, October 21, 1893, 9.

164. *Outlook*, November 18, 1893, 876.

CONCLUSION: SCIENCE, HISTORICALLY SPEAKING

1. Audrey Moore, "The Relations between Science and Religion," *Quarterly Review* 159 (1885): 380.

2. Moore, "Relations," 363.

3. The full list is found in Charles H. Smith and Megan Derr, eds., *Alfred Russel Wallace's 1886–1887 Travel Diary: The North American Lecture Tour* (Manchester: SIRI Scientific Press, 2013), 167–68.

4. Thomas Augst, "Humanist Enterprise in the Marketplace of Culture," in *Cosmopolitan Lyceum: Lecture Culture and the Globe in Nineteenth-Century America*, ed. Tom F. Wright (Amherst: University of Massachusetts Press, 2013), 235.

5. Marcel Chotkowski LaFollette, *Science on the Air: Popularizers and Personalities on Radio and Early Television* (Chicago: University of Chicago Press, 2008).

6. For a helpful summary of this literature, see David A. Kirkby, "Film, Radio and Television," in *A Companion to the History of Science*, ed. Bernard Lightman (Chichester: Wiley Blackwell, 2016), 428–41.

7. Tim Boon, "1962: 'What Manner of Men?': Meeting Scientists through Television," *Public Understanding of Science* 28, no. 3 (April 2019): 372–78.

8. For a recent exploration of Brian Cox's rhetorical performances, see Tina Skouen, "Brian Cox and the New Enlightenment," *Notes and Records of the Royal Society* 74, no. 1 (March 2020): 149–78.

BIBLIOGRAPHY

MANUSCRIPT COLLECTIONS

British Library, Manuscript Collections, Add MS 63092.

The Darwin Correspondence Project, https://www.darwinproject.ac.uk.

The Huntington Library, San Marino, California, William Jones Rhees Papers.

Imperial College London, Records and Archives, Thomas Henry Huxley Collection.

National Library of Scotland, Papers of Professor Henry Drummond, Acc. 5890.

Natural History Museum Library and Archives.

Royal Institution of Great Britain, Papers of John Tyndall, RI MS JT.

Smithsonian Institution Archives, Joseph Henry Collection.

University of Toronto Discover Archives, https://discoverarchives.library.utoronto.ca/index.

PRIMARY AND SECONDARY SOURCES

Adams, Amanda. *Performing Authorship in the Nineteenth-Century Transatlantic Lecture Tour*. Farnham: Ashgate, 2014.

"Alfred Russel Wallace." *Mining and Scientific Press* 54 (1887): 397.

Allen, Grant. *The Lower Slopes*. London: Stone and Kimball, 1894.

Allen, Grant. *A Mortal Coil: A Novel*. London: Chatto and Windus, 1888.

"The American Piracy of 'The Ascent of Man.'" *Bookman* 6 (1894): 109–10.

Anderson, Erin Lorraine. "Nature on Display: The Wagner Free Institute of Science, 1855–1900." PhD diss., University of Delaware, 2020.

Anderson, Oliver. "Hansard's Hazards: An Illustration from Recent Interpretation of Married Woman's Property Law and the 1857 Divorce Act." *English Historical Review* 112, no. 449 (November 1997): 1202–15.

Andrews, Malcolm. *Charles Dickens and His Performing Selves*. Oxford: Oxford University Press, 2006.

Annual Report of the Provost to the Trustees of the Peabody Institute, 2 (1869).

Annual Report of the Provost to the Trustees of the Peabody Institute, 10 (1877).

Annual Report of the Provost to the Trustees of the Peabody Institute, 20 (1887).

Applegate, Debby. *The Most Famous Man in America: A Biography of Henry Ward Beecher*. New York: Doubleday, 2006.

Atwater, Isaac. *History of the City of Minneapolis*. New York: Munsell, 1893.

Augst, Thomas. *The Clerk's Tale: Young Men and Moral Life in Nineteenth-Century America*. Chicago: University of Chicago Press, 2003.

Augst, Thomas. "Composing the Moral Senses: Emerson and the Politics of Character in Nineteenth-Century America." *Political Theory* 27, no. 1 (February 1999): 85–120.

Augst, Thomas. "Humanist Enterprise in the Marketplace of Culture." In *Cosmopolitan Lyceum: Lecture Culture and the Globe in Nineteenth-Century America*, edited by Tom F. Wright, 223–40. Amherst: University of Massachusetts Press, 2013.

Ball, W. Valentine. *Letters and Reminiscences of Sir Robert Ball*. London: Cassell, 1915.

Barrows, Samuel J., ed. *Science and Immortality: The* Christian Register *Symposium*. Boston: George H. Ellis, 1887.

Barton, Ruth. "Sunday Lecture Societies: Naturalistic Scientists, Unitarians and Secularists Unite against Sabbatarian Legislation." In Dawson and Lightman, *Victorian Scientific Naturalism*, 189–219.

Beccaloni, G. W., ed. *Wallace Letters Online*. Accessed June 6, 2021. https://www.nhm.ac.uk/research-curation/scientific-resources/collections/library-collections/wallace-letters-online/index.html.

Bell, Trudy E. "Mitchel, Ormsby MacKnight." In *Biographical Dictionary of Astronomers*, edited by Thomas Hockey, vol. 2, 790. New York: Springer, 2007.

Bergland, Renée. *Maria Mitchell and the Sexing of Science: An Astronomer among the American Romantics*. Boston: Beacon, 2008.

Bigelow, William Reed. "Harvard's Better Self." *New England Magazine* 9 (1890): 504–11.

"Biographical Sketch of Professor Tyndall." *Appleton's Journal* 2 (1869): 339–41.

Bogle, Lori Lyn. "Pandering to the Crowd: The American Governing Elite's Changing Views on Mass Media and Publicity during the Nineteenth Century." *Journalism History* 43, no. 2 (2017): 62–74.

Boon, Tim. "1962: 'What Manner of Men?': Meeting Scientists through Television." *Public Understanding of Science* 28, no. 3 (April 2019): 372–78.

Bosco, Ronald A., and Joel Myerson, eds. *The Selected Lectures of Ralph Waldo Emerson*. Athens: University of Georgia Press, 2005.

Boyer, John W. *The University of Chicago: A History*. Chicago: University of Chicago Press, 2015.

Bremner, G. A., and Jonathan Conlin. "Consolations of Amero-Teutonism: E. A. Freeman's Tour of the United States, 1881–82." In *Making History: Edward Augustus Freeman and Victorian Cultural Politics*, edited by G. A. Bremner and Jonathan Conlin, 101–8. Oxford: Oxford University Press, 2015.

Bridgman, Howard A. "Henry Drummond." *New England Magazine* 14 (1893): 725–29.

Broadus, John. A. *A Treatise on the Preparation and Delivery of Sermons*. Philadelphia: Smith, English, 1871.

Brooke, John H. "The History of Science and Religion: Some Evangelical Dimensions." In *Evangelicals and Science in Historical Perspective*, edited by David N. Livingstone, D. G. Hart and Mark Noll, 17–42. Oxford: Oxford University Press, 1999.

Brooker, Jeremy. "A Lecture on Locust Street: Morton, Tyndall, Pepper, and the Construction of Scientific Identity." In *Science Museums in Transition: Cultures of Display in Nineteenth-Century Britain and America*, edited by Carin Berkowitz and Bernard Lightman, 111–38. Pittsburgh: University of Pittsburgh Press, 2017.

Brown, Daniel. *The Poetry of Victorian Scientists: Style, Science and Nonsense*. Cambridge, UK: Cambridge University Press, 2013.

Brownell, William C. "English Lecturers in America." *Galaxy* 20 (1875): 62–72.

Buckingham, James Silk. *An Improved Syllabus of Mr Buckingham's Lectures on the Oriental World*. London: Hurst, Chance, 1830.

Bush, Martin. "The Proctor-Parkes Incident: Politics, Protestants and Popular Astronomy in Australia in 1880." *Historical Records of Australian Science* 28, no. 1 (2017): 26–36.

Bynum, W. F. and Caroline Overy, eds. *Michael Foster and Thomas Henry Huxley Correspondence, Medical History Supplement, No. 28*. London: Welcome Trust Centre for the History of Medicine, 2009.

Cahan, David. *Helmholtz: A Life in Science*. Chicago: University of Chicago Press, 2018.

Cahan, David. "Helmholtz in Gilded Age America: The International Electrical Congress of 1893 and the Relations of Science and Technology." *Annals of Science* 67, no. 1 (2010): 1–38.

Cantor, Geoffrey, and Sally Shuttleworth, eds. *Science Serialized: Representations of the Sciences in Nineteenth-Century Periodicals*. Cambridge, MA: MIT Press, 2004.

Cantor, Geoffrey, Gowan Dawson, Graeme Gooday, Richard Noakes, Sally Shuttleworth, and Jonathan R. Topham, eds. *Science in the Nineteenth-Century Periodical: Reading the Magazine of Nature*. Cambridge, UK: Cambridge University Press, 2004.

Chalaby, Jean. *The Invention of Journalism*. London: Macmillan, 1998.

Chalmers, Mitchell P. "Professor Tyndall." *New Review* 10 (1894): 77–85.

Charteris, Evan. *The Life and Letters of Sir Edmund Gosse*. London: William Heinemann, 1931.

Clarke, William. *Academic Charisma and the Origins of the Research University*. Chicago: University of Chicago Press, 2006.

Collini, Stefan. "The Idea of 'Character' in Victorian Political Thought." *Transactions of the Royal Historical Society* 35 (1985): 29–50.

Collini, Stefan. *Public Moralists: Political Thought and Intellectual Life in Britain, 1860–1930*. Oxford: Clarendon Press, 1991.

Cope, Emily Murphy. "'Inspiration of Delivery': John A. Broadus and the Evangelical Underpinnings of Extemporaneous Orator." *Rhetoric Society Quarterly* 45, no. 4 (2015): 279–99.

Cranfield, Jonathan. *Twentieth-Century Victorian: Arthur Conan Doyle and the* Strand Magazine, *1891–1930*. Edinburgh: Edinburgh University Press, 2016.

Crowe, Michael J. *The Extraterrestrial Life Debate 1750–1900: The Idea of a Plurality of Worlds from Kant to Lowell*. Cambridge, UK: Cambridge University Press, 1988.

Darnton, Robert. "What Is the History of Books?" *Daedalus* 111, no. 3 (Summer 1982): 65–83.

Darwin, Charles. *The Descent of Man*, vol. 2. London: John Murray, 1871.

Dawson, Gowan. *Darwin, Literature and Victorian Respectability*. Cambridge, UK: Cambridge University Press, 2007.

Dawson, Gowan, and Bernard Lightman, eds. *Victorian Scientific Naturalism: Community, Identity, Continuity*. Chicago: University of Chicago Press, 2014.

Dawson, Gowan, Bernard Lightman, Sally Shuttleworth, and Jonathan R. Topham, eds. *Science Periodicals in Nineteenth-Century Britain: Constructing Scientific Communities*. Chicago: University of Chicago Press, 2020.

Dawson, J. William. *The Geological History of Plants*. New York: D. Appleton, 1888.

Dell, John Henry. *The Dawning Grey*. London: Simpkin Marshall, 1885.

Desmond, Adrian. *Huxley: From Devil's Disciple to Evolution's High Priest*. Reading, MA: Addison-Wesley, 1997.

Dixon, Thomas. *The Invention of Altruism: Making Moral Meanings in Victorian Britain*. Oxford: Oxford University Press, 2008.

Dott, Robert. "Lyell in America: His Lectures, Field Work and Mutual Influences, 1841–1853." *Earth Sciences History* 15, no. 2 (1996): 101–40.

Doyle, Arthur Conan. "The Voice of Science." *Strand Magazine* 1 (1891): 312–17.

Drummond, Henry. *The Ascent of Man*. London: Hodder and Stoughton, 1894.

Drummond, Henry. "Christianity and the Evolution of Society." In *Christianity Practically Applied: The Discussions of the International Christian Conference*, vol. 2, edited by Evangelical Alliance for the United States, 466–71. New York: Baker and Taylor, 1894.

Drummond, Henry. "Evolution and Christianity." In *The World's Parliament of Religions*, vol. 2, edited by John Henry Barrows, 1316–35. Chicago: Parliament Publishing, 1893.

Drummond, Henry. *The Evolution of Man*. Edited by William Templeton. Philadelphia: Henry Altemus, 1893.

Drummond, Henry. *A Life for a Life and Other Addresses*. New York: Fleming H. Revell, 1897.

Drummond. Henry, *Love: The Supreme Gift*. New York: Fleming H. Revell, 1897.

Drummond, Henry. "Mr. Moody: Some Impressions and Facts." *McClure's Magazine* 4 (1894–1895): 55–69.

Drummond, Henry. *Natural Law in the Spiritual World*. New York: James Pott, 1884.

Drummond, Henry. *The New Evangelism and Other Addresses*. New York: Dodd, Mead, 1899.

Drummond, Henry. *Sermons Rolled Away and Other Addresses to Young Men Delivered in America*. New York, James Pott, 1899.

Dryer, J. L. E., and H. H. Turner. *History of the Royal Astronomical Society*. London: Royal Astronomical Society, 1923.

Eastman, Carolyn. "Placing Platform Culture in Nineteenth-Century American Life." In Ray and Stob, *Thinking Together*, 187–202.

Elderkin, John. *A Brief History of the Lotus Club*. New York: Lotus Club, 1895.

Ellison, Robert H. *The Victorian Pulpit: Spoken and Written Sermons in Nineteenth-Century Britain*. Selinsgrove, PA: Susquehanna University Press, 1998.

Emerson, Ralph Waldo. "Character." In *Collected Works of Ralph Waldo Emerson*, vol. 3, 53–57. Cambridge, MA: Harvard University Press, 1983.

Emerson, Ralph Waldo. "Eloquence." In *Collected Works of Ralph Waldo Emerson* vol. 7, 30–51. Cambridge, MA: Harvard University Press, 2007.

Emerson, Ralph Waldo. "Eloquence." In *Collected Works of Ralph Waldo Emerson* vol. 8, 59–71. Cambridge, MA: Harvard University Press, 2010.

Escott, Thomas H. S. "The House of Commons: Its Personnel and Its Oratory." *Fraser's Magazine* 10 (1874): 504–17.

The Evangelical Alliance for the United States. *Christianity Practically Applied: The Discussions of the International Christian Conference*. 2 vols. New York: Baker and Taylor, 1894.

Evensen, Bruce J. *God's Man for the Gilded Age: D. L. Moody and the Rise of Modern Mass Evangelism*. Oxford: Oxford University Press, 2003.

Findlay, James F. *Dwight L. Moody: American Evangelist 1837–1899*. Eugene: Wipf and Stock, 2007.

Finnegan, Diarmid A. "Daniel William Cahill (1796–1864) and the Rhetorical Geography of Science and Religion." In Kember, Plunkett and Sullivan, *Popular Exhibitions*, 97–114.

Finnegan, Diarmid A. "Eve and Evolution: Christian Responses to the First Woman Question, 1860–1900." *Journal of the History of Ideas*, 75, no. 2 (April 2014): 281–303.

Finnegan, Diarmid A. "Finding a Scientific Voice: Performing Science, Space and Speech in the Nineteenth Century." *Transactions of the Institute of British Geographers* 42, no. 2 (June 2017): 192–205.

Finnegan, Diarmid A. "Lectures." In Lightman, *Companion to the History of Science*, 414–27.

Finnegan, Diarmid A. "Placing Science in an Age of Oratory: Spaces of Scientific Speech in Mid-Victorian Edinburgh." In *Geographies of Nineteenth-Century Science*, edited by David N. Livingstone and Charles W. J. Withers, 153–77. Chicago: University of Chicago Press, 2011.

Finnerty, Páiric, and Rod Rosenquist. "Transatlantic Celebrity: European Fame in Nineteenth-Century America." *Comparative American Studies: An International Journal* 14, no. 1 (2016): 1–6.

Fisher, Allan. *Fleming H. Revell Company: The First 125 Years, 1870–1995*. Grand Rapids: Revell, 1995.

Fiske, John. *Edward Livingston Youmans: Interpreter of Science for the People*. New York: D. Appleton, 1894.

Fiske, John. *Outlines of Cosmic Philosophy*. 2 vols. London: Macmillan, 1874.

Forgan, Sophie. "Listening and Learning: Audiences and their Roles in Nineteenth-Century Britain." In *Participating in the Knowledge Society: Researchers beyond the University Walls*, edited by Ruth Finnegan, 65–78. Basingstoke: Palgrave Macmillan, 2005.

Franklin, Fabian. *The Life of Daniel Coit Gilman*. New York: Dodd, Mead, 1910.

Freeberg, Ernest. *The Age of Edison: Electric Light and the Invention of Modern America*. New York: Penguin, 2013.

Friedman, David. *Wilde in America: Oscar Wilde and the Invention of Modern Celebrity*. New York: W. W. Norton, 2015.

Fyfe, Aileen, and Bernard Lightman, eds. *Science in the Marketplace: Nineteenth-Century Sites and Experiences*. Chicago: University of Chicago Press, 2007.

Gange, David. *Dialogues with the Dead: Egyptology in British Culture and Religion, 1822–1922*. Oxford: Oxford University Press, 2013.

Ganter, Granville. "Women's Entrepreneurial Lecturing in the Early National Period." In Ray and Stob, *Thinking Together*, 41–55.

Gilman, Daniel C. *Address at the Inauguration of Daniel C. Gilman*. Baltimore: John Murphy, 1876.

Gilman, Daniel C. *The Launching of a University*. New York: Dodd, Mead, 1906.

Gladden, Washington. "The New Evolution." *McClure's Magazine* 3 (1894): 235–42.

Gloege, Timothy E. W. *Guaranteed Pure: The Moody Bible Institute, Business, and the Making of Modern Evangelicalism*. Chapel Hill: University of North Carolina Press, 2015.

Goffman, Erving. *Forms of Talk*. Philadelphia: University of Pennsylvania Press, 1981.

Gooday, Graeme. "Ethnicity, Expertise and Authority: The Cases of Lewis Howard Latimer, William Preece and John Tyndall." In *Scientists' Expertise as Performance*, edited by Joris Vandendriessche, Evert Peeters and Kaat Wils, 15–30. London: Pickering and Chatto, 2015.

Goodman, Susan, and Carl Dawson. *William Dean Howells: A Writer's Life*. Berkeley: University of California Press, 2005.

Goss, Charles F. *Echoes from the Pulpit and Platform*. Hartford, CT: A. D. Worthington, 1900.

Gundry, Stanley. *Love Them In: The Life and Theology of D. L. Moody*. Chicago: Moody Press, 1999.

Gustafson, Sandra M. *Eloquence Is Power: Oratory and Performance in Early America*. Chapel Hill: University of North Carolina Press, 2000.

Haley, Andrew P. *Restaurants and the Rise of the American Middle Class, 1880–1920*. Chapel Hill: University of North Carolina Press, 2011.

Hall, Thomas C. *John Hall: Pastor and Preacher*. New York: Fleming H. Revell, 1901.

Hamlin, Kimberly A. *From Eve to Evolution: Darwin, Science and Women's Rights in Gilded Age America*. Chicago: University of Chicago Press, 2014.

Hanes, Susan R. *Wilkie Collins's American Tour, 1873–74*. London: Pickering and Chatto, 2008.

Hart, Daryl. "Faith and Learning in the Age of the University: The Academic Ministry of Daniel Coit Gilman." In *The Secularization of the Academy*, edited by George M. Marsden and Bradley J. Longfield, 107–45. Oxford: Oxford University Press, 1991.

Harvey, David. "Globalization and 'the Spatial Fix.'" *Geographische Revue* 2 (2001): 23–30.

Hays, Jo N. "The Rise and Fall of Dionysius Lardner." *Annals of Science* 38, no. 5 (1981): 527–42.

Helmstadter, R. J. "Wild, Joseph." In *Dictionary of Canadian Biography*, vol. 13. Toronto: University of Toronto/Université Laval, 2003.

Henson, Louise, Geoffrey Cantor, Gowan Dawson, Richard Noakes, Sally Shuttleworth, and Jonathan R. Topham, eds. *Culture and Science in the Nineteenth-Century Media*. Farnham: Ashgate, 2004.

Hesketh, Ian. "Technologies of the Scientific Self: John Tyndall and His Journals." *Isis* 110, no. 3 (September 2019): 460–82.

Hewitt, Martin. "Beyond Scientific Spectacle: Image and Word in Nineteenth-Century Popular Lecturing." In Kember, Plunkett and Sullivan, *Popular Exhibitions*, 79–96.

Higginson, Thomas Wentworth. "The American Lecture-System." *Macmillan Magazine* 18 (1868): 48–56.

Hoegaerts, Josephine. "Speaking Like Intelligent Men: Vocal Articulations of Authority in the House of Commons in the Nineteenth Century." *Radical History Review* 121, no. 1 (January 2015): 123–44.

"How Fast Can People Talk?" *Shorthand Review* 5 (1893): 50–1.

"How Voice Reveals Character." *Review of Reviews* 9 (1893): 156.

Howard, George W. *The Monumental City*. Baltimore: J. D. Ehler, 1873–1876.

Howard, Jill. "'Physics and Fashion': John Tyndall and His Audiences in Mid-Victorian Britain." *Studies in History and Philosophy of Science* 35, no. 4 (2004): 729–58.

Huxley, Leonard. *Life and Letters of Thomas Henry Huxley*. 2 vols. New York: D. Appleton, 1900.

Huxley, Thomas H. *Collected Essays*. London: Macmillan, 1893–94.

Huxley, Thomas H. *Discourses Biological and Geological*. London: Macmillan, 1894.

Huxley, Thomas H. "Professor Tyndall." *Nineteenth Century* 35 (1894): 1–11.

Huxley, Thomas H. "The School Boards: What They Can Do, and What They May Do." *Contemporary Review* 16 (1870): 1–15.

Jensen, J. Vernon. *Thomas Henry Huxley: Communicating for Science*. Newark: University of Delaware Press, 1991.

Jensen, J. Vernon. "Thomas Henry Huxley's Lecture Tour of the United States, 1876." *Notes and Records of the Royal Society of London* 42, no. 2 (July 1988): 181–95.

Jordan, Philip D. *The Evangelical Alliance for the United States of America, 1847–1900*. New York: E. Mellen, 1982.

Joyce, Patrick. *Democratic Subjects: The Self and the Social in Nineteenth Century England*. Cambridge, UK: Cambridge University Press, 1994.

Kaalund, Nanna Katrine Lüders. "Of Rocks and 'Men': The Cosmogony of John William Dawson." In *Historicizing Humans: Deep Time, Evolution and Race in Nineteenth-Century British Sciences*, edited by Efram Sera Shriar, 44–67. Pittsburgh: University of Pittsburgh Press, 2018.

Kember, Joe, John Plunkett and Jill A. Sullivan, eds. *Popular Exhibitions, Science and Showmanship, 1840–1910*. London: Pickering and Chatto, 2012.

Kendall, Phebe Mitchell. *Maria Mitchell: Life, Letters and Journals*. Boston: Lee and Shepard, 1896.

Kidd, Benjamin, *Social Evolution*. London: Macmillan, 1894.

Kirkby, David A. "Film, Radio and Television." In Lightman, *Companion to the History of Science*, 418–41.

Knight, David. "Scientific Lectures: A History of Performance." *Interdisciplinary Science Reviews* 27, no. 3 (2002): 217–24.

Knott, Cargill Gilston. *The Life and Scientific Works of Peter Guthrie Tait*. Cambridge, UK: Cambridge University Press, 1911.

LaFollette, Marcel Chotkowski. *Science on the Air: Popularizers and Personalities on Radio and Early Television*. Chicago: University of Chicago Press, 2008.

Lee, Michael. "Higher Criticism and Higher Education at the University of Chicago: William Rainey Harper's Vision of Religion in the Research University." *History of Education Quarterly* 48, no. 4 (November 2008): 508–33.

Lennox, Cuthbert [John H. Napier]. *The Practical Life of Henry Drummond*. New York: James Pott, 1901.

Lightman, Bernard, ed. *A Companion to the History of Science*. Chichester: John Wiley and Sons, 2016.

Lightman, Bernard. "Darwin and the Popularization of Evolution." *Notes and Records of the Royal Society* 64, no. 1 (March 2010): 5–24.

Lightman, Bernard. "Interpreting Agnosticism as a Nonconformist Sect: T. H. Huxley's 'New Reformation.'" In *Science and Dissent in England, 1688–1945*, edited by Paul Wood, 197–214. Aldershot: Ashgate, 2004.

Lightman, Bernard. "Lecturing in the Spatial Economy of Science." In Fyfe and Lightman, *Science in the Marketplace*, 93–132.

Lightman, Bernard. "Scientists as Materialists in the Periodical Press: Tyndall's Belfast Address." In Cantor and Shuttleworth, *Science Serialized*, 199–237.

Lightman, Bernard. "The Theology of the Victorian Scientific Naturalists." In *Science without God: Rethinking the History of Scientific Naturalism*, edited by Peter Harrison and Jon H. Roberts, 234–54. Oxford: Oxford University Press, 2019.

Lightman, Bernard. *Victorian Popularizers of Science: Designing Nature for New Audiences*. Chicago: University of Chicago Press, 2007.

Lightman, Bernard, and Michael Reidy, eds. *The Age of Naturalism: Tyndall and his Contemporaries*. Pittsburgh: University of Pittsburgh Press, 2014.

"Literary News," *Atlantic Monthly* 34 (1874): 363–64.

Livingstone, David N. *Dealing with Darwin: Place, Politics and Rhetoric in Religious Engagements with Evolution*. Baltimore: Johns Hopkins University Press, 2014.

Livingstone, David N. "Science, Site and Speech: Scientific Knowledge and the Spaces of Rhetoric." *History of the Human Sciences* 20, no. 2 (May 2007): 71–98.

Logan, Shirley Wilson. *Liberating Language: Sites of Rhetorical Education in Nineteenth-Century Black America*. Carbondale: SIU Press, 2008.

Lucier, Paul. "The Professional and the Scientist in Nineteenth-Century America." *Isis* 100, no. 4 (December 2009): 699–732.

Lyons, Sherrie. "The Many Influences Shaping Wallace's Views of Human Evolution." In *An Alfred Russel Wallace Companion*, edited by Charles H. Smith, James T. Costa and David A. Collard, 167–90. Chicago: University of Chicago Press, 2019.

Macdonald, Frederic W. *The Life of William Morley Punshon, LL.D.* London: Hodder and Stoughton, 1887.

Manna, Salvatore John. "A Brothers' Reunion: Evolution's Champion Alfred Russel Wallace and Forty-Niner John Wallace." *California History* 85, no. 4 (2008): 4–25, 70–71.

Marchant, James, ed. *Alfred Russel Wallace Letters and Reminiscences*, 2 vols. London: Cassell, 1916.

Marden, Orison Swett. *How They Succeeded*. Boston: Lothrop, 1901.

Marsden, George. *The Soul of the American University*. Oxford: Oxford University Press, 1994.

Martin, Janette. "Popular Political Oratory and Itinerant Lecturing in Yorkshire and the North East in an Age of Chartism, 1837–60." PhD diss., University of York, 2010.

Martyn, W. Carlos. *John B. Gough: The Apostle of Cold Water*. New York: Funk and Wagnalls, 1894.

Massachusetts Institute of Technology. *The Massachusetts Institute of Technology, a Brief Account of Its Foundation, Character, and Equipment*. Boston: MIT Press, 1895.

Maxwell, James Clerk. "To the Chief Musician on Nubla [sic]: A Tyndallic Ode." *Nature* 4 (1871): 291.

McClure, Samuel S. (with Willa Cather). "My Autobiography." *McClure's Magazine* 42 (1914): 85–95.

McCosh, James. *The Development Hypothesis: Is It Sufficient?* New York: Robert Carter and Brothers, 1876.

McGee, Anita N. "The Women's Anthropological Society of America." *Science* 13 (1889): 241.

McIlvaine, Joshua Hall. *Elocution: The Source and Elements of Its Power*. New York: Charles Scribner, 1871.

McKivigan, John R. *Forgotten Firebrand: James Redpath and the Making of Nineteenth-Century America*. Ithaca: Cornell University Press, 2008.

McKivigan, John R. "'A New Vocation before Me': Frederick Douglass's Post-Civil War Lyceum Career." *Howard Journal of Communication* 29, no. 3 (2018): 269–81.

McMullin, Ernan. "Darwin and the Other Christian Tradition." *Zygon* 46, no. 2 (June 2011): 291–316.

Meisel, Joseph S. *Public Speech and the Culture of Public Life in the Age of Gladstone.* New York: Columbia University Press, 2001.

Mendelssohn, Michèle. *Making Oscar Wilde.* Oxford: Oxford University Press, 2018.

Milner, Richard. "Charles Darwin and Associates, Ghostbusters." *Scientific American* 275, no. 4 (October 1996): 96–101.

"Modern Skepticism." *Scribner's Monthly* 6 (1873): 424–32.

Moody, Dwight L. "A Tribute," in Drummond, *A Life for a Life,* 7–11.

Moore, Aubrey. "The Relations between Science and Religion." *Quarterly Review* 159 (1885): 360–86.

Moore, James. "Evangelicals and Evolution: Henry Drummond, Herbert Spencer and the Naturalisation of the Spiritual World." *Scottish Journal of Theology* 38, no. 3 (August 1985): 383–418.

Morus, Iwan R. "Seeing and Believing science." *Isis* 97, no. 1 (March 2006): 101–10.

Morus, Iwan R. "Worlds of Wonder: Sensation and the Victorian Scientific Performance." *Isis* 101, no. 4 (December 2010): 806–16.

Mott, Frank Luther. *A History of American Magazines, 1865–1885.* Cambridge, MA: Harvard University Press, 1938.

"Mr. Proctor on Star Drift and Nebulae," *Scientific Opinion* 3 (1870): 417.

Mullin, Robert Bruce. "Science, Miracles and the Prayer-Gauge Debate." In *When Science and Christianity Meet,* edited by David C. Lindberg and Ron Numbers, 203–24. Chicago: University of Chicago Press, 2003.

Murdoch, James E. *Orthophony, or, the Cultivation of the Voice, in Elocution,* 7th ed. Boston: Ticknor, Reed and Fields, 1851.

Nall, Joshua. *News from Mars: Mass Media and the Forging of a New Astronomy, 1860–1910.* Pittsburgh: University of Pittsburgh Press, 2019.

Nichol, John P. *Views of Astronomy.* New York: Greeley and McElrath, 1848.

"Notes." *Nature* 7 (1872): 150.

O'Neill, Bonnie Carr. "'The Best of Me Is There': Emerson as Lecturer and Celebrity." *American Literature* 80, no. 4 (December 2008): 739–67.

O'Neill, Bonnie Carr. *Literary Celebrity and Public Life in Nineteenth-Century United States.* Athens: University of Georgia Press, 2017.

Pandora, Katherine. "Popular Science in National and Transnational Perspective: Suggestions from the American Context." *Isis* 100, no. 2 (June 2009): 346–58.

Picker, John M. *Victorian Soundscapes.* Oxford: Oxford University Press, 2003.

Pollock, Juliet. "Michael Faraday." *Saint Paul's Magazine* 6 (1870): 292–303.

Pond, James B. *Eccentricities of Genius.* London: Chatto and Windus, 1901.

Proceedings at the Farewell Banquet to Professor Tyndall. New York: D. Appleton, 1873.

Proctor, Richard A. "Astronomy in America." *Popular Science Review* 15 (1876): 351–63.

Proctor, Richard A. "Colours of the Double Stars." *Cornhill Magazine* 8 (1863): 679–87.

Proctor, Richard A. "Editorial Gossip." *Knowledge* 3 (1883): 155.

Proctor, Richard A. "Editorial Gossip." *Knowledge* 4 (1883): 292.

Proctor, Richard A. "Editorial Gossip." *Knowledge* 4 (1883): 318.

Proctor, Richard A. "Gossip." *Knowledge* 8 (1885): 228.

Proctor, Richard A. "Gossip." *Knowledge* 8 (1886): 195.

Proctor, Richard A. "Lectures and the London Papers." *Knowledge* 3 (1883): 217.

Proctor, Richard A. "The Lecture System in America and England." *English Mechanic and the World of Science* 24 (1876): 1–2.

Proctor, Richard A. "Lecturing Notes." *Knowledge* 3 (1883): 40

Proctor, Richard A. "Letter to the Editors." *English Mechanic and World of Science*, 22 (1875–1876): 220.

Proctor, Richard A. "Letters to the Editors." *English Mechanic and World of Science* 20 (1874–1875): 473.

Proctor, Richard A. "Old and New Astronomy: An Autobiographical Sketch." *Knowledge* 11 (1888): 112–13.

Proctor, Richard A. *Our Place among the Infinities*. New York: D. Appleton, 1876.

Proctor, Richard A. "Our Two Brains." *Knowledge* 5 (1884): 309–10.

Proctor, Richard A. "The Past and Future of Our Earth." *Contemporary Review* 25 (1874): 74–92.

Proctor, Richard A. *Pleasant Ways in Science*. New York: R. Worthington, 1879.

Proctor, Richard A. *Six Lectures on Astronomy*. New York: Truth Seeker, 1876.

Proctor, Richard A. "Sydney and Sunday Lecturing." *Knowledge* 3 (1883): 57.

Proctor, Richard A. "The Type Writer." *Knowledge* 2 (1882): 402.

Proctor, Richard A. "To the Editor of the *Atlantic Monthly*." *Atlantic Monthly* 34 (1874): 750–51.

Proctor, Richard A. "Waking at Will." *English Mechanic and World of Science* 20 (1875): 551.

"Professor Drummond's Booklets." *Wesleyan-Methodist Magazine* 114 (1891): 713.

"Professor Huxley at Oxford." *Speaker* 7 (1893): 596.

"Professor Tyndall's Deed of Trust," *Popular Science Monthly* 3 (1873): 100.

"Professor Tyndall's Lectures." *Galaxy* 14 (1872): 561–2.

"Professor Tyndall's Lectures in New York." *Galaxy* 15 (1873): 273.

Quincey, Thomas De. "System of the Heavens, as Revealed by Lord Rosse's Telescope," *Tait's Edinburgh Magazine* 13 (1846): 564–78.

Radick, Gregory. *The Simian Tongue: The Long Debate about Animal Language*. Chicago: University of Chicago Press, 2007.

Randel, William Peirce. "Huxley in America." *Proceedings of the American Philosophical Society* 114, no. 2 (April 1970): 73–99.

Rankin, Jeremiah, and Ruth Barton. "Tyndall, Lewes and Popular Representations of Scientific Authority in Victorian Britain." In *The Age of Scientific Naturalism*, edited by Bernard Lightman and Michael S. Reidy, 51–70. London: Pickering and Chatto, 2014.

Ratcliff, Jessica. *The Transit of Venus Enterprise in Victorian Britain*. London: Pickering and Chatto, 2008.

Ray, Angela G. *Lyceum and Public Culture in the Nineteenth-Century United States*. East Lansing: Michigan State University Press, 2005.

Ray, Angela G. "What Hath She Wrought: Women's Rights and the Nineteenth-Century Lyceum." *Rhetoric and Public Affairs* 9, no. 2 (Summer 2006): 183–213.

Ray, Angela G., and Paul Stob, eds. *Thinking Together: Lecturing, Learning and Difference in the Long Nineteenth Century*. Philadelphia: Pennsylvania University Press, 2018.

Reidy, Michael S. "Evolutionary Naturalism on High: The Victorian Sequester the Alps." In Dawson and Lightman, *Victorian Scientific Naturalism*, 55–78.

"Richard Anthony Proctor." *Scribner's Monthly* 7 (1873): 172–75.

Rieser, Andrew C. *The Chautauqua Moment: Protestants, Progressives and the Culture of Modern Liberalism*. New York: Columbia University Press, 2003.

Robert, Dana L. *Occupy until I Come: Arthur T. Pierson and the Evangelization of the World*. Grand Rapids: William B. Eerdmans, 2003.

Robertson, William. *The Life and Times of Right Hon. John Bright*. London: Cassell, 1883.

Ross, Sydney. "Scientist: The Story of a Word." *Annals of Science* 18, no. 2 (1962): 65–85.

Rossiter, Margaret W. "Benjamin Silliman and the Lowell Institute: The Popularization of Science in Nineteenth-Century America." *New England Quarterly* 44, no. 4 (December 1971): 602–26.

Rossiter, Margaret W. *Women Scientists in America: Struggles and Strategies to 1940*. Baltimore: Johns Hopkins University Press, 1982.

Rothenberg, Marc, ed. *The Papers of Joseph Henry*. Vol. 11, *January 1866–May 1878*. Sagamore Beach, MA: Watson Publishing, 2007.

Rowe, Kenneth E. "Building Monumental Methodist Cathedrals in America's Capital City, 1850–1950." *Methodist History* 50, no. 3 (April 2012): 171–78.

Rush, James. *Philosophy of the Human Voice*. Philadelphia: n.p., 1827.

Schaffer, Simon. "Transport Phenomenon: Space and Visibility in Victorian Physics." *Early Popular Visual Culture* 10, no. 1 (2012): 71–91.

Schott, Kenneth Ronald. "An Analysis of Henry Drummond and His Rhetoric of Reconciliation." PhD diss., Ohio State University, 1972.

Schroeder, Janice. "Speaking Volumes: Victorian Feminism and the Appeal of Public Discussion." *Nineteenth-Century Contexts* 25, no. 2 (2003): 97–117.

Schudson, Michael. "Questioning Authority: A History of the News Interview in American Journalism, 1860s–1930s." *Media, Culture and Society* 16, no. 4 (October 1994): 565–87.

Scott, Donald. "The Popular Lecture and the Creation of a Public in Mid-Nineteenth-Century America." *Journal of American History* 66, no. 4 (March 1980): 791–809.

Secord, James. "How Scientific Conversation Became Shop Talk." *Transactions of the Royal Historical Society* 17 (2007): 129–56.

Silliman, Robert H. "The Hamlet Affair: Charles Lyell and the North Americans." *Isis* 86, no. 4 (December 1995): 541–61.

Simpson, James Young. *Henry Drummond*. New York: Charles Scribner's Sons, 1901.

"Sketch of R. A. Proctor." *Popular Science Monthly* 4 (1874): 487.

Skouen, Tina. "Brian Cox and the New Enlightenment." *Notes and Records of the Royal Society* 74, no. 1 (March 2020): 149–78.

Slotten, Ross A. *Heretic in Darwin's Court: The Life of Alfred Russel Wallace*. New York: Columbia University Press, 2004.

Smalley, George W. "Mr. Huxley." *Scribner's Magazine* 18 (1895): 514–24.

Smith, Charles H. "Wallace, Spiritualism and Beyond: 'Change' or 'No Change.'" In *Natural Selection and Beyond: The Intellectual Legacy of Alfred Russel Wallace*, edited by Charles H. Smith and George Beccaloni, 391–424. Oxford: Oxford University Press, 2008.

Smith, Charles H., and Megan Derr, eds. *Alfred Russel Wallace's 1886–1887 Travel Diary: The North American Lecture Tour*. Manchester: SIRI Scientific Press, 2013.

Smith, George Smith. *The Life of Henry Drummond*. New York: Doubleday and McClure, 1898.

Smith, Harriette K. *History of the Lowell Institute*. Boston: Lamson, Wolffe, 1898.

Smith, James M. "Thomas Henry Huxley in Nashville: Part II." *Tennessee Historical Quarterly* 33, no. 3 (Fall 1974): 322–41.

Smythe, Ted Curtis. *The Gilded Age Press, 1865–1900*. Westport, CT: Praeger, 2003.

Spurgeon, Charles H. *Lectures to My Students*. London: Passmore and Alabaster, 1875.

Stewart, DeLisle. "Vaughan, Daniel." In *American Dictionary of National Biography*, vol. 19, edited by Dumas Malone, 235–36. New York: Charles Scribner's Sons, 1936.

Strong-Boag, Veronica. *Liberal Hearts and Coronets: The Lives and Times of Ishbel Marjoribanks Gordon and John Campbell Gordon, the Aberdeens*. Toronto: University of Toronto Press, 2015.

Summerville, James. "Albert Roberts: Journalist of the New South: Part II." *Tennessee Historical Quarterly* 42, no. 2 (Summer 1983): 179–202.

"The Sun." *English Mechanic and World of Science* 18 (1873–1874): 528.

"Table Talk." *Appletons' Journal* 8 (1872): 247–48.

Tetrault, Lisa. "The Incorporation of American Feminism: Suffragists and the Postbellum Lyceum." *Journal of American History* 96, no. 4 (March 2010): 1027–56.

Theerman, Paul H. "Dionysius Lardner's American Tour." In *Experiencing Nature*, edited by Paul H. Theerman and Karen Hunger Parshall, 211–36. Dordrecht: Kluwer Academic, 1997.

Thompson, Jane Smeal, and Helen Thompson. *Silvanus Phillips Thompson: His Life and Letters*. London: T. Fisher Unwin, 1920.

Tribune Popular Science. Boston: Henry L. Shepard, 1874.

Turner, Frank M. *Between Science and Religion: The Reaction to Scientific Naturalism in Victorian England.* New Haven, CT: Yale University Press, 1974.

Turner, Frank M. "Rainfall, Plagues and the Prince of Wales: A Chapter in the Conflict of Religion and Science." *Journal of British Studies* 13, no. 2 (May 1974): 46–65.

Tyler, Moses Coit. "How They Manage Their Lectures in England." *Putnam's Magazine* 13 (1869): 98–106.

Tyndall, John. "The Constitution of the Universe." *Fortnightly Review* 3 (1865): 129–44.

Tyndall, John. *The Forms of Water in Clouds and Rivers, Glaciers and Ice.* 4th ed. New York: D. Appleton, 1877.

Tyndall, John. *Fragments of Science for Unscientific People.* New York: D. Appleton, 1871.

Tyndall, John. *Lectures on Light.* New York: D. Appleton, 1873.

Tyndall, John. "*Life and Letters of Faraday.*" *Academy* 1, no. 8 (1870): 204–7.

Tyndall, John. "My Schools and Schoolmasters." *Popular Science Monthly* 26 (1885): 333–35.

Tyndall, John. *New Fragments.* New York: D. Appleton, 1892.

Tyndall, John. *Notes of a Course of Nine Lectures on Light.* London: Longmans, Green, 1870.

Tyndall, John. "On a New Series of Chemical Reactions Produced by Light." *Proceedings of the Royal Society* 17 (1868): 92–102.

Tyndall, John. "Personal Recollections of Thomas Carlyle." *Fortnightly Review* 48 (1890): 28.

Tyndall, John. "Science and Religion." *Popular Science Monthly* 2 (1872): 79–82.

Tyndall, John, and Henry Thompson. "The 'Prayer for the Sick.' Hints Towards a Serious Attempt to Estimate Its Value." *Contemporary Review* 20 (1872): 205–10.

Ungureanu, James C. "Edward L. Youmans and the 'Peacemakers' in the *Popular Science Monthly.*" *Fides et Historia* 51, no. 2 (Summer–Fall 2019): 31–32.

Ungureanu, James C. "Science, Religion and the 'New Reformation' of the Nineteenth Century." *Science & Christian Belief* 31, no. 1 (2019): 41–61.

Ungureanu, James C. *Science, Religion and the Protestant Tradition: Retracing the Origins of Conflict.* Pittsburgh: University of Pittsburgh Press, 2019.

van Krieken, Robert. *Celebrity Society.* London: Routledge, 2012.

"The Voice and Character." *Bow Bells* 5 (1866): 89.

"The Voice as an Index of Character." *Musical Standard* 29 (1885): 192.

Wallace, Alfred Russel. *Darwinism.* London: Macmillan, 1889.

Wallace, Alfred Russel. "The Depression of Trade: Its Causes and Its Remedies." 1886. Online at https://people.wku.edu/charles.smith/wallace/S387.htm.

Wallace, Alfred Russel. "An Essay, on the Best Method of Conducting the Kington Mechanics' Institution." In *The History of Kington*, edited by Richard Parry, 66–70. Kington: Charles Humphreys, 1845.

Wallace, Alfred Russel. *My Life: A Record of Events and Opinions.* 2 vols. London: Chapman and Hall, 1905.

Wallace, Alfred Russel. "Sir Charles Lyell on Geological Climates and the Origin of Species." *Quarterly Review* 126 (1869): 359–94.

Ward, Wilfrid. "Thomas Henry Huxley: A reminiscence." *Nineteenth Century* 40 (1896): 274–92.

Warren, Arthur. "Henry Drummond." *Woman at Home* 2 (1894): 1–16.

Wells, Kentwood D. "Dionysius Lardner: Popular Science Showman of the 1840s." *Magic Lantern Gazette* 29, no. 1 (Spring 2017): 3–18.

White, Paul. *Thomas Huxley: Making the "Man of Science."* Cambridge, UK: Cambridge University Press, 2003.

Wilson, Daniel. "Proceedings at the Dinner." *Proceedings of the American Philosophical Society* 27 (1889): 31–34.

Wilson, Elizabeth. "Wandering Stars: The Impact of British Evangelists in Australia, 1870–1900." PhD diss., University of Tasmania, 2011.

Wilson, R. Jackson. "Emerson as Lecturer." In *Cambridge Companion to Ralph Waldo Emerson*, edited by Joel Porte, 76–96. Cambridge, UK: Cambridge University Press, 1999.

Winters, Alison. *Mesmerized: Powers of Mind in Victorian Britain.* Chicago: University of Chicago Press, 1998.

Withers, Charles W. J. *Zero Degrees: Geographies of the Prime Meridian.* Cambridge, MA: Harvard University Press, 2017.

Wright, Tom F. *Lecturing the Atlantic: Speech, Print and an Anglo-American Commons, 1830–1870.* Oxford: Oxford University Press, 2017.

Wright, Tom F. "Listening to Emerson's 'England' at Clinton Hall." *Journal of American Studies* 46, no. 3 (August 2012): 641–62.

Wright, Tom F. "The Transatlantic Larynx in Wartime: John Gough's London Voices." In *Transatlantic Traffic and (Mis)Translations*, edited by Robin Peel and Daniel Maudlin, 45–62. Durham: University of New Hampshire Press, 2013.

Yates, John. "Professor Tyndall." *Christian Union* 6 (1872): 343–44.

Youmans, Edward L. "A Correction: Letter from Prof. Tyndall." *Popular Science Monthly* 3 (1873): 241–43.

Youmans, Edward L. "Prof. Huxley." *Popular Science Monthly* 9 (1876): 621–22.

Youmans, Edward L. "Professor Tyndall." *Popular Science Monthly* 1 (1872): 751–52.

Youmans, Eliza A. "Tyndall's American Visit." *Popular Science Monthly* 44 (1894): 502–14.

Young, Jeremy C. *The Age of Charisma: Leaders, Followers and Emotions in American Society, 1870–1940.* Cambridge, UK: Cambridge University Press, 2017.

Zimmerman, Sarah. *The Romantic Literary Lecture in Britain.* Oxford: Oxford University Press, 2019.

INDEX

Academy of Music (Baltimore), 80

Academy of Music (Brooklyn), 50, 111

Academy of Music (New York), 49, 57

Academy of Music (Sioux City), 156

acoustics, 19, 34, 103, 148–49, 157

Adams, John Quincy, 19

Agassiz, Alexander, 150

Agassiz, Louis, 16, 50, 58, 111, 114, 150

agnosticism, 23, 95, 120, 123, 143, 161,
189, 198, 213

Airy, George Biddell, 104–8, 110

Albany Evening Journal, 114

Alexander, James Bradun, 194

Alger, Rev. William, 201

Allen, Grant, 162

American Association for the Advancement of Science, 74, 178

American Geographical Society, 151

American Literary Bureau, 107, 113

Amherst College, 178, 191

Angelus (Millet), 195, 198, 202

Anglo-American relations, 27, 41, 44,
56–57, 106

Anthropological Society of London, 135

Anthropological Society of Washington,
151

Appleton Chapel (Harvard), 184, 186,
189, 191

Appleton's Journal, 40, 44

Arcturus (Toronto), 154

Arnold, Matthew, 97–98, 182

astrology, 114, 117

Association Hall (New York), 111, 112,
129

astronomy: advances in, 94, 101, 106,
115, 121, 126, 212–13; amateur, 107,
122; American, 19, 24, 27, 105–6,
108, 110, 112; British, 103–5, 110;
as news, 98; professional 95, 99,
107; and religion, 113–14, 117–18,
122–25, 127; and self-improvement,
96, 99; teaching of, 95–96, 133

Atlantic Monthly, 106

atheism, 82, 95, 107, 114, 116, 217

Attenborough, David, 217

audiences: behavior of, 13, 19, 32, 48,
93, 141–42, 182; command of, 10,
49, 69, 142, 167, 209; composition of, XI, 39, 144, 151, 170, 174,
179–80, 193, 198, 250n3; attitudes
towards, XI, 3, 36–38, 46, 48, 67,
104; emotions of, 9–10, 18, 31, 37,
64, 81–82, 185; expectations of,
10–11, 47, 55–56, 82, 103, 118,
211; and gender, 24, 38, 185, 191;
influence of speaker on, 39, 41, 43;